Proceedings

Second Conference on Tall Buildings in Seismic Regions

May 16 and 17, 1991
Los Angeles, California

The Los Angeles Tall Buildings Structural Design Council and the Council on Tall Buildings and Urban Habitat assumes no responsibility of the statements made in the pages of these Proceedings. Any opinions expressed are those of the individual authors. Interested readers should contact the individual authors for necessary clarification. The material contained herein reflects reproduction and reduction from original materials submitted by the individual authors.

Council Report 903.409

ISBN 0-939493-07-1

SECOND CONFERENCE ON TALL BUILDINGS IN SEISMIC REGIONS

Current and Future Trends in Design and Construction of Tall Buildings for Wind and Seismic Effects

Organized by:

The Los Angeles Tall Buildings Structural Design Council
and
The Council on Tall Buildings and Urban Habitat

May 16 and 17, 1991

Los Angeles Hilton and Towers
Los Angeles, California

Recent earthquakes around the world continue to focus the need for dialog between developers, architects, engineers and builders, for construction of all types. The Second Conference on Tall Buildings in Seismic Regions was organized to enhance the understanding of many of the broad and particular issues of development, planning, design and construction of tall buildings in the 1990's, particularly in seismic regions.

The Conference is the 52nd Regional Conference and follows a Council on Tall Buildings and Urban Habitat week-long conference held in Hong Kong in November of 1990. The Conference is only the second of these sessions to be held in the seismically sensitive west coast region. The initial west coast regional conference was held in Los Angeles in February 1988, and emphasized the combined theme of seismicity and tall buildings. In the intervening two years, the Los Angeles Tall Buildings Structural Design Council has held one-day sessions as a forum to discuss structural design issues in a strong seismic environment.

The general sessions focus on the process of development of tall building projects from inception through the development process, including the design and construction of selected tall buildings.

The plenary sessions are lead by well known developers, architects and engineers and consist of teams representing the development process, the architectural design, the engineering development and construction issues. The presentations and discussions overview the general and particular issues in current development processes, architectural design and state-of-the-art of engineering, as well as the future visions for development and design of tall and very tall buildings.

Specialty sessions emphasize particular topics in greater detail including wind and seismicity, research, innovations, applications, analysis and design techniques, design and control of tall/very tall buildings, cladding and other architectural and engineering issues.

LOS ANGELES TALL BUILDINGS STRUCTURAL DESIGN COUNCIL

With the emergence of Los Angeles as the key center for business and commerce, the significant trend of tall buildings in the downtown area is coincident with rising land costs. The need for vitality in the city has led toward development of many hotels and high-rise residential buildings to complement the business, commercial, and convention space.

Since many issues in seismic areas relate to the structural aspects of tall buildings, the Los Angeles Tall Buildings Structural Design Council was formed in 1988 to provide a forum for the discussion of issues relating to the design of tall buildings in the Los Angeles region.

The annual meeting of the Council presents a program for engineers, architects, contractors, and building officials that includes research reports on areas of emerging importance, case studies of current structural designs, and consensus documents by the membership on contemporary design issues.

Its members seek to advance state-of-the-art structural design through interaction with other professional organizations, building departments, and university researchers as well as recognize significant contributions to the structural design of tall buildings and promote scholarship in the field.

The student scholarship program has recognized an outstanding student each year and provides financial support and on the job experience in a design office to assist a student whose career objective is structural engineering of major projects.

The Council is a nonprofit California corporation whose members are those individuals who have demonstrated exceptional professional accomplishments in the structural design of tall buildings.

Dr. Gregg Brandow (President)
Brandow & Johnston Associates

Dr. Lauren D. Carpenter
Wong Hobach Lau

Dr. Robert E. Englekirk
Robert Englekirk, Inc.

Robert N. Harder
LA City Dept. of Building & Safety

Dr. Gary C. Hart
UCLA/Englekirk & Hart, Inc.

Richard E. Holguin
LA City Dept. of Building & Safety

Dr. George W. Housner
California Institute of Technology

Roy G. Johnston
Brandow & Johnston Associates

John A. Martin
John A. Martin & Associates

John A. Martin, Jr.
John A. Martin & Associates

Clarkson W. Pinkham
S.B. Barnes Associates

Donald R. Strand
Brandow & Johnston Associates

Edward J. Teal
Seismic Engineering Assoc., Ltd.

Nabih Youssef (Exec. Secretary/Treasurer)
Nabih Youssef & Associates

Los Angeles Tall Buildings Structural Design Council
660 South Figueroa Street, Suite 1660
Los Angeles, California 90017
(213) 688-3014
FAX: (213) 688-3018

Council on Tall Buildings and Urban Habitat

Council Contributors

Contributing Participants

Table of Contents

Proceedings of the Second Conference on Tall Buildings in Seismic Regions
55th Regional Conference
May 16 and 17, 1991, Los Angeles, California

Comparing the Behavior of 40 Story Structures Designed According to US and Japanese Codes

Navin Amin[1], Kirk Martini[2], Hamed Fatehi[3], Tomio Kurata[4]

Abstract

The paper is a case study of differences in structural behavior resulting from US and Japanese design codes. The subject of the study is a 40 story steel frame structure for a hotel located in San Diego, originally designed according to the 1988 Uniform Building Code (UBC). For the purpose of this study, the structural members of the primary frame were redesigned by a Japanese designer according to the Japanese Building Standard Law (JBSL). The behavior of the two designs is compared using static and dynamic non-linear analysis of simplified two dimensional models. The building code requirements are compared in order to explain differences in behavior. The study does not address aspects of structural detailing

1. Introduction

The subject of this study is a 40 story steel frame structure for a hotel to be constructed in San Diego. After the design was completed according to the 1988 UBC, a visiting Japanese engineer, who was involved with the project, redesigned the members of the primary superstructure according to Japanese structural codes in order to compare US and Japanese design methods and results. The following discussion outlines differences in structural behavior. The configuration of the structure and the basis for the two designs is described in section 2. Section 3 describes the comparison of structural behavior based on static and dynamic non-linear analysis. Section 4 provides a summary and conclusions.

2. Configuration and Design

Figures 1 through 4 show the configuration of this dual braced-moment frame structure. In the transverse direction there are three types of frame: the heavily braced frames on lines D and F carry most of the transverse lateral loads; the outrigger frames on lines B, C, G, and H provide additional overturning resistance; and the braced frames on lines A and I provide torsional resistance. In the longitudinal direction, the braced-moment frames on lines 2 and 3 carry most of the longitudinal lateral loads, with the moment frames on lines 1 and 4 providing additional lateral resistance. Both designs were performed with three dimensional elastic analysis, considering static wind and seismic loads as well as dynamic response spectra using modal analysis. The Japanese design used zone coefficients for lateral loads which corresponded to site conditions in San Diego. Reference [1] describes the design procedures in detail.

In both designs, certain key factors were chosen at the discretion of the designer according to common practice rather than specific code requirements. The US design limited interstory drift due to wind loads to h/400; the UBC does not specify a drift limit for wind, so this value was chosen by the designer. In the

---

[1] Chief Structural Engineer, and Associate Partner
  Skidmore, Owings and Merrill, San Francisco
[2] Engineer, Skidmore, Owings and Merrill, San Francisco
  Ph.D. Candidate, University of California Berkeley
[3] Project Engineer
  Skidmore, Owings and Merrill, San Francisco
[4] Visiting Engineer
  Sumitomo Construction, Tokyo

case of the Japanese design, the designer chose to increase the base shear coefficients from code minimum values. In the longitudinal direction, the base shear coefficient was increased from 4.1% to 5%; in the transverse direction, the coefficient was increased from 6.2% to 7%. For both designs, these choices had a significant impact on the resulting structures and the study illustrates that comparisons of structural design methods must include aspects of common practice in addition to code requirements.

3. Comparing behavior

3.1 Modelling Assumptions

The behavior of the two designs was compared using two dimensional non-linear analysis models based on the ANSR-III program [2]. For both the longitudinal and transverse directions, half of the lateral load resisting frames were aligned in a single plane using slaving links for the horizontal displacements at the floor levels. Only half of the frames are included in each direction because of symmetry. This configuration neglects torsional effects and biaxial effects at columns which participate in both transverse and longitudinal frames, however it was considered adequate for the purposes of this study.

This simplified arrangement was chosen because of the complexity involved in three dimensional non-linear analysis of large models. The behavior of these two dimensional models was compared with the three dimensional models by checking the corresponding periods of vibration (figure 5). The periods of the two dimensional models are within 6% of the corresponding modes of the three dimensional models.

All members were modelled using lumped plasticity beam-column elements [3]. The yield strength of compression members was reduced to reflect the buckling strength. This is not an ideal arrangement since it does not correctly model load reversal on buckled members and also results in reduced tension strength, which can lead to premature yielding of tension braces, however this approach was considered a reasonable compromise. The analyses did not include geometric stiffness (P-delta) effects for computational reasons.

3.2 Static Behavior

Figures 6 and 7 show the static load deflection curves for the longitudinal and transverse directions respectively. In both figures, the deflection quantity is the lateral deflection at the top of the structure, and the force quantity is the base shear in kips. The load is given in kips rather than in terms of a percentage of gravity because the two designs have different masses. The JSBL includes a portion of the live load as reactive seismic mass, so that this mass is approximately 12% greater for the Japanese design. The lateral load for these diagrams is applied to the structure in a pattern according to the Japanese code. This pattern is roughly similar to the UBC pattern of a triangular distribution with an additional top force.

The curves illustrate that the Japanese design has approximately 20% more strength and stiffness in both directions; the Japanese design uses approximately 11% more steel to achieve this extra strength and stiffness. The difference arises from differences in the strength and stiffness requirements for lateral loads. Figure 8 compares the strength and stiffness requirements for two designs. Required strength is measured in terms of a "strength index", defined as the base shear divided by the allowable stress increase factor for short term loading. Required stiffness is measured in terms of a similar index, defined as the base shear divided by the allowable drift. For the purpose of comparison, the values in the figure are normalized in each case so that the US seismic value is equal to 1.0.

The strength and stiffness indices are used in order to reflect aspects of demand and capacity in a single quantity. The Japanese design uses larger base shears to represent structural demands, but it also uses larger values for allowable short term stress and for allowable drifts representing structural capacities. For strength, the long term allowable stresses are nearly the same for US and Japanese codes, but the Japanese use a short term increase of 1.5 rather than the value of 1.33 used in the UBC. For stiffness, the Japanese codes used a maximum drift of h/200, while the US design used a value of h/333 for seismic loads, as specified by the UBC, and a value of h/400 for wind design, as selected by the designer. The strength and stiffness indices give a more balanced view of the design criteria than base shears alone.

Comparing the values in figure 8 reveals that the Japanese design used much more severe criteria for seismic strength, 2.4 times the US value in the transverse direction, and 1.7 times larger in the longitudinal direction. The strength indices for wind strength are much closer, with the Japanese values being about 25% greater in each direction. In the transverse, direction this difference in wind strength criteria largely determined the difference in strength and stiffness between the two structures, In addition, the US design used a much larger index for wind stiffness, approximately 1.4 times the Japanese value in both directions. This requirement for wind stiffness governed the size of many members in the US design and contributed to the relatively small difference between the two designs, compared to the difference which the seismic requirements alone would indicate.

The strength indices above do not account for many aspects of lateral load design. In particular, there are important aspects of the US design which the index does not reflect. First, the UBC requires that columns have adequate capacity to resist $3*R_w/8$ times the seismic axial force. Also, at the discretion of the designer, the members were designed to have sufficient ultimate capacity for a response spectrum representing a maximum probable earthquake. These provisions, along with the requirements for wind strength and stiffness, contributed to the transverse seismic overstrength of the US design with respect to basic UBC loads.

3.3 Dynamic Behavior

3.3.1 Vibration Characteristics

Figure 9 shows the first three modal periods and participation factors obtained from three dimensional elastic analysis of the two designs; in the figure, the X direction is longitudinal and the Y direction is transverse. Despite the 20% difference in stiffness observed above, the modal frequencies are very close. This is because, as stated above, the Japanese design used 12% more mass since it included a portion of the live load. Therefore, the increased stiffness of the Japanese design is largely offset by the increased mass, resulting in only slightly shorter modal periods.

Although the structures were designed considering different mass levels, it was considered more reasonable to use equal masses when comparing dynamic behavior. Thus, for the two dimensional non-linear time history analysis discussed below, the US mass level was used for both designs; this reduced the longitudinal period of vibration of the Japanese design from 4.75 seconds to 4.52 seconds, and reduced the transverse period from 3.29 to 3.12 seconds.

3.3.2 Time History Analysis

Using the non-linear models, each design was subjected to a version of the Loma Prieta earthquake motion measured in San Francisco. The ground motion was scaled to a maximum velocity of 40 cm/sec. (15.7 in/sec), with a resulting maximum peak acceleration of 0.27g. In contrast to the US, where maximum peak acceleration is a common measure of ground motion severity, the Japanese typically use maximum velocity on the basis that it is a better indication of damage potential. In Japan a ground motion scaled to 40 cm/sec would typically be used for evaluating ultimate limit states for a structure located in a region with seismicity similar to that of San Diego.

Figure 10 compares the envelopes of horizontal displacement at the floor levels and figure 11 compares envelopes of interstory drifts in response to the ground motion. In the transverse direction, the general level of drifts and displacements is noticeably lower for the Japanese design, however in the longitudinal direction, there is a smaller difference between the response of the two designs. An explanation for this behavior can be found in the response spectrum for this ground motion (figure 12). In the transverse direction, the two designs are in the velocity region of the spectrum, so that the reduced period of the Japanese design results in smaller displacements. However in the longitudinal direction, the two designs are close to the displacement region of the spectrum, so that displacements do not vary with vibrational period.

The analysis of these structures for this particular ground motion is not sufficient to draw general conclusions, but the results indicate that the additional strength and stiffness of the Japanese design may not be a significant benefit for long period structures. Complete study of the question requires a wide variety of structures and ground motions, particularly including ground motions with significant frequency content in the long period range. However, such analysis was beyond the scope of this study.

4. Conclusions

The Japanese design used significantly more severe requirements for seismic strength and stiffness than the US design, however the US design used more severe requirements for wind stiffness. The net result was an approximately 20% difference in static strength and stiffness.. It is important to note that these results may not apply to other configurations, particularly for structures with less oblong plan shape where deflections due to wind load are less dominant. In addition, since the structure is located in San Diego, the designs were based on a seismicity level corresponding to UBC seismic zone 3. In a more severe seismic zone, wind loads would be less dominant. In such cases, the more severe Japanese requirements for seismic strength would probably lead to much greater differences in strength, stiffness, and total steel weight.

In addition to differences in code provisions, each design reflected differences in common practice and the attitudes and judgement of its designer. The selection of wind drift criteria for the US design, and the adjustment of seismic base shears for the Japanese design had a significant effect on the final results. Any comparison of design methods must consider such aspects.

When the two designs were subjected to a ground motion derived from the Loma Prieta earthquake, the higher strength and stiffness of the Japanese design resulted in smaller drifts and displacements in the transverse direction, but had less influence on the longitudinal response. The results indicate that adding strength and stiffness to long period structures may result in higher internal forces without limiting drifts and displacements. Further study is required to fully address this issue.

5. References

1. N. Amin, H. Fatehi, T. Kurata. A Comparison of Design Procedures and Results of a 40 Story Steel Building in San Diego using UBC-88 and Japanese Code. 4th US-Japan Workshop, ATC, 1990.

2. C.V. Oughourlian, G.H. Powell. ANSR-III: General Purpose Computer Program for Nonlinear Structural Analysis. University of California EERC, UCB/EERC-82/21, November 1982.

3. P.F. Chen, G.H. Powell, Generalized Plastic Hinge Concepts for 3D Beam-Column Elements. University of California EERC, UCB/EERC-82/20, November 1982.

Figure 1: Perspective

Typical Floor
Framing

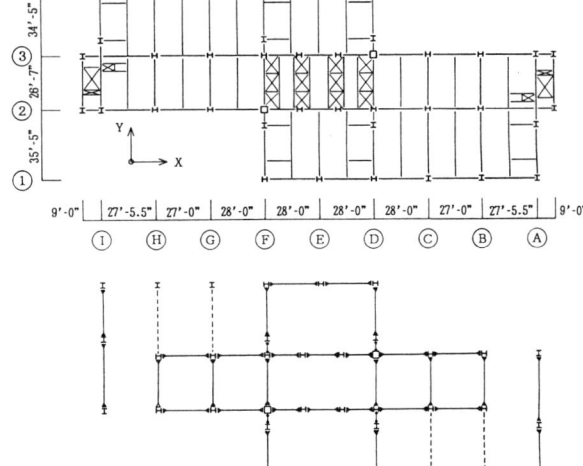

Lateral Resisting
Frame Configuration

Figure 2: Plans

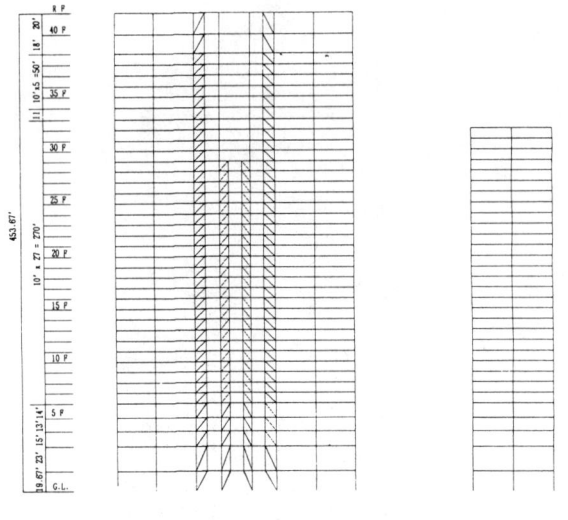

Lines 2 and 3   Lines 1 and 4

Figure 3: Longitudinal Frame Elevations

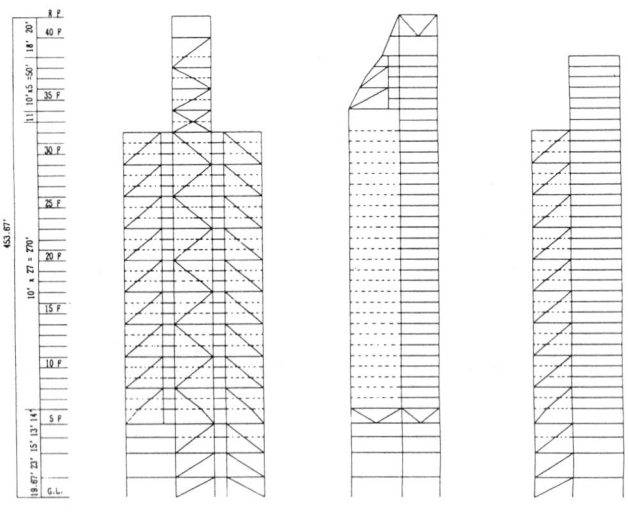

Lines F and D   Lines B, C,   Lines I and A
                  H and G

Figure 4: Transverse Frame Elevations

		3d Model	2d Model
US	transverse	3.23 sec.	3.43 sec.
	longitudinal	4.83	4.98
Japan	transverse	3.12	3.29
	longitudinal	4.67	4.75

Figure 5: Comparison of Vibration Periods of 2D and 3D Models

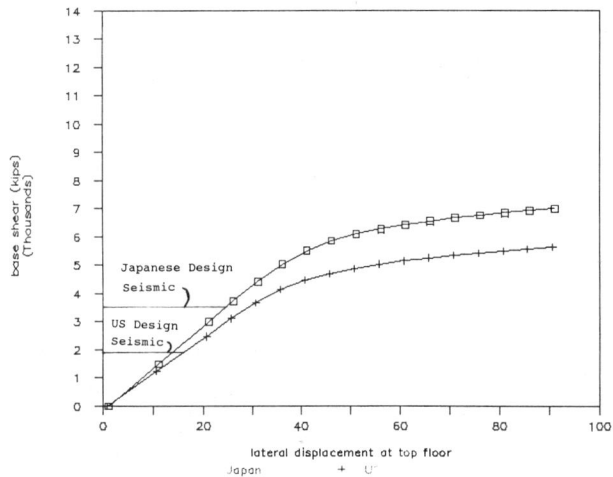

Figure 6: Static Load-Deflection Curve for Longitudinal Direction

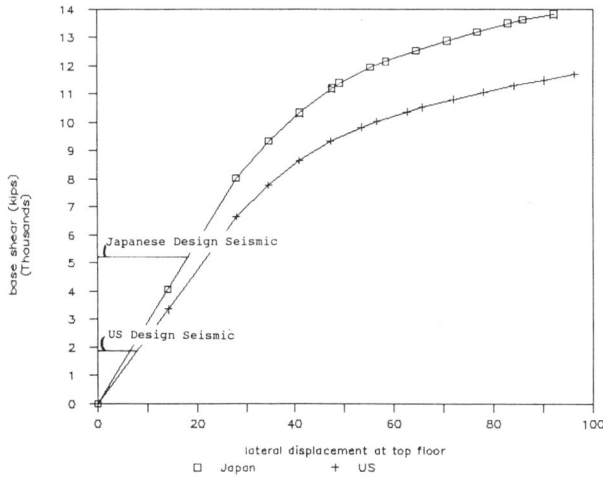

Figure 7: Static Load-Deflection Curve for Transverse Direction

7

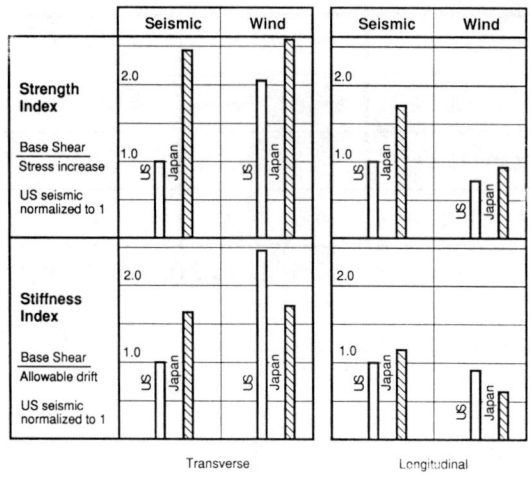

Figure 8: Comparison of Lateral Strength and Stiffness Criteria

			1st. MODE	2nd. MODE	3rd. MODE
U · S · A	PERIOD (sec)		4.832	4.217	3.235
	PARTICIPATION FACTOR	X - TRANS.	11.48	0.02	0.38
		Y - TRANS.	-0.29	0.14	11.05
		Z - ROTAT.	-2.47	8928.65	-119.45
J A P A N	PERIOD (sec)		4.672	4.175	3.115
	PARTICIPATION FACTOR	X - TRANS.	12.20	-0.25	-0.45
		Y - TRANS.	0.33	0.20	11.63
		Z - ROTAT.	168.8	9401.2	-191.1

Figure 9: Comparison of 3D Elastic Vibration Characteristics

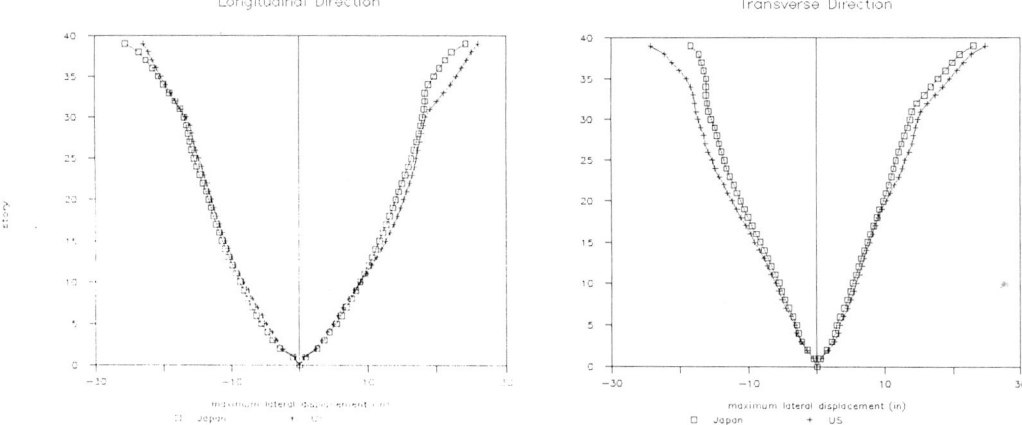

Figure 10: Comparison of Displacement Envelopes

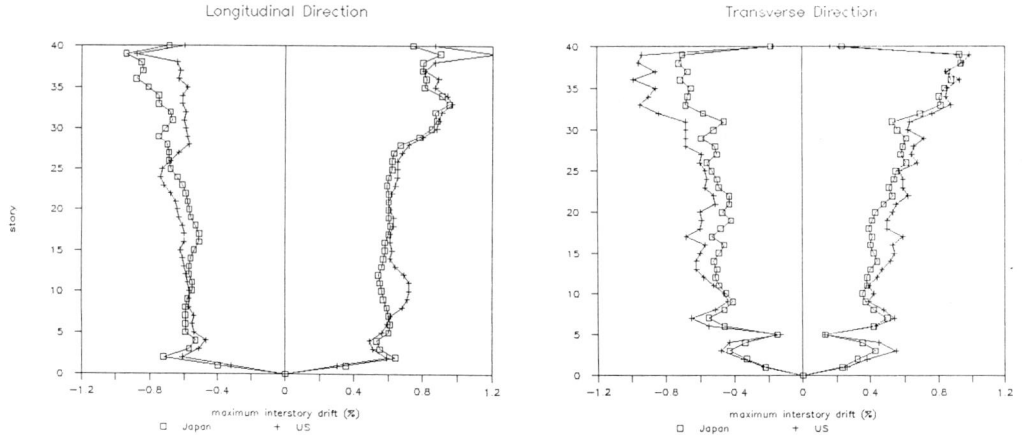

Figure 11: Comparison of Drift Envelopes

9

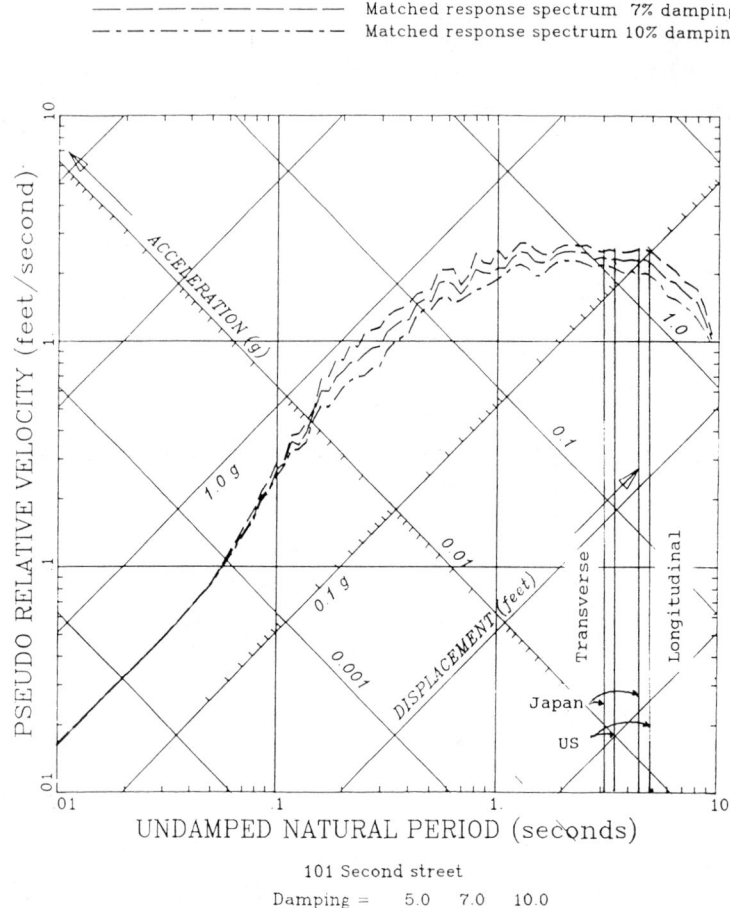

Figure 12: Response Spectrum for Ground Motion Derived from the Loma Prieta Earthquake

10

Proceedings of the Second Conference on Tall Buildings in Seismic Regions
55th Regional Conference
May 16 and 17, 1991, Los Angeles, California

SEISMIC RESPONSE ANALYSIS OF THE PACIFIC PARK PLAZA BUILDING

J. C. Anderson, V. V. Bertero and E. Miranda

INTRODUCTION

The Pacific Park Plaza Building at 6363 Christie Avenue in
Emeryville, California, is located on the east side of San
Francisco Bay between Oakland and Berkeley. The 30 story
apartment building was built in 1983 and has a plan shown in
Fig. 1 which consists of three equally spaced wings
extending outward a distance of 112 feet from a central
core. The foundation which is underlain by a layer of soft,
silty clay known locally as bay mud, consists of a five foot
thick base mat supported on nine hundred, 14 inch by 14 inch
prestressed concrete piles, 60 to 70 feet long. The lateral
force system consists entirely of ductile, moment resistant
reinforced concrete frames with the exception of two story
shear walls located at the base of each wing. At the time of
its construction, this building was the tallest reinforced
concrete structure in California. It remains th tallest
structure of its type in northern California but has been
eclipsed by a 32 story concrete ductile frame in Burbank,
California. This building was also the first structure on
the west coast to make selective use of high strength
concrete. Minimum concrete strength was specified as 5000
psi, however, concrete having a compressive strength of 6500
psi was used in the columns of the lower floors and in two
story levels at the twentieth floor level.

Forced and ambient vibration tests were conducted on the
structure in August of 1983 by researchers at the University
of California, Berkeley (1). This data forms a basis for
evaluating the mechanical characteristics of the structure
in the as built condition. In order to monitor the dynamic
response of the building under earthquake ground motions, it
was instrumented by the United States Geological Survey
(USGS) with 21 accelerometers which were deployed throughout
the builidng at the base, 13th floor, 21st floor and roof
levels (2) as shown in Fig. 1. In addition, free field
recording stations were located on the north and south sides
of the building.

Since construction, the building has experienced base
motions from several bay area earthquakes, however, by far
the most significant of these are the ones due to the Loma
Prieta Earthquake of October 17, 1989. This earthquake was
centered northeast of Santa Cruz, approximately 60 miles
southwest of the building site and had a surface wave

 (JCA)   Department of Civil Engineering, University of
           Southern California, Los Angeles, CA 90089
(VVB,EM) Department of Civil Engineering, University of
           California, Berkeley, CA. 94720

magnitude of 7.1. Major damage occurred within 3 miles of the building, including the collapse of the Cypress Street Viaduct. Peak accelerations recorded in the building were in the east-west direction and varied from 0.21g at the base to 0.48g at the roof. Inspection of the building following the earthquake showed no visible structural or nonstructural damage although the five level parking structure next to the building experienced flexural cracks in the floor system and shear failure of two columns in the first level.

ANALYTICAL MODELS

Detailed three dimensional finite element models of the structure were developed in order to evaluate the ability of current computer programs to reproduce the recorded response, given the input base motion. These models will also be used to evaluate the effect of the earthquake on the individual members of the structure and to evaluate the developed inertia forces relative to the specified building code requirements. In the current phase of the study, critical comparisons will be made between calculated accelerations and corresponding recorded values.

Several computer programs are available on a commercial basis which can be used to evaluate the elastic, dynamic response of the Pacific Park Plaza. The SAP90 (3) program was selected for use in this study for the following reasons:

1. An extended memory version, SAP90 Plus, is available which has extended memory capability and permits modeling the structure in detail for computation on a personal computer (IBM PS/2 with 8MB of memory and 120MB disk).

2. The program allows the user to plot the time history response of any node point. This feature was crucial for comparison of the calculated results with the recorded response.

3. The program permits consideration of a discontinuous, rigid floor diaphragm through the use of a master node-slave node option. This feature was essential for modeling the discontinuous floor diaphragm at the mezzanine level. It also greatly reduces the amount of modeling required to represent the in-plane effect of the floor diaphragm and also reduces the time required to perform a dynamic analysis.

4. The program has a postprocessor which can be used to perform a capacity check of the frame members. This will be used in later studies to evaluate the lateral force capacity of the structure and to compare it to code design provisions.

The completed SAP90 model of the structure consisting of 2,362 nodes, 5,700 elements and 6,816 degrees of freedom, is shown in an isometric view in Fig. 2.

MODELING CONSIDERATIONS FOR REINFORCED CONCRETE

Reinforced concrete is a nonhomogenous material which is normally placed monolithically. This results in the following modeling considerations which need to be considered when working with reinforced concrete structural systems:

1. <u>Finite Width Joints.</u> Due to monolithic construction and the overall size of the beam and column members, the clear span of the beams and columns can be reduced significantly thereby stiffening the structure. This condition is considered in the program by the inclusion of rigid offsets on the ends of the frame elements. There is no bending or shear deformations within the rigid offset which extends from the joint to the face of the support. It is possible that the use of rigid offsets which are equal to the full dimension of the beam-column intersection may stiffen the structure too much since deformations do occur in the joint region. This is accounted for in the program by the inclusion of a rigid joint reduction factor which reduces the length of the offset and thereby approximates the effect of the deformation that occurs in the joint region. Analyses based on the centerline to centerline dimensions are identified as having zero width joints.

2. <u>Effective Beam Section.</u> Initially, monolithic slab and beam construction results in a tee section for the beams with the flange having the slab thickness and extending a specified distance (nominally equal to eight times the slab thickness) on either side of the web. Under service loads, microcracking occurs in the concrete. This causes sections under negative moment to act as rectangular sections and sections under positive moment to continue to act as tee sections. Furthermore, as cracking occurs, the section properties are reduced from those of the gross section to those of the cracked, transformed section used in working stress analysis with the actual section property somewhere in between depending on the amount of cracking.

These modeling considerations are incorporated in six different SAP90 models of the Pacific Park Plaza Building which are identified in the following manner:

PPLAZA3 - Zero width joints, rectangular beams, gross
          section properties
PPLAZA4 - Finite width joints, rectangular beams, gross
          section properties
PPLAZA5 - Finite width joints, tee beams, gross section
          properties
PPLAZA6 - Finite width joints, tee beams, gross section
          properties, vertical foundation springs
PPLAZA7 - Finite width joints, average cracked transformed
          section properties considering tee section at
          center and rectangular section at supports
PPLAZA8 - Finite width joints with a 25% reduction factor,
          average cracked transformed section properties

MODAL ANALYSES

The results of modal analyses of the various building models
listed above are summarized in Table 1. Here it can be seen
that in the initial condition, model PPLAZA5 which uses
finite width joints and tee sections for the beams and
girders gives a very close approximation to the results
measured in the ambient vibration test. This might be
expected since the vertical loading is largely self weight
and the lateral deformations are quite small. The deformed
plan shape for the first three modes of this model are shown
in Figs. 3, 4 and 5. The translational mode in the north-
south direction is shown in Fig. 3. Here it can be seen that
this mode has some torsional deformation due to the
nonsymmetry due to the mezzanine slab in the west wing. The
translational mode in the east-west direction is shown in
Fig. 4. Here the structure is almost symmetrical about the
east-west axis and the displaced shape is almost pure
translation. The third mode, shown in Fig. 5 is a torsional
mode which is readily apparant from the displaced shape.
Models identified as PPLAZA4, PPLAZA6 and PPLAZA3 represent
modelling characteristics which result in increasing periods
of vibration, however, none of these gives a good estimate
of the recorded values. Models PPLAZA7 and PPLAZA8
incorporate estimates of the moment of inertia of the
cracked transformed sections. This was done by calculating
the moment of inertia of the cracked transformed section at
the two ends of the member and at midspan. The average value
used in the computations was obtained by averaging the two
end values (negative moment) and then averaging this value
with the midspan value (positive moment). As can be seen
from the table, this results in a structural model that is
still a little stiffer than indicated by the recorded data.
As a final adjustment, the rigid zones at the beam to column
intersections were reduced by 25% to account for some
flexibility in the joint region. The final model, PPLAZA8,
gives a close approximation to the recorded periods.

Table 1. Comparison of Computed and Recorded Modal Periods

Model	Mode 1 N-S	Mode 2 E-W	Mode 3 Torsion	Mode 4 N-S	Mode 5 E-W	Mode 6 Torsion
AMBIENT TEST	1.77	1.69	1.68	0.60	0.60	0.59
PPLAZA5	1.78	1.77	1.69	0.61	0.60	0.54
PPLAZA4	2.09	2.06	1.17	----	----	----
PPLAZA6	2.37	2.33	2.23	0.62	0.61	0.56
PPLAZA3	2.45	2.43	2.24	0.85	0.85	0.75
PPLAZA7	2.60	2.54	2.29	0.89	0.83	0.75
PPLAZA8	2.69	2.63	2.36	0.93	0.87	0.78
RECORDED	2.69	2.59	----	1.07	0.89	----

DYNAMIC RESPONSE ANALYSIS

Using the accelerations recorded on the ground floor in the
north wing as input, the time history response of the
PPLAZA8 model was evaluated. In this analysis, accelerations
recorded in the north-south and east-west directions were
applied simultaneously to the model and the dynamic response
calculated using the modal time history approach. Floor
response spectra were also generated from both the recorded
and calculated motions in order to better compare the
results. The damping in the structure was assumed to be 5%
in all modes. The roof spectra in the east-west direction is
shown in Fig. 6. Here it can be seen that there is a good
match between the periods of the recorded and calculated
values. It is also of interest to note that the peak
response occurs at a period of approximately 0.8 seconds
which is the second mode of vibration. The peak due to the
first mode occurs at about 2.6 seconds but is much smaller.
The corresponding acceleration time histories at the roof
level are shown in Fig. 7. Floor spectra at the 21st level
are shown in Fig. 8 and the corresponding time history
response is shown in Fig. 9. In both cases the comparison is
quite good. The floor spectra obtained at the 13th level is
shown in Fig. 10 and the corresponding time history response
is shown in Fig. 11. As in the previous cases the
camparisons are quite good.

CONCLUSIONS

The data presented represents a summary of the results obtained during the first phase of the continuing study of the Pacific Park Plaza Building. Future studies will consider earthquake demand versus structure capacity, critical comparisons with building code requirements and nonlinear response analyses. Based on the initial elastic dynamic response analyses, the following conclusions are made:

1. The initial dynamic characteristics of the building are modeled very accurately by considering the finite width of the beam to column joints and by using effective tee sections for the beams and girders.

2. Since the structure was built and occupied, it has experienced several earthquakes, the most significant of which was the Loma Prieta Earthquake of October, 1989. This has caused the fundamental period of the structure to increase by 52% with a resulting reduction in the initial stiffness of 128%. This behavior is represented by the changing of the fundamental period of the building from 1.78 seconds when built to 2.69 seconds currently. This condition can be approximated in the mathematical model of this structure by using an average stiffness of cracked, transformed sections for the beams and girders and allowing a 25% reduction in the beam to column joint stiffness.

3. The acceleration response at the top of the building is predominately due to the second mode of vibration. This type of dynamic behavior is due to a combination of the building height, lateral stiffness and characteristics of the base motion.

REFERENCES

1. Stephen, R.M., Wilson, E.L., and Stander, N., "Dynamic Properties of a Thirty-Story Condominium Tower Building," UCB/EERC-85/03, University of California, Berkeley, April, 1985.

2. Maley, R., et.al., " U.S. Geological Survey Strong-Motion Records from the Northern California (Loma Prieta) Earthquake of October 17, 1989," USGS Open-file Report 89-568, Oct. 1989.

3. Wilson, E.L. and Habibullah, A., "SAP90 User's Manual", Computers & Structures, Inc., Berkeley, CA., July, 1989.

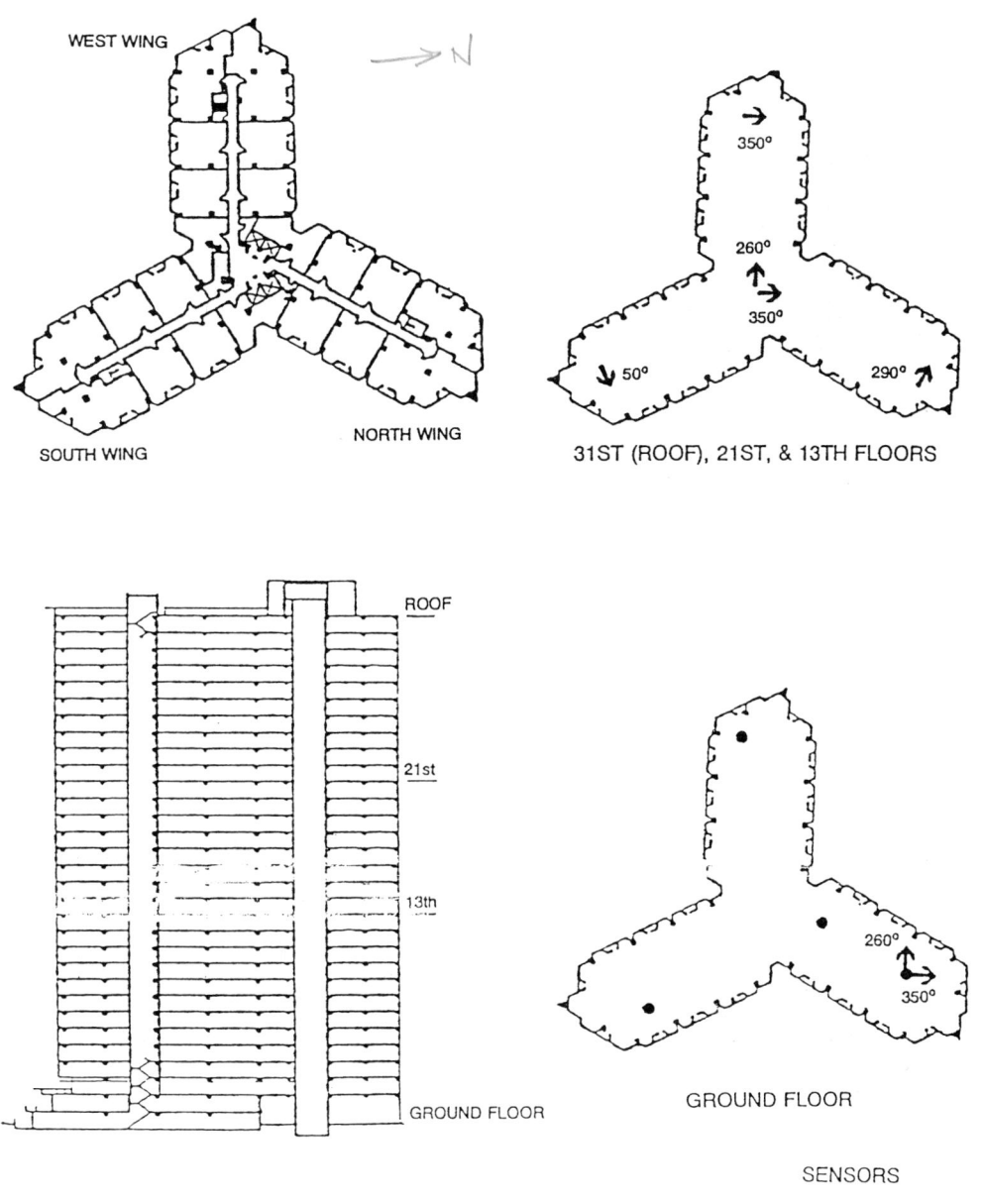

WEST WING

N

SOUTH WING

NORTH WING

31ST (ROOF), 21ST, & 13TH FLOORS

350°

260°

350°

50°

290°

ROOF

21st

13th

GROUND FLOOR

GROUND FLOOR

260°

350°

SENSORS

→ HORIZONTAL
● VERTICAL

Figure 1. Building Plan and Instrument Location

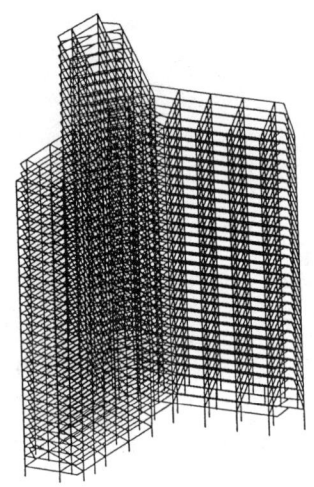

Figure 2. Computer Model

Figure 3. Mode 1, North-South

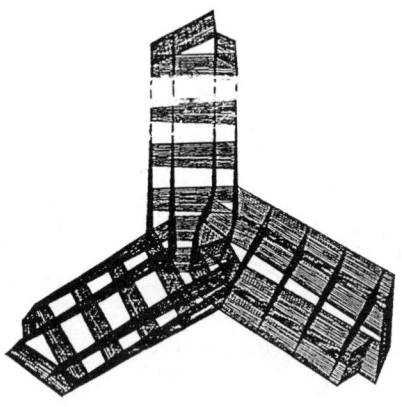

Figure 4 Mode 2, East-West

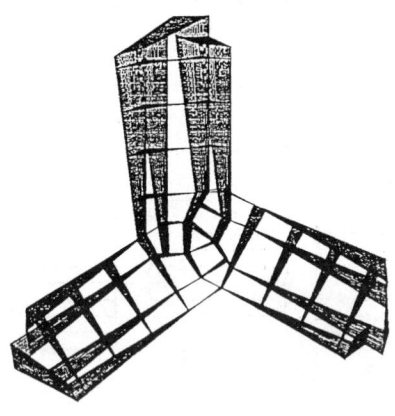

Figure 5. Mode 3, Torsion

PACIFIC PLAZA ROOF

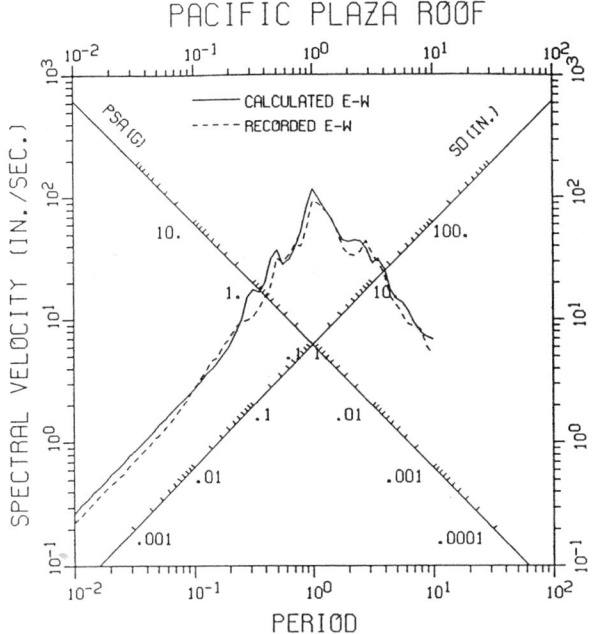

Figure 6. Roof Spectra

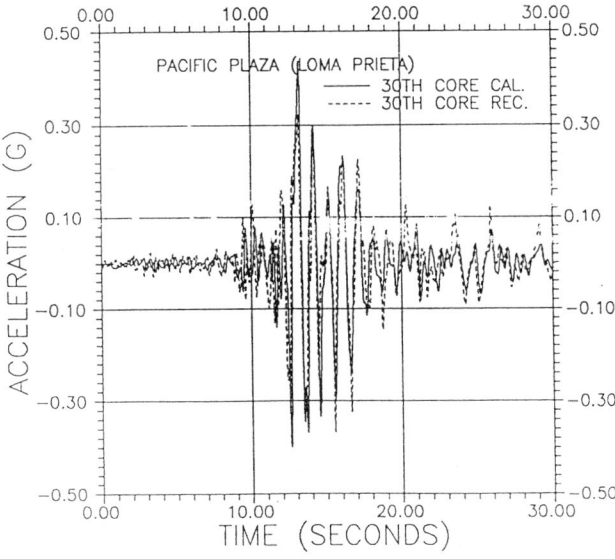

Figure 7. Roof Accelerations

19

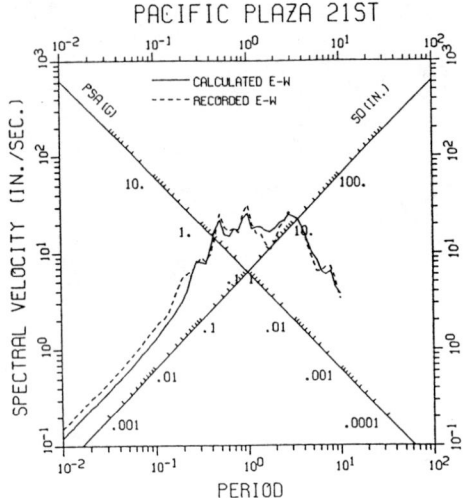

PACIFIC PLAZA 21ST

Figure 8. 21st Floor Spectra

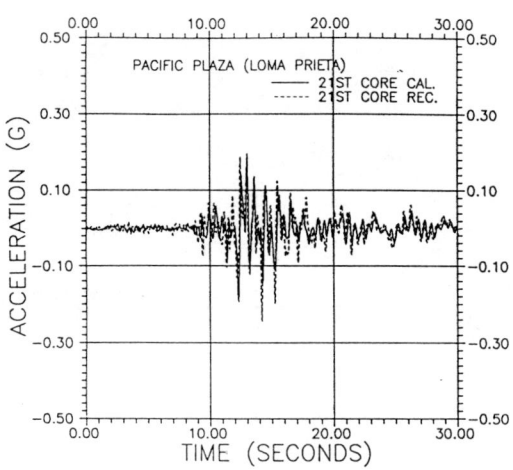

Figure 9. 21st Floor
Accelerations

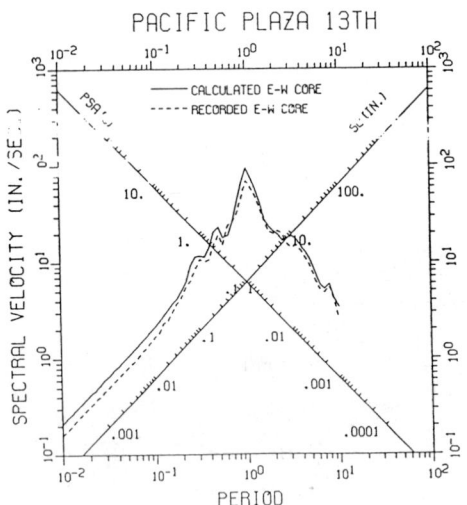

PACIFIC PLAZA 13TH

Figure 10. 13th Floor Spectra

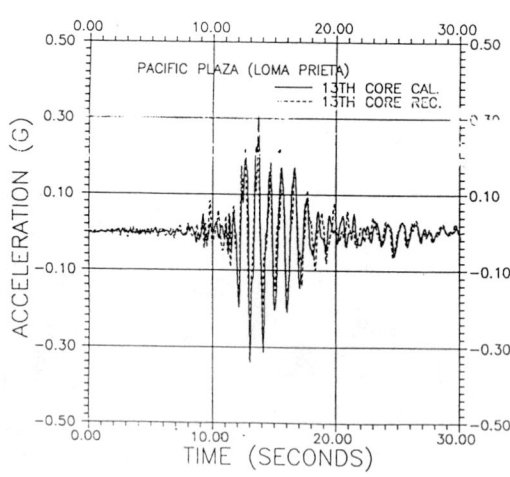

Figure 11. 13th Floor
Accelerations

20

Proceedings of the Second Conference on Tall Buildings in Seismic Regions
55th Regional Conference
May 16 and 17, 1991, Los Angeles, California

Cyclic Behavior and Design of Steel Bolted Top and Bottom Plate Moment Connections

By

Abolhassan Astaneh and James D. Harriott
Department of Civil Engineering and
The Earthquake Engineering Research Center
University of California at Berkeley

1. INTRODUCTION

Steel structures are divided into three types; rigid or fully restrained (FR), semi-rigid or partially restrained (PR) and simple. Current seismic design practice and seismic codes such as Uniform Building Code[7] recognize moment frames with rigid connections as well as braced frames with simple connections. The codes have sufficient information regarding loading and resistance of these well known steel structural systems. However, information on seimic design of steel structures with semi-rigid connections is very limited.

The objective of this study was to investigate cyclic behavior of three types of steel beam-to-column connections under severe seismic loads. By using the information collected during experiments, seismic design procedures for these connections are developed and proposed.

2. EXPERIMENTAL RESEARCH

Details of beam-to-column connection assemblages that were tested are shown in Figure 1. Three full size specimens were fabricated and tested at the University of California Berkeley under cyclic loading conditions that represented the effects of strong earthquakes. Each specimen consisted of a 7 feet long W18x50 beam connected to a 3 feet long column by top and bottom flange plates and a shear connections. In all specimens the top and bottom plates were the same. The only difference among specimens was the mechanism of shear transfer.

In Specimen A, web connection was a structural tee. The flange of the tee was welded to the flange of the column by two vertical fillet weld line. The stem of the tee was bolted to the beam web by five high strength bolts as shown in Figure 1(a).

Specimen B did not have web connection. To transfer shear from beam to column, in this connection, a vertical stiffener was used under bottom flange. The stiffener was welded to column flange as well as to the bottom plate, Figure 1(b).

Specimen C shown in Figure 1(c) had a single plate

shear connection. The plate was welded to the column flange and bolted to the beam web by five bolts. In recent years, considerable research is done on single tee and single plate shear connections subjected to static loading and new design procedures have been developed[1,2]

The loading sequence that was used in the tests is shown in Figure 2. An attempt is made to ensure that this cyclic loading is similar to cyclic loading that was used in 1970"s by Professor Popov and his research associates to study cyclic behavior of fully welded moment connections[6]

Figure 3 shows a sketch of the test set-up used in these studies. The set-up has been used at the University of California at Berkeley earlier in studies of steel moment connections[5,6] In recent years, the set-up is modernized by adding state of the art electronic data acquisition and processing systems.

Test specimens were placed in the set-up and were subjected to cyclic moment-rotation applications following the sequence shown in Figure 2.

3. EXPERIMENTAL RESULTS

Test results that were collected were in the form of observations made on the behavior of each specimen as well as quantitative data. Complete results of this study can be found in Reference 3.

3.1. Behavior of Specimen A with Tee Web Connection

The moment versus rotation hysteresis loops for this specimen are given in Figure 4. Up to four completed cycles of load the connection appeared to be completely elastic. There were no signs of slip, either visible or audible. However, the connection was not as stiff rotationally as a perfectly rigid model would predict. The familiar flexibility coefficient of $L^3/3EI$ would predict 31.92 kips force per a tip deflection of 0.1372 inch, while the actual load was only 15.7 kips. One reason for the flexibility of connection was that even though slip was not visible, the processed data indicated that small amounts of slip were occurring. Also, due to flexibility of web tee connection, not all the beam web was participating in bending. In other words, near the column face, within the connection, the assumption of plane sections remaining plane was not true.

During sixth cycle, first inelastic deformations were observed. Some local yielding was evident in the flange plates between the column face and the first row of bolts. As Figure 5 shows, the yield lines appeared in 45 degree patterns around each bolt hole. Some opening of tee was evident. At moment of 1098 k-in and a rotation of 0.002 radians the opening was 1/16 inch. The first significant slip occurred with a loud bang at a moment of 2246 k-in and rotation of 0.0034 radians.

During 14th cycle massive yielding occurred in the top plate at a moment of 5411 k-in and a rotation of 0.0354 radians. The tee opening was about 1/4 inch. During the 15th cycle massive tension yielding occurred in the bottom plate while the top plate experienced minor local buckling. Failure occurred at the maximum moment of about 5500 k-in and a maximum rotation of about .05 radians. The failure occurred due to fracture of net section of plate. Before fracture plate showed considerable necking through the net section.

3.2. Behavior of Specimen B with Stiffened Seat Connection

This specimen did not have any web connection. However, a 3/4 inch thick triangular plate was welded to the column flange and to the bottom flange plate much as in a stiffened seat connection as shown in Figure 1(b). The moment versus rotation hysteresis loops of Specimen B are shown in Figure 6. Up to five completed cycles of load the connection appeared to be completely elastic with no signs of visible or audible slip. As in Test A, this specimen also was more flexible than a theoretical fixed connection. During the test, there were some signs of stiffener plate buckling out of plane when it was under compression and becoming straight as load reversed and tension force was developed in the stiffener. As loading continued, the processed data indicated that small amount of slippage was taking place during these initial cycles.

During ninth cycle at a moment of 3489 k-in and a rotation of 0.0026 radians, a shock (large enough to knock off dial gages) accompanied by a bang went through the specimen probably due to bolt slip. The measurements indicated that slippage has occurred eight times during previous cycles and there was no significant slip of top plate that might have caused the big shock. However, the measurements on the bottom plate indicated a major slip in this plate. During these cycles, gradual yielding was developing in the vicinity of bolt holes of first net section of plates.

During the eleventh cycle, a crack in the nominally bottom plate was detected at the base of the fillet weld joining the stiffener to the plate. During the twelfth cycle the crack widened to approximately 0.04 inch but did not propagate. During thirteenth cycle there were large shocks to the specimen from bolt slip in the positive loading. During fourteenth cycle yielding propagated around the bolt holes of the bottom plate.

During fifteenth cycle failure occurred at a moment of 5800 k-in and a rotation of about 0.045 radians. The failure was due to fracture across the first net section of the bottom flange plate that had stiffener plate. Figure 7 shows this specimen after failure.

3.3. Behavior of Specimen C with Single Plate Web Connection

Figure 8 shows hysteresis loops for this specimen. Up to six completed cycles of load the connection appeared to be elastic. There were no signs of slip, visible audible or from measurements. Much like Tests A and B, the connection in Test C was not as stiff as a perfectly fixed model would predict. Again, the plane sections-remain-plane hypothesis was likely invalid at the column face and adjacent areas.

During the positive loading portion of the seventh cycle a slip occurred between the nominally bottom plate and shim. This produced a 1/16 inch gap in the whitewash. During the eighth cycle again a 1/16 inch slip occurred. During ninth cycle a slight amount of yielding became apparent in the bottom plate.

During tenth, eleventh and twelfth cycles, the behavior was more or less similar to the behavior during ninth cycle. During cycles 13, 14 and 15 more yielding occurred near the bolt holes of first net section of plates. Failure occurred during positive loading of fifteenth cycle by tensile fracture of net section of nominally top plate. The moment at the failure was about 6600 k-in while rotation was about 0.047 radian. Figure 9 shows Specimen C after failure.

4. DISCUSSION OF RESULTS AND SEISMIC DESIGN RECOMMENDATIONS

The experimental studies summarized here indicated that the connections that were tested should be considered semi-rigid for large seismic effects. The main source of semi-rigidity was gradual and sometimes sudden slips in the top and bottom plates. The hysteresis behavior of connections as shown in Figures 4,6 and 8, could be modeled realistically by considering a bi-linear moment-rotation model.

The most important finding of the study was that these connections were very ductile and could survive at least 14 large inelastic cycles of rotation. Based on the findings of this study as well as other information available in the literature on the subject, seismic design procedures were developed and proposed for the type of connections that were studied. The main consideration in developing these design procedures was to ensure that connection will have sufficient ductility to survive a strong earthquake with no fracture and no local buckling. In order to achieve this objective, possible limit states of failure of connection components were identified. Then, these limit states were prioritized such that ductile limit states such as yielding of plates or flanges occur before fracture of the net sections and fracture of net sections occur before fracture of the welds or bolts. After prioritizing limit states, appropriate equations and concepts were developed to enable designers to address each limit state. The ultimate goal of proposed design procedures was to ensure that behavior of connections will be as ductile and as desirable as what was observed during testing. The proposed design procedures are outlined in Reference 3 in details. The failure modes are defined and appropriate equations are proposed to enable the designer to consider each failure mode properly.

5. CONCLUSIONS

Based on the data obtained from the tests as well as observations during the tests, the following conclusions and commentary were formulated. More detailed conclusions can be found in Reference 3.

a. Connections behaved in a very ductile manner without any significant local buckling until at least 14 cycles of severe inelastic rotations were applied. The connections tolerated easily cyclic rotations in the excess of .02 radian. This rotation roughly corresponds to a drift index of about 2% in the building.

b. No fracture of Net section occurred until rotations reached at least 0.03 radians.

c. Generally, cyclic behavior of these connection was better than commonly used fully welded moment connections. The fully welded connections usually experience premature failure of locally buckled area of the beam flange near the welds or fracture of the heat affected zone when subjected to large seismic effects.

d. Connections showed flexibility whenever they passed through slip load level. It does not appear that this flexibility which made the connection semi-rigid, necessarily is harmful to seismic performance of the structures. In fact, based on research data on several other projects conducted to study seismic behavior of semi-rigid frames, it is shown that in some cases the flexibility of semi-rigid connections can be beneficial in reducing inertia forces without significant increase in drift[4].

6. REFERENCES

1. Astaneh, A., S.M. Call and K. M. McMullin, "Design of Single Plate Shear Connections," Engineering Journal, AISC, 1st. Quarter, 1989.

2. Astaneh, A. and M. N. Nader,"Design of Tee Framing Shear Connections," Engineering Journal, AISC, 1st. Quarter, 1989.

3. Harriott, J.D. and A. Astaneh-Asl, "Cyclic Behavior of Steel Top-and-Bottom-Plate Moment Connections", EERC Report 90-19Earthquake Engineering Research Center, University of California at Berkeley, August 1990.

4. Nader M.N. and A. Astaneh-Asl, "Experimental Studies of a Single Story Steel Structure with Fixed, Semi-rigid and Flexible Connections", EERC Report 90-19Earthquake Engineering Research Center, University of California at Berkeley, August 1990.

5. Popov, E.P. and R.B. Pinkney, "Cyclic Yield Reversal in Steel Building Connections," _J. of the Structural Division,_, ASCE, ST3, March 1969.

6. Popov, E.P. and R. Stephen, "Cyclic Loading of Full-Size Steel Connections,"_EERC Report 70-3,_ Earthquake Engineering Research Center, University of California at Berkeley, June 1970.

7. "_Uniform Building Code_", International Conference of Building Officials, Whittier, 1988.

7. ACKNOWLEDGMENTS

The project was sponsored in part by the A. C. Martin and Associates of Los Angeles. The support of Nabih Youssef of Nabih Youssef and Associates and the assistance of Roy Stephen, laboratory manager, throughout this project is sincerely appreciated.

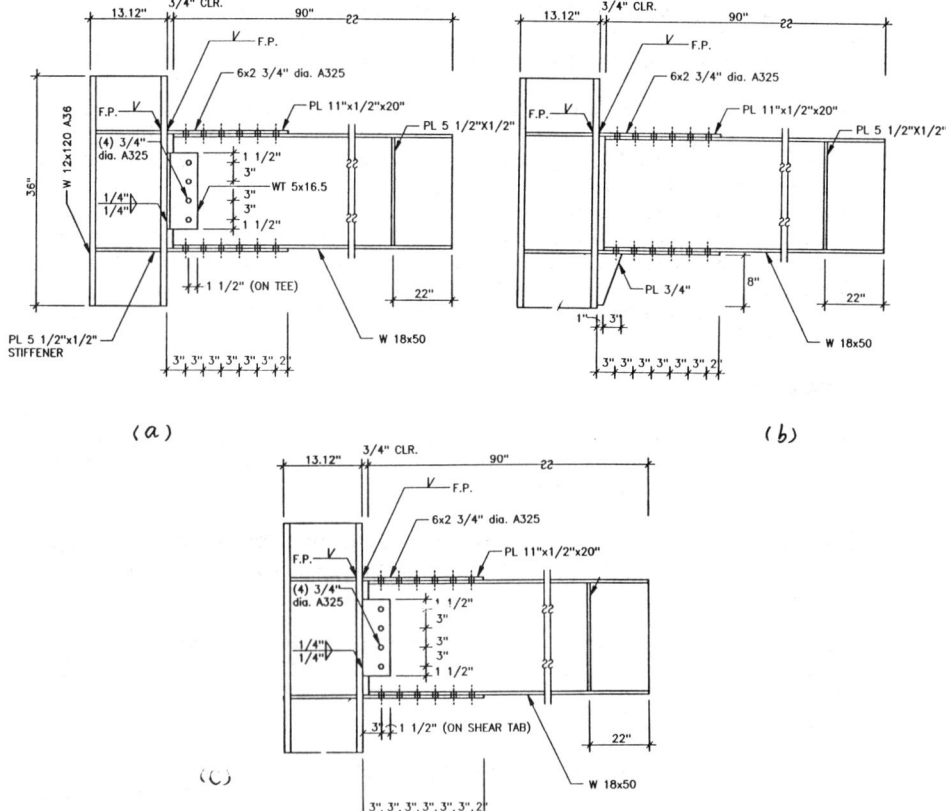

Figure 1. Details of Test Specimens

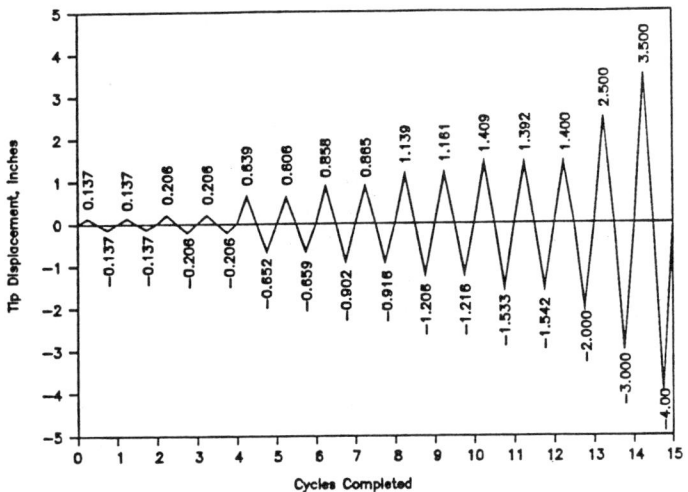

Figure 2. Cyclic Displacement History

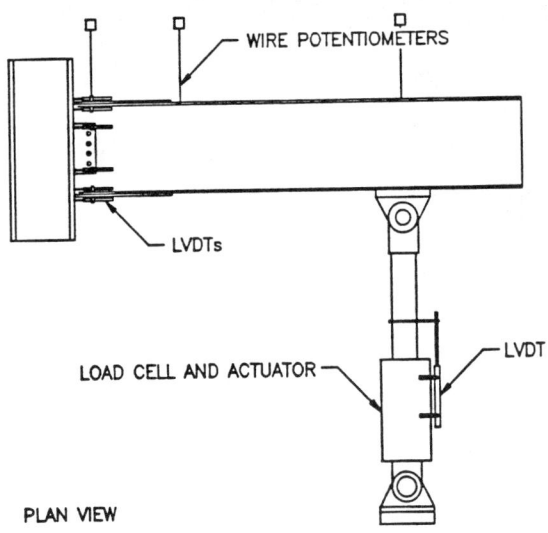

Figure 3. Test Set-up

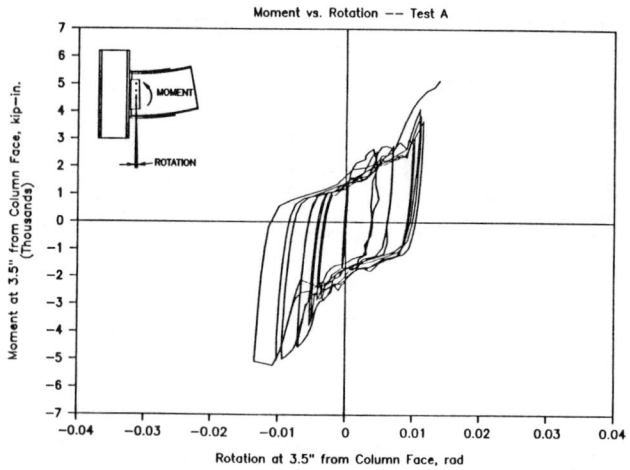

Figure 4. Moment-Rotation Hysteresis   Loops for Specimen A

Figure 5. Failure of Specimen A

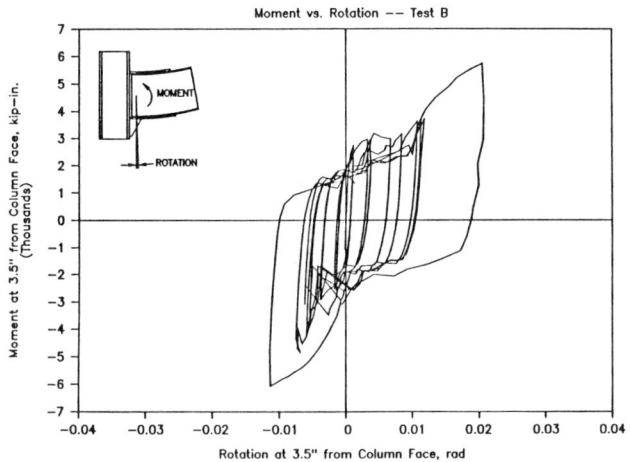

Figure 6. Moment-Rotation Hysteresis Loops for Specimen B

Figure 7. Failure of Specimen B

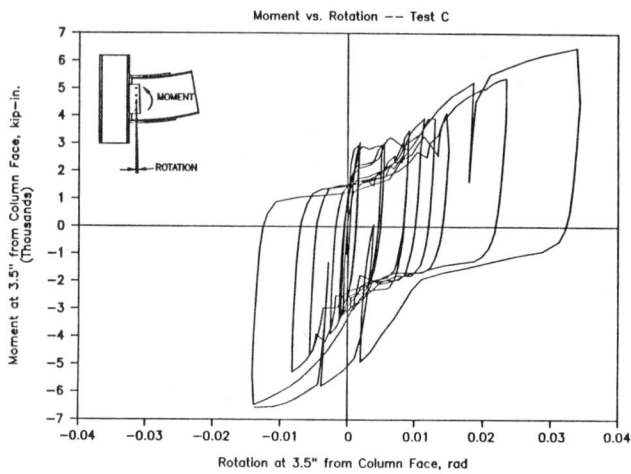

Figure 8. Moment-Rotation Hysteresis   Loops for Specimen C

Figure 9. Failure of Specimen C

Proceedings of the Second Conference on Tall Buildings in Seismic Regions
55th Regional Conference
May 16 and 17, 1991, Los Angeles, California

Studies of a 49-story Instrumented Steel Structure Shaken During the Loma Prieta Earthquake

By

Abolhassan Astaneh, Cheng-Chi Chen, and David Bonowitz
Department of Civil Engineering and
Earthquake Engineering Research Center
University of California at Berkeley

1. INTRODUCTION

The magnitude 7.1 Loma Prieta earthquake of October 17, 1989 has created valuable data on actual seismic performance of many major structures designed and constructed according to current structural engineering and design technology and practices. The California Strong Motion Instrumentation Program (SMIP) of Division of Mines and Geology of California had many strong motion recording stations throughout the greater San Francisco Bay area and was able to collect vast amount of data on ground motion as well as response of structures. One of the major structures that has been instrumented by the SMIP and was shaken during the Loma Prieta earthquake is a 49 story steel structure in san Francisco shown in Figure 1. The strong motion instruments performed well during the earthquake and more than 120 seconds of structural response were recorded. This paper discusses the instrumentation, presents the recorded response and provides general comments on the performance of the structure.

2. RESEARCH PROJECT

Currently, the authors are conducting a study of seismic performance of this building and the complete results will be published in the future. The study has various phases as follows:

1. Collect data with regard to geometry, material properties, non-structural elements, equipment, dead and live load at the time of earthquake as well as during regular service condition. The information is used in building realistic computer models of the structural system and its components.

2. Collect data with regard to any damage that might have occurred during the earthquake. At the time of this writing, no damage is reported. The damage information, if any, could be used to calibrate the computer model as well as to evaluate the results of the analyses.

3. Obtain strong motion records from SMIP and process the data which included using a special software[1].

4. Analyze the strong motion data processed in step 3 above.

5. Construct a realistic computer model of the structural system. Major components of structure such as beams, columns, foundations, eccentric bracings, floor diaphragms and connections, particularly panel zones, need to be modeled carefully and realistically.

6. The computer model of the structure is being subjected to the base excitations recorded in the building and the response of the structure to the Loma Prieta is being studied. The intent of this phase is to calibrate the computer model and to study the effect of current modeling methods on the actual response.
   In this phase of analysis, the structural analysis software ETABS-90PLUS[3] is being used. The program is currently used in the elastic, 3-dimensional analyses of the building structures by many structural engineers. One of the reasons in using this program is to investigate the power and short-comings of the current state of the art software for seismic analysis.

7. Concurrent with Step 6 above, to study non-linear response, computer models of the E-W and N-S frames are being built and will be subjected to ever-increasing levels of accelerations recorded during the Loma Prieta as well as other ground motions that can be established for the site of the building.

8. Using information obtained in above phases, study the code provisions used in design of the building[7] and investigate the adequacy and accuracy of the provisions in establishing the demand and supply side of general equation of design. The study includes force as well as displacement considerations to obtain an understanding of the damage potential of future earthquakes. Accordingly, design oriented recommendations will be formulated and proposed.
   In this phase, also recommendations will be made to SMIP with regard to refinement of existing instrumentation based on the findings of the research.

3. THE BUILDING STRUCTURE AND INSTRUMENTATION

The building is located in downtown San Francisco. It was designed[4] according to UBC-76 and its construction was completed in 1979. Figure 2 shows typical framing plan and the elevations. The floors consist of 2.5 inch concrete over a 3 inch deep metal deck supported by steel beams and girders. Shear studs are used to attach the floors to steel framing. The structural system consists of moment resisting frames in both E-W and N-S direction. However, to increase lateral stiffness in N-S direction which is the narrow direction of the building, four bays along axes 4, 5, 6 and 7 are braced using eccentric braces. The super-structure is supported by a 5 feet thick reinforced concrete mat foundation which in turn is supported on 150 to 200 feet deep composite steel/concrete piles.

The strong motion instrumentation of the building consists of 18 channels of accelerographs installed at various levels and directions as shown in Figure 3.

4. PRELIMINARY RESULTS OF THE STUDIES

The data that was recovered by SMIP[6] were processed and currently are being analyzed. The analysis so far has revealed certain dynamic characteristics of the building as discussed in the following.

4.1. Accelerations

Figure 4 shows plots of E-W and N-S component of acceleration time histories recorded at the 44th floor and at the basement B level. The maximum peak horizontal accelerations at the basement level were 0.159g and 0.101g in the E-W and N-S directions respectively. The corresponding values for the 44th floor were 0.403g and 0.478g respectively. The ratios of maximum peak acceleration of 44th floor to basement B were 2.53 in E-W direction and 4.73 in the N-S direction.

4.2. Periods

The preliminary analysis of the strong motion data revealed the approximate values of periods of vibration as given in Table 1. The values of periods calculated by using several methods used in design offices are also given in Table 1.

4.3. Damping Ratio

The preliminary analysis of the recorded data indicates that the damping during the earthquake was about 1.7 to 2.0 percent of critical damping.

4.4. Deflected Shape of the Structure

Figure 5 shows displacement time histories for 44th floor (one floor below the roof) in N-S and E-W direction. As figure indicates the displacement at the east and west end of the building plotted on the top plot were very close indicating small and negligible torsional effects. The displacement histories shown in Figure 5 have two distinct regions: (a) the region from start of the record to about 30 seconds while earthquake was present and; (b) the displacement history recorded after 30 seconds. During the initial phase, in the N-S direction, a period of vibration

of about 1.8 seconds is superposed on the fundamental mode
of about 5.0 seconds. The 1.8 second period represents the
vibration of top 6 unbraced floors in the N-S direction.

Figure 6 shows deflected shape of the structure in N-S
and E-W direction in an animated Plot. Figure 6(a) shows
the deflected shape during the time interval from 13th
second to 23th seconds of the record. Notice that during
this time interval ground motion was present. As Figure 6(a)
indicates, the N-S direction, the upper part of the
structure was experiencing large dispacements relative to
the lower floors. Notice in Figure 2 that the upper ④ ⑥
stories were smaller in the area than the other floors and
in addition, upper ④ floors did not have bracing. In the E-W
direction, apparently the vibration was initially dominated
by second mode but gradually the first mode became the most
dominant mode of vibration.

Figure 6(b) shows the deflected shape of the structure
in N-S and E-W direction during a time interval of from 39.5
to 42 seconds. at this time, the earthquake had stopped and
the vibration is mostly free vibration of the structure.
Notice that the structure during this time interval was
vibrating in its first mode in both N-S and E-W directions.

Figure 7 shows a plan view of displacement of a point
on 44th and 39th floors. Figure 7(a) is plotted using data
recorded during time interval of 13 to 23 second. During
this time the earthquake was ongoing and the motion is
relatively chaotic with several lower modes being present.
on the other hand, Figure 7(b) shows same particle motions
during a later time interval during 30th through 80th
seconds of the motion. The study of plots indicated that in
this case the motion is quite regular. The floors move for
several cycles in a diagonal N-E to S-W direction. However,
gradually, the motion shift almost 90 degrees to other
diagonal direction in direction of the N-W and S-E. This
behavior was very consistent after earthquake had stopped.
The exact cause and ramifications of this behavior is
currently under study.

4.5. Drift Ratios

Figure 8 shows average drift ratios for the N-S
direction and Figure 9 shows drift ratio time history for E-
W direction. As Figure 8 indicates, relatively large drift
ratios were developed in N-S direction between 39th and 44th
floors with a maximum of about 0.0055. These floors were
smaller in plan and did not have eccentric bracing while the
lower floors were substantially larger and had eccentric
braces. It appears that due to sudden change of stiffness
and other dynamic characteristics of the building above the
39th floor, large drift values are developed. Unlike the
top floors, the average drift ratios for floors between 16th
and 39th was very small with a maximum of less than 0.0018.

As Figure 9 indicates, in the E-W direction the
variation of drift ratio was much smoother than the N-S
direction.

5. SUMMARY

Since at the time of this writing, the research is still ongoing, final conclusions cannot be formulates. However, from analyses that are done so far the following observations can be made.

1. The recorded data provide very valuable information on actual seismic behavior of this modern steel high rise.

2. The analysis of the recorded data indicated that addition of eccentric braces in narrow direction of the structure has resulted in reducing drift and shortening the fundamental period of vibration.

3. The irregularity that was created by sudden decrease of plan area and discontinuity of the bracings above the 38th floor resulted in significant change and increase in displacement response above the 38th floor. Further focus on this aspect of the behavior is expected to shed more light on the effect of irregularities on overall seismic response.

4. The recorded data indicates that the structure performed satisfactorily during the Loma Prieta earthquake.

6. REFERENCES

1. Boroschek, R. "ANIM-2D", Software, Department of Civil Engineering, University of California at Berkeley, 1989.

2. "CSMIP Strong Motion Records from the Santa Cruz Mountains (Loma Prieta) California Earthquake of October 17, 1989), California Division of Mines, 1989

3. "ETABS", Software and Manual, Structures & Computers Inc., Berkeley, California, 1990.

4. Merovich, A.T., J. P. Nicoletti and E. Hartle, "Eccentric Bracing in Tall Buildings", J. of Structural Division, ASCE, Vol. 108, ST9, September,1982.

5. Shen, J-H. and A. Astaneh, "Seismic Response Evaluation of an Instrumented Six Story Steel Building", Research Report,UCB/EERC-90/20, Earthquake Engineering Research Center, University of California at Berkeley, 1990

6. "Plots of the Processed Data for San Francisco 47-Story Office Building from the Santa Cruz Mountains (Loma Prieta) Earthquake of October 17, 1989", CSMIP, 1990.

7. "Uniform Building Code", 1976 & 1988 Editions, International Conference of Building Officials, Whittier, CA. 1976.

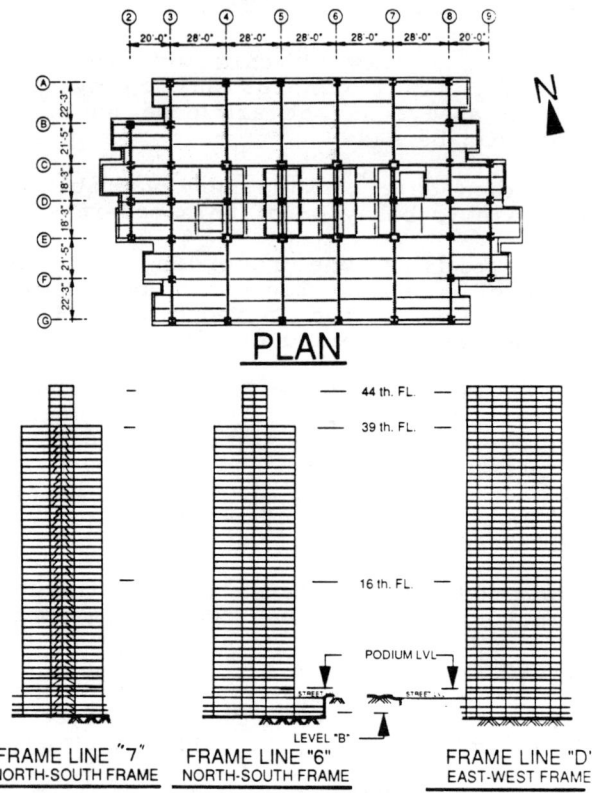

Figure 2. Typical Framing Plan, East-West and North-South Frames

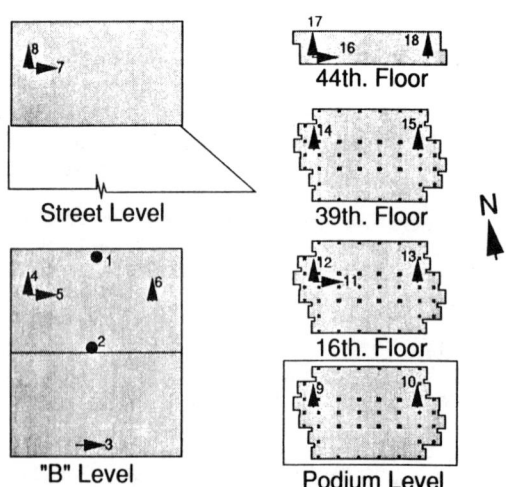

Figure 3. Plan of SMIP Instrumentation

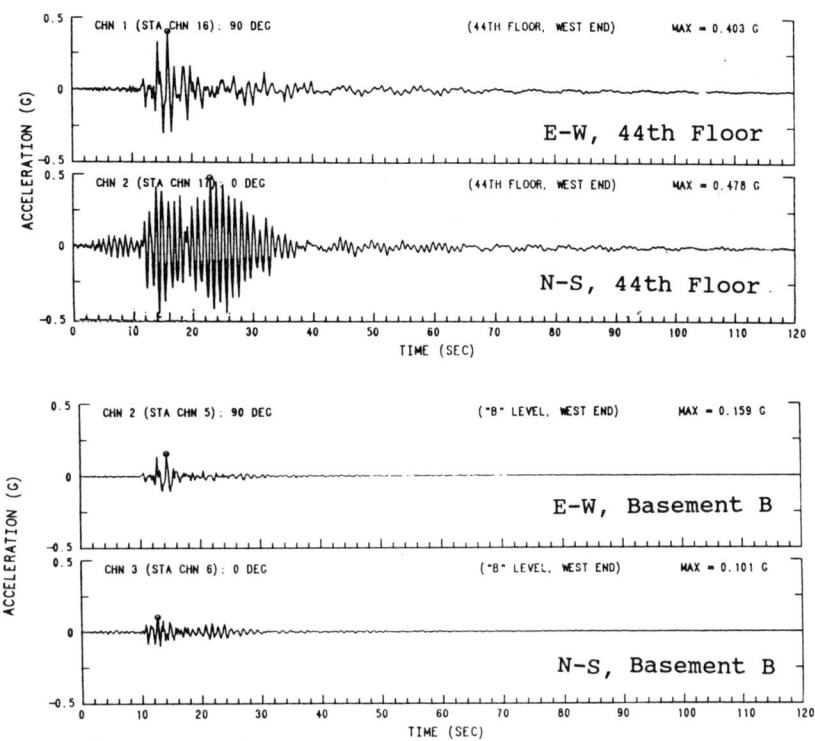

Figure 4. Accelerates Recorded at 44th Floor and Basement

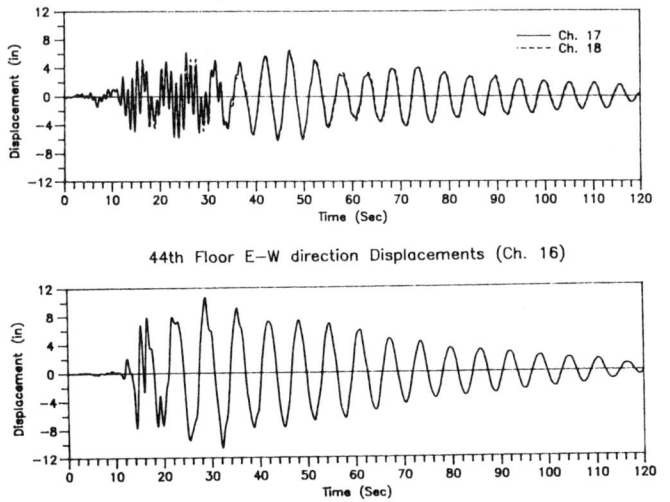

Figure 5. Displacement Time Histories of 44th Floor

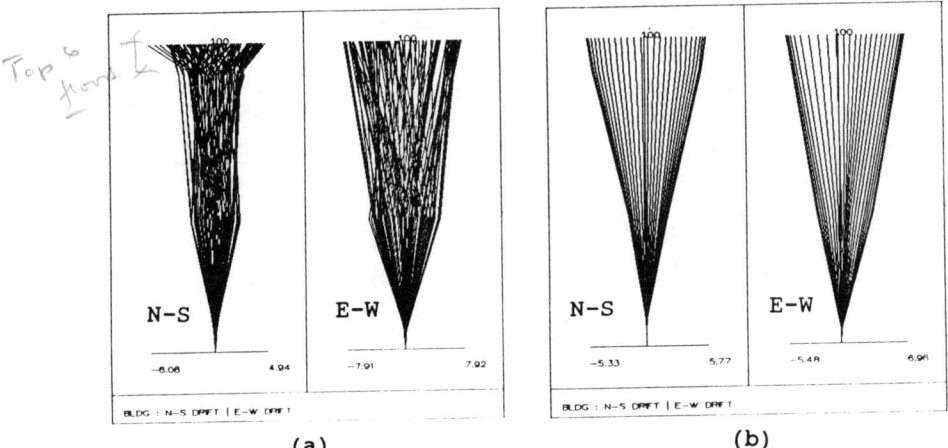

Figure 6. Animated Deflected Shape of the Structure
      (a) During 13 through 23 Seconds of Response
      (b) During 23 through 39.5 Seconds of Response

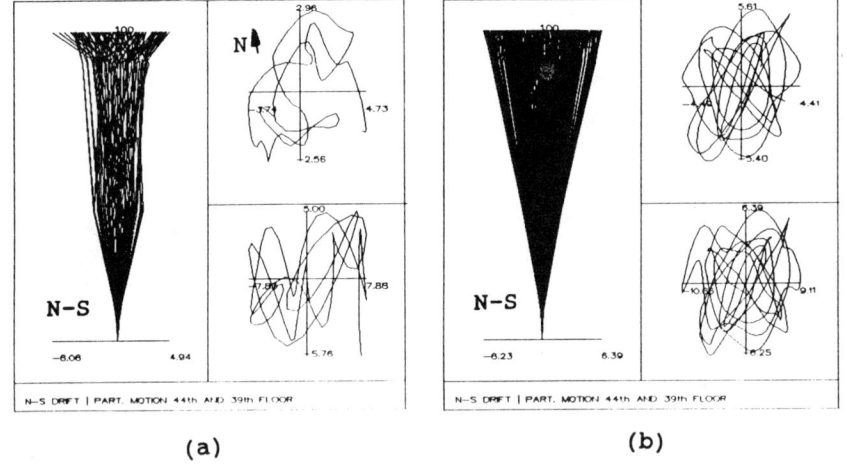

Figure 7. Plan View Plots of Movements of Floors

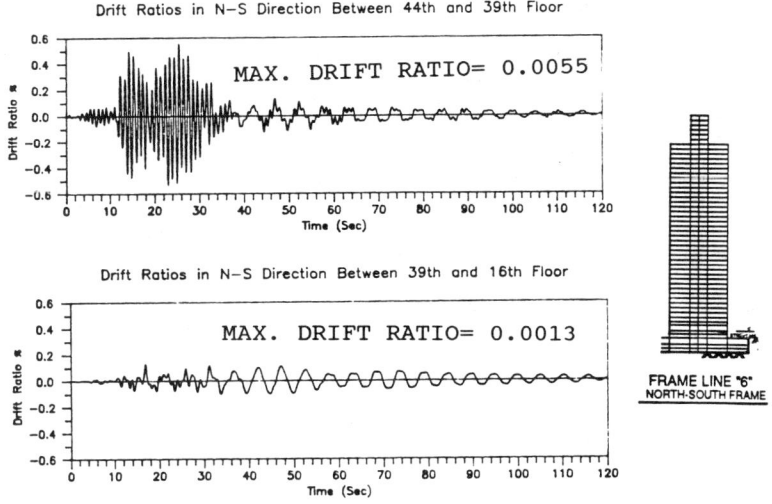

Figure 8. Average Drift Ratios for the N-S Direction

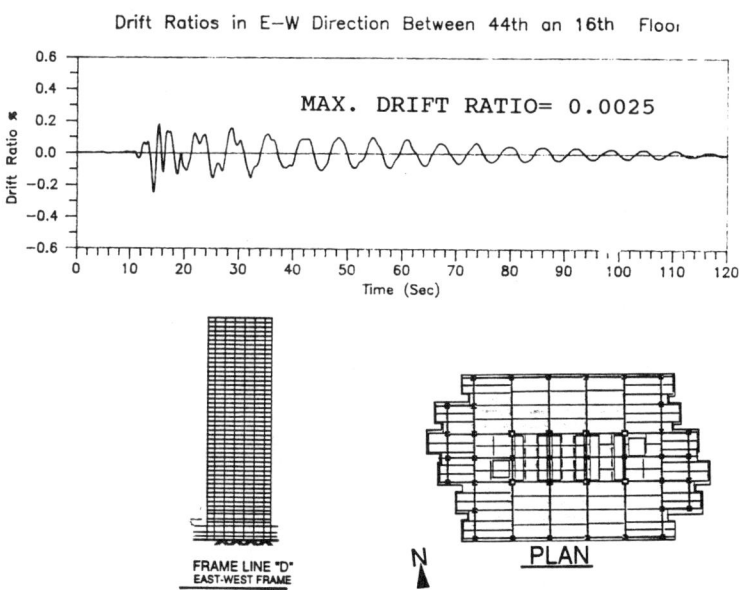

Figure 9. Average Drift Ratios for the E-W Direction

7. ACKNOWLEDGMENTS

The project is sponsored by the Strong Motion Instrumentation Program (SMIP) of the Division of Mines and Geology of the California Department of Conservation. The efforts of SMIP in collecting and partially processing the data, particularly, the efforts of A. Shakal and M. Huang are acknowledged. The authors are thankful to Structures & Computers Inc for generously donating a copy of ETABS program to the project. The authors also thank R. Boroschek for the use of his program ANIM-2D[1]

TABLE 1. Periods Extracted from Strong Motion Records and Calculated by Current Practice

Mode of Vibration (1)	Period in Seconds Based on:			
	SMIP Data (2)	UBC-76 [a] (3)	UBC-88 [b] (4)	N/10 (5)
First Mode (E-W)	6.5	4.7	4.1	4.7
Second Mode (E-W)	2.0	4.7/3 [c]	4.1/3	4.7/3
Third Mode (E-W)	1.3	4.7/5 [c]	4.1/5	4.7/5
First Mode (N-S)	5.0	4.0	3.5	4.7
Second Mode (N-S)	1.8	4.0/3	3.5/3	4.7/3
Third Mode (N-S)	1.0	4.0/5	3.5/5	4.7/5

a: Uniform Building Code, 1976 was used in original design.
b: Uniform Building Code, 1988 is the current code.
c: The second and third modes are approximately calculated as $T_1/3$ and $T_1/5$ respectively.

Figure 1. Downtown San Francisco and Building Under Study

Proceedings of the Second Conference on Tall Buildings in Seismic Regions
55th Regional Conference
May 16 and 17, 1991, Los Angeles, California

Earthquake-and Wind-resistant Design
of the Tokyo Marine Building

Shigeru Ban[1], Seiichi Muramatsu[2]

INTRODUCTION

The Tokyo Marine Building is a 27-story office building located in the Osaka Business Park district, being developed just east of Osaka Castle, Japan. It rises to a height of about 390 feet above ground. The building provides about 740,000 square feet area for office space, retail space and parking space. The building was designed and constructed by Kajima Corporation. The construction work was completed in Dec.1990.

Architecturally, the building was conceived to fit into the environment of the Osaka Business Park, and reflects the image of the client, Tokyo Marine. As the base of operations in western Japan for Tokyo Marine, the building was designed to have high-tech capabilities, to appropriately reflect its prestige by its external appearance, and also to be attractive to tenants as a prestigious office space.

The building has a rectangular plan to fit into the rectangular site surrounded by two high-rise building on both longer sides. The exterior facade exposing columns and beams outside the building brings to mind the simple lines of traditional Japanese wood-frame details and gives a clear identity as well. A picture of the building is shown in Figure 1.

The lateral force resisting system of the building consists of the framed columns interconnected with long span beams(10.8m,21m). They allow for open public area at the plaza level of the building and column free space across the 21m. width, with a 2.7m ceiling height on the office floors.

This paper describes the characteristics of this system, earthquake resistant design, wind vibration design and the fire protection of the external steel work.

[1] Chief Engineer,Structural Engineering Department,Kajima Corporation,Tokyo,Japan
[2] Senior Engineer,Structural Engineering Department,Kajima Corporation,Tokyo,Japan

STRUCTURAL FRAMING

The building is framed in structural steel, which is moment-resisting frame of columns(500mmx500mm Box shape) and beams(900mm depth H-shape). The framed columns are split into a frame shape. Each framed column consists of four vertical members joined by short beams(2.7m) to create a three-dimensional structure. The combination of the short beams in between the framed columns elements and long span beams create an unusual static characteristics: migration of column loads, short span beams loaded more lightly in bending than the long spans, and an unusual failure hinge mechanism for the extreme earthquake check. Also the building is supported by 14 pilotis framed columns 12.9m high from ground to 4th floor(Figure 2). The four vertical column elements are connected by X shaped steel box section horizontal bracing in place of the short beams.

A. Migration of Column Loads

Isolating a typical framed column element, elastic analysis using the program "KANSAS" was carried out to investigate whether the overall frame action affected the axial load distribution between the inner and outer column members. Figure 5 shows the simple model used for a typical bay. The sequence of construction was taken into account in a simplified modelling. The flow of load from the inner column elements to the outer ones was demonstrated (see Figure 5 and 6 for results). In spite of the large difference in tributary areas, overall deformations of the framed column almost eliminated the imbalance of axial load, as shown in Figure 7 where the scale is normalized by dividing the load by a quarter of the total, to give a measure of the extent to which load is evenly shared among all four columns.

B. Short Span Beam Stresses During Earthquake Loading

In a medium rise building, short span beams yield at their supports under relatively low load levels due to concentration of bending stiffness. However in this case the effect of axial deformations of the framed columns was the dominant mode of behavior and so the short beams attracted much less bending moment than if axial deformations were absent (refer to Figure 8, which is comparing the effects of ignoring the axial deformations in a static analysis under earthquake load).

C. Plastic Characteristics of the Structure

An incrementally increasing static analysis on an elastic-plastic model ("pushover analysis") was carried out to obtain the skeleton curves of story shear-deformation characteristics for the structure. From this analysis, the formation of plastic hinges at the supports of long span beams, with the short span beams remaining elastic even at high earthquake load levels was notable. The development of plastic hinges and the resulting mechanism for the structure are shown in Figure 9.

THE EFFECT OF VERTICAL COMPONENT OF EARTHQUAKE MOTION

The structure was planned with an unusually long main span of 21m across the building width. In such cases the vertical component of an earthquake can have a significant effect on stresses in the beams, and so this was investigated by modelling a typical bay as a two dimensional frame structure (refer to Figure 10). Dynamic analysis was carried out using the time history of four earthquake records, inputting their lateral and vertical components simultaneously.

As a preliminary step the modes of vibration of the structure were obtained, from which it was noted that the fifth mode, which is the first mode in the vertical direction, involved axial stretching and compression of the columns, with all the beams vibrating together. (refer to Figure 10). Only at higher modes did the beam vibration become more complex.

Analysis was carried out at two levels of earthquake input. The following results were obtained:

i) At 25cm/sec peak input level, the long span beams developed large bending moments, particularly higher up in the building. It was verified that even including these stresses, the beams remained within the short term allowable stresses (the elastic limit, according to Japanese design criteria).

ii) At 40cm/sec peak input level, due to the effect of the vertical component of the earthquake, plastic hinges at the ends of long span beams developed early on, but as shown in Figure 11, which plots the loops of time history for story shear versus story lateral drift for the TAFT record at the 15th story, the structure remained elastic in its overall behavior, and did not degrade its overall dynamic stiffness characteristics. The reason for this is considered to be that the period of vibration vertically is so much shorter than horizontally that the structure was able to recover its elastic characteristics. From this result it was concluded that the use of horizontal earthquake time history records only for the main response analysis was appropriate.

DESIGN WIND FORCES AND PREDICTION OF WIND VIBRATION

Although the building is not of irregular shape, wind tunnel testing was carried out to investigate the effect of the exposed column frames on surface roughness and effective frontal area of the building. It was also desirable to predict the vibration behavior of the building under wind loading, since its period is relatively long.

Using the pressure coefficients from the wind tunnel test, the wind pressure distribution up the building was obtained using the vertical distribution of wind velocity as stated in the AIJ Recommendations for Design Loads for Buildings. The base shear produced by this wind load is 91% of "design" earthquake base shear (equivalent to the level 1 ground motion 25cm/sec peak input).

As part of the wind tunnel test, using a device called a manifold, simultaneous measurements were made of wind pressure vibrations at each story. These measured records were input to an analytical model with lumped mass parameters (refer to Figure 12), to perform a time history analysis. The wind load was applied from two directions, and the time history of wind pressure were normalized to correspond to the 10 minutes average windspeed of 15m/sec (Return Period 5 years). Figure 13 shows the result at the 25th floor level, in terms of acceleration. It was predicted that although the upper levels would experience a peak acceleration of around 4 Gal, this would not cause any discomfort.

DESIGN FOR FIRE RESISTANCE

It was considered that since the exposed frame was outside the main building glazing line, it would not be subject to the same intensity of fire as a normal frame, and therefore it could be fire-protected to an approriately lesser degree. As shown in Figure 14, a simulation analysis was performed on the heat experienced by the external frame when subjected to flames discharging from inside the building.

The results of this simulation were that the beams would be heated to 273 degrees on the outside of their aluminum cladding, and the columns to a maximum of 281 degrees centigrade. Since the steel temperature would be lower than the cladding, and according to the Ministry of Construction Bulletin No.2999 the critical temperature for steel in a fire may be taken as 350 degrees average (with a maximum of 450 degrees), the conclusion was drawn that no fire protection was needed at all for the external frame.

Upon examination by the Building Center Fire Safety and Protection Committee, 1cm of fire resistant cladding material was finally agreed upon, in order to obtain the Minister of Construction's approval. Compared with the normal 5cm thickness required for 3 hour's protection or 2cm for one hour's protection, this was a substantial relaxation.

ACKNOWLEDGMENTS

This project was realized with the close cooperation of Kajima's Architectural Design Division, Construction Division of Osaka Branch Office, Kobori Reserch Institute, and Kajima Reserch Institute, who overcame many challenges in carrying out the design. Thanks are due to all concerned.

FIGURE 1   TOKYO MARINE PLAZA

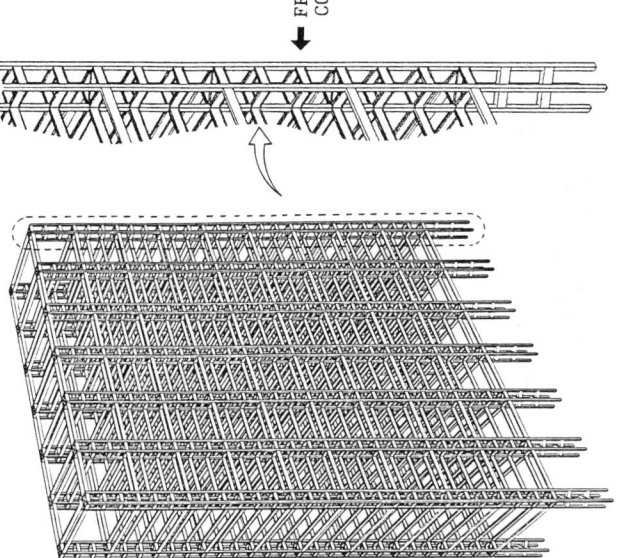

FRAMED COLUMN

FIGURE 3   FRAMING PERSPECTIVE

FIGURE 2   PILOTIS COLUMNS

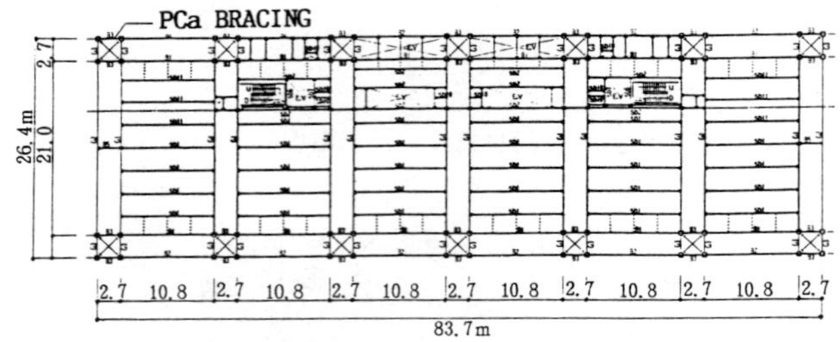

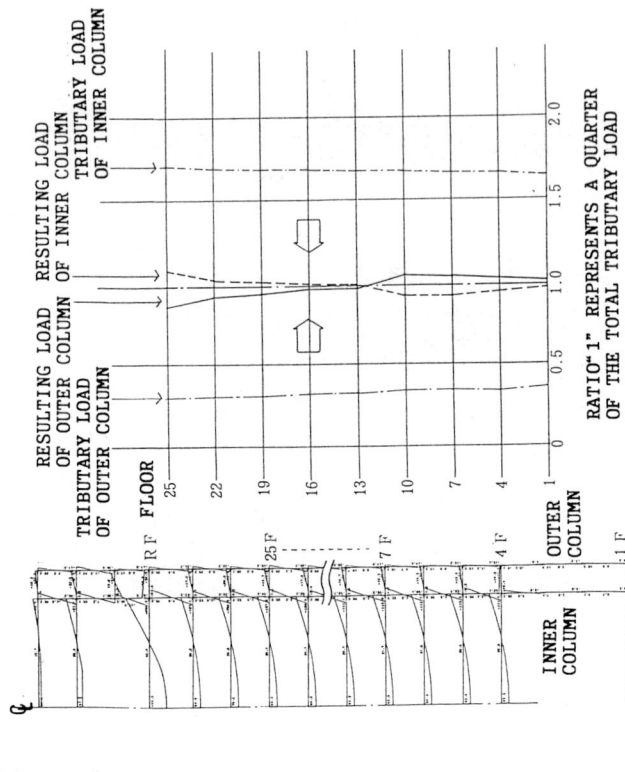

RESULTING LOAD OF OUTER COLUMN

TRIBUTARY LOAD OF OUTER COLUMN

RESULTING LOAD OF INNER COLUMN

RESULTING LOAD OF INNER COLUMN

TRIBUTARY LOAD OF INNER COLUMN

RATIO "1" REPRESENTS A QUARTER OF THE TOTAL TRIBUTARY LOAD

FLOOR

25
22
19
16
13
10
7
4
1

0   0.5   1.0   1.5   2.0

R F
25 F
7 F
4 F
1 F

OUTER COLUMN

INNER COLUMN

FIGURE 6   MOMENT DIAGRAM   FIGURE 7   AXIAL LOAD DISTIBUTION

PCa BRACING

2.7 10.8 2.7 10.8 2.7 10.8 2.7 10.8 2.7 10.8 2.7 10.8 2.7 10.8 2.7

83.7m

2.7 21.0 2.7
26.4m

FIGURE 4   FRAMING PLAN FOR TYPICAL FLOOR

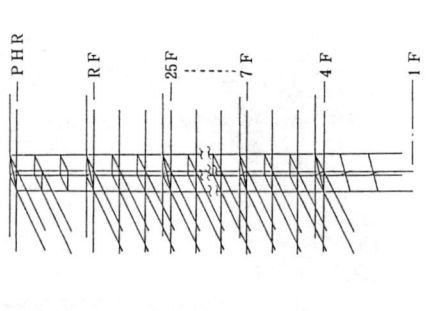

PHR
R F
25 F
7 F
4 F
1 F

FIGURE 5   ANALYTICAL MODEL FOR VERTICAL LOADS

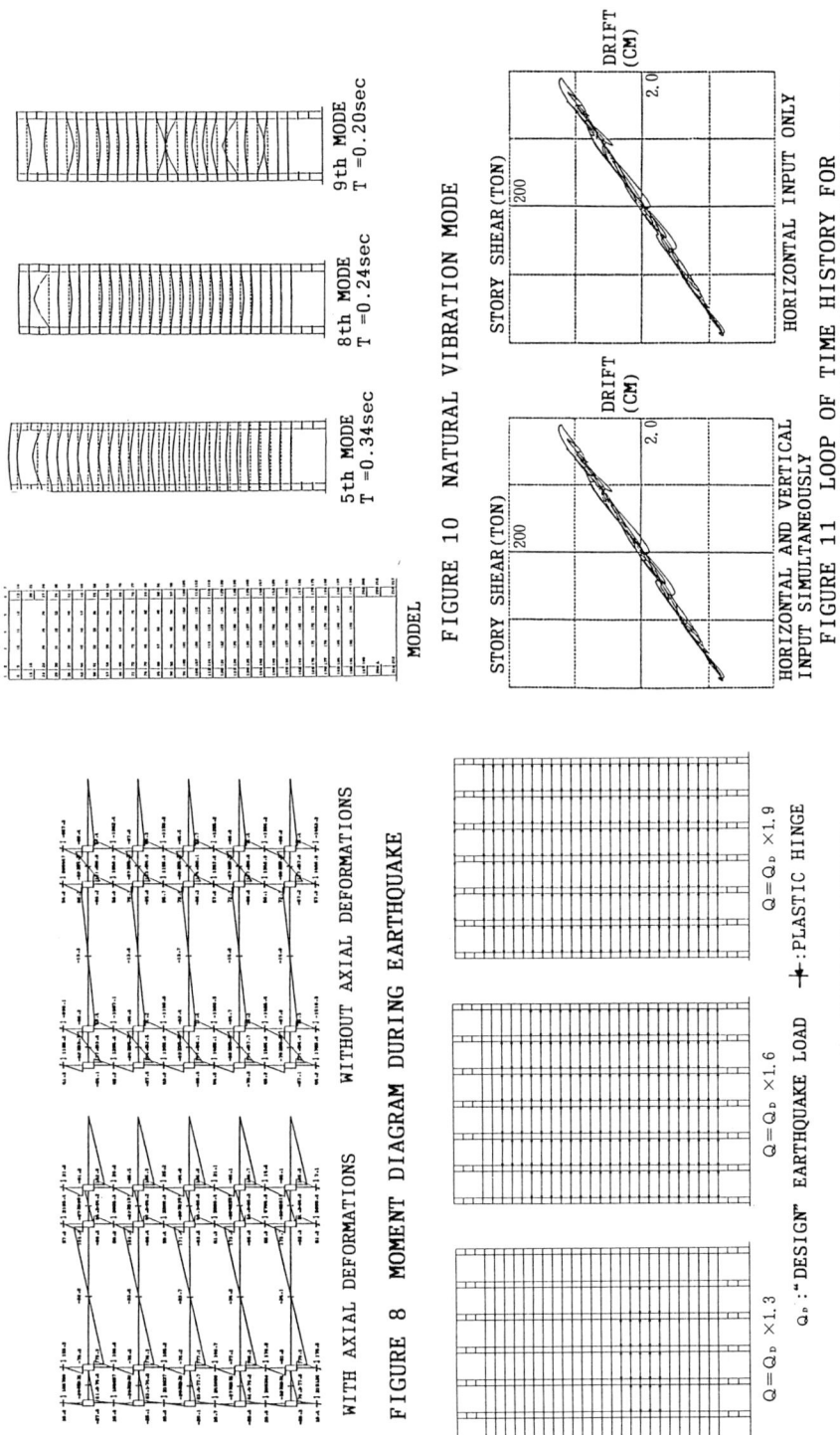

9th MODE
T=0.20sec

8th MODE
T=0.24sec

5th MODE
T=0.34sec

MODEL

FIGURE 10 NATURAL VIBRATION MODE

STORY SHEAR(TON)

200

DRIFT
(CM)

2.0

HORIZONTAL AND VERTICAL
INPUT SIMULTANEOUSLY

STORY SHEAR(TON)

200

DRIFT
(CM)

2.0

HORIZONTAL INPUT ONLY

FIGURE 11 LOOP OF TIME HISTORY FOR
STORY SHEAR VERSUS STORY DRIFT

WITH AXIAL DEFORMATIONS

WITHOUT AXIAL DEFORMATIONS

FIGURE 8 MOMENT DIAGRAM DURING EARTHQUAKE

14

13

12

11

10

9

Q=Q$_D$×1.3

Q=Q$_D$×1.6

Q=Q$_D$×1.9

Q$_D$:"DESIGN" EARTHQUAKE LOAD   ━┿━:PLASTIC HINGE

FIGURE 9 DEVELOPMENT OF PLASTIC HINGES

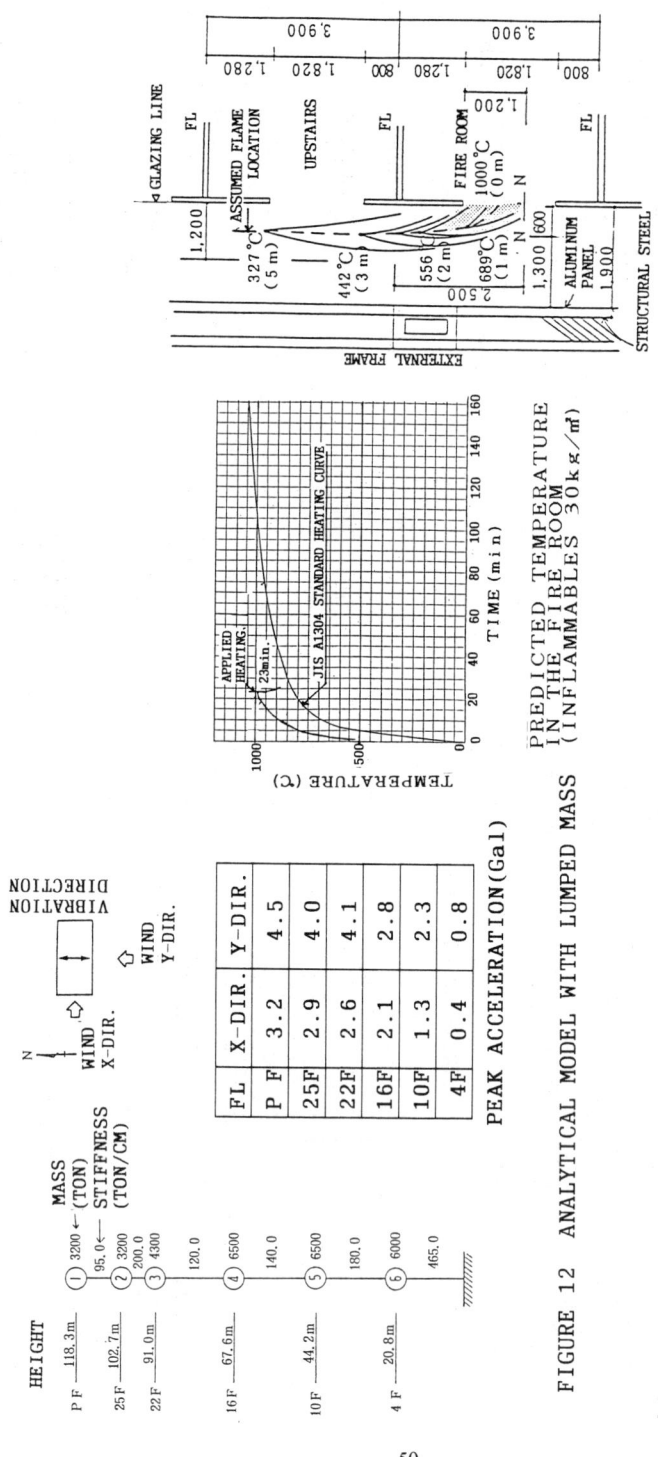

FIGURE 14 PREDICTION OF DISCHARGING FLAME AND TEMPERATURE DISTRIBUTION

PREDICTED TEMPERATURE IN THE FIRE ROOM (INFLAMMABLES 30 kg/m²)

FIGURE 12 ANALYTICAL MODEL WITH LUMPED MASS

FIGURE 13 TIME HISTORY OF RESPONSE ACCELERATION AT 25TH FLOOR

PEAK ACCELERATION (Gal)

FL	X-DIR.	Y-DIR.
P F	3.2	4.5
25F	2.9	4.0
22F	2.6	4.1
16F	2.1	2.8
10F	1.3	2.3
4F	0.4	0.8

Proceedings of the Second Conference on Tall Buildings in Seismic Regions
55th Regional Conference
May 16 and 17, 1991, Los Angeles, California

FREE SPANNING DUCTILE VIERENDEEL FRAME:
A SYSTEM FOR HIGH-RISE BUILDINGS
IN SEISMIC AREAS

by

P. V. Banavalkar, Ph.D., P.E.
Chief Structural Engineer/Executive Vice President
CBM ENGINEERS, INC., Houston

INTRODUCTION

In a seismic area a choice of the structural system is dictated by its ductility, energy absorption capacity and redundancy. The vertical continuity both in the overturning and the lateral shear resistance is a desirable feature to ensure a good structural performance of a building in a seismic event. The modern architecture with tall slender towers exhibiting multiple step backs, mixed-use facilities and subterranean parking under the footprint of the tower creates necessity for transfer girders in a conventionally designed closely spaced ductile tubular system.

The author developed a cost effective free spanning ductile vierendeel system placed within the uninterrupted spine of the building. The ductility to the system is mainly provided by the yielding of the vertical stubs in the vierendeel girders. The performance of this system in terms of its energy absorption capacity, redundancy, overall safety, and the criteria of resistance versus demand is compared here with a conventionally designed ductile moment resisting frame for an identical structure. Along with this comparison, the paper presents the special details and the design criteria required for the system. The application of this structural system for major high-rise buildings is also discussed.

COMPOSITION, STRUCTURAL BEHAVIOR, AND DETAILS

COMPOSITION

The system consists of five to six stories tall free spanning vierendeel girder (fig. 1a) anchored by the main columns which are continuous to the foundation. The vertical stubs are rigidly connected to the continuous horizontal beams which in turn are rigidly connected to the main columns. Each vierendeel girder is stacked vertically in such a way that the hinge connections between vertical stubs is capable of transmitting only horizontal shears (fig. 1b). These hinges insure that all the vertical gravity load is accumulated in the main columns only and the vertical stubs act only as a shear membrane. Furthermore, the location of stubs not necessarily placed symmetrically with respect to the main columns can be reasonably adjusted within a vierendeel girder to suit the occupancy of the floor such as, hotel, office or parking garage.

STRUCTURAL BEHAVIOR

Though difficult to quantify, the lateral resistance of the system can be partitioned into two parts (fig. 2).

1) The resistance provided by super-frame without the stubs (frame action).

2) The shear resistance provided by the stubs and the beam assembly (vierendeel action).

The vertical load carrying capacity of the vierendeel is quite well known.

DETAILS

To achieve desired level of ductility following details are required:

1) The vertical stubs are rigidly connected to the beam in such a way to develop full plastic moment capacity of the stubs and the corresponding shear. The panel zone of the beam is stiffened accordingly. The connection of the stubs to the foundation also follows the same principles.

2) The beam to main column connection is a conventional ductile frame connection. The shear connection of the beam to the main columns is designed to develop the shear due to full plastification of the stubs as well as the plastification of the beam in addition to the supported gravity load (fig. 3).

3) As discussed before, the hinge connection in the vertical stubs is designed to transmit only horizontal forces.

4) All joints at the intersection of stubs and the beam should be braced laterally by floor beams.

STRUCTURAL PERFORMANCE OF THE SYSTEM

The structural performance of the ductile vierendeel frame DVF (RII) is compared with the a conventional DMRSF (RI) designed in accordance with the 1988 UBC code. The 17 stories tall RI frame (fig. 4a) consists of 5 bay (20'-0" o.c.) 100 ft. long ductile frame, whereas the DVF (RII) consists of 60'-0" single span vierendeel with vertical stubs at 20'-0" o.c. (fig. 4b). Both frames have identical floor heights and similar base condition. It may be stated that RI is an as-built structure for a building in California. The relevant properties of both structures, RI and RII, are given in Table I.

The dynamic analysis of both structures was performed for S-3 response spectra of UBC 1988.

The structural performance of both structures (RI and RII) was evaluated by the following criteria:

1) Inelastic response characteristics.

2) Redundancy of the structure in vertical load carrying capacity.

INELASTIC RESPONSE

1) Ultimate Shear Capacity

Because of the strong column-weak beam concept, the ultimate lateral shear capacity for frame RI is limited by the plastification of the beams, whereas in RII structure the same is limited by the plastification of the stubs as well as the beams. As shown in Table II, the ultimate shear capacity of structure RII is greater than that of RI structure over the majority of the floors. Structure RII is slightly softer in the lowest three floors of the building

2) Elastic Energy Absorption Capacity

The total energy stored in the RI structure is calculated at the time of plastification of beams. Whereas the total energy stored in RII structure is determined at the plastification of both the vertical stubs as well as the beam. As anticipated RII structure, because of its higher stiffness, indicates larger energy absorption capacity, particularly in the ductile element, such as the beams and the vertical stubs (see Table III). The additional input energy from the seismic motion will have to be absorbed by ductile beam-column joint.

3) Hierarchy of Plastic Hinge Formation and Load Deflection Curve

To qualitatively establish the hierarchy of plastic hinge during degration of the structure (fig. 5), the ratios of response generated bending moments in the flexural members due to response spectra S-3 were compared to the plastic moment capacities of the members.

As anticipated for RI structure, the plastification of the beams Sb-3 is immediately followed by the plastification of beams Sb-2 and Sb-1.

For the structure RII, at the load level approximately 2.5 times smaller than the response generated by S-3 spectra the vertical stubs form plastic hinges at the junction of the beams. The onset of the plastification of the stubs provides ductility along with the increase in structural damping.

The lateral load deflection (P-Δ) characteristic at the 9th floor, which is representative of the rest of the structure, is shown in fig. 6.

For RI structure, the ultimate lateral shear capacity at the 9th floor is only 1.076 (1142/1061) times the shear capacity at the initial set of plastification of the beam Sb-3. As opposed to this, RII structure after the plastification of the stubs gains approximately 32% of its ultimate shear capacity during the plastification of the beam. The energy absorption capacity, which is represented by the area under P-Δ curve for the RII structure (fig. 6), is superior to that of RI structure.

The overturning resistance of the structure during the process of degradation by the plastification is also very important. After the plastification of the beams Sb-2 and Sb-3, the fundamental period of the RI frame increases to 5.84 seconds as opposed to the initial period of 3.27 seconds. The RI structure reduces itself to two single bay frames of 20'-0" span with the overturning stiffness of the building being reduced by a dramatic factor of 17.50 times the original (fig. 5).

After the plastification of the stubs, the fundamental period of the RII structure also increases to 5.56 seconds/cycle from 3.14 seconds/cycle initial period. However, the overturning stiffness and resistance of RII structure remains unaltered during this excursion, a very decisive advantage over conventional RI structure.

4) Resistance versus Demand

The inelastic response of the structure is further examined by the requirement of Resistance versus Demand, a litmus test for a desirable ductile behavior (fig. 7). As shown in the figure, the plastification of the stubs starts approximately at 40% of the response shear 3,724 kips determined by the spectra S-3, a measure of the maximum expected earthquake. After the plastification of the stubs, both the ultimate lateral shear capacity and the damping of the structure continues to increase, achieving the ultimate shear capacity of 1,567 kips. The increase in period of the building to 5.56 seconds along with the increase damping reduces the shear demand of the structure to 1,141 kips, realistically represented by an equivalent spectra for the increased damping.

The desired performance of the structure for Resistance matching or exceeding the Demand is well documented by this behavior.

REDUNDANCY IN VERTICAL LOAD CARRYING CAPACITY (FIG. 8)

It is essential that a system can perform inelastically without jeopardizing its vertical load carrying capacity. The performance of the conventional RI structure is very well documented. For the RII structure, the following criteria needs to be followed:

1) Prevention of Three Hinge Mechanism in a Girder

   The continuous main beam supporting the floor with rigidly attached vertical stubs is designed in such a way that even after full plastification of the stubs and the formation of the plastic moment hinges at the beam column connections there is still a minimum of 1.20 factor of safety against the support of the full gravity load. This criterion protects formation of the unstable three hinge mechanism in the beam and ensures stable vertical load carrying capacity of the structure. Since the design of multistory structures is generally determined by the stiffness criteria, the criterion of providing safe load carrying capacity in the beam does not impose any additional cost penalty on the RII structure over a conventionally designed strong column-weak beam ductile frame.

2) Main Column Capacity

   The main columns supporting the vierendeel should be checked for the axial load due to plastification of the beam and the stubs in addition to the gravity load. This criterion is found to be far in excess of the ⅜ Rw amplification by the present UBC code.

APPLICATION TO HIGH-RISE BUILDINGS

In high-rise buildings, columns heavily loaded by gravity loads and interlinked by a shear membrane in the form of ductile vierendeel girder or bracing can provide very efficient cost effective overturning resistance to the structure.

As shown in fig. 9a, an office building with underground parking always has a main shear transfer diaphragm at its base. The ductile vierendeel can be devised along the perimeter of the building with main columns providing continuity in the overturning resistance to the foundation. The system can also be applied in the core of a tower where the multiple door openings to the core make the application of eccentric bracing impossible (fig. 9b). The system has been applied to already constructed 53 stories tall Gas Company tower in Los Angeles and is applied to two other major office buildings as well.

The system presents an ideal solution for multi-use complexes such as a tall slender hotel on top of a office building with a Porte Cochere at its base, super-posed on an underground parking garage (fig. 10). A supplemental stiffening at the Porte Cochere level may be required to eliminate soft story. The notching of the beam to give enough head room clearance in the corridors of the hotel requires special attention in sizing the member to prevent the formation of three hinge mechanism (fig. 8).

CONCLUSION

Designed with special design criteria, it is shown that the ductile free-spanning vierendeel system is equivalent to a conventional ductile moment resisting frame in terms of ductility, energy absorption capacity, ultimate shear capacity, and demand versus resistance performance. Because of this structural behavior, when designed for code level forces, Rw = 12 should be used.

The proposed system is superior to the conventional system in terms of its overturning resistance and load deflection (P-Δ) characteristics. The proposed system has two ductile components, namely the beam and the vertical stubs. The onset of first yielding in the vertical stubs acting as a fuse is a unique feature of the system.

The system can be applied efficiently to multi-use hotel tower and office buildings with underground parking.

Heavily loaded columns supporting the free-spanning vierendeels provide flexibility in the underground parking spaces and at the same time minimize the foundation and structural cost by eliminating uplifting loads in the columns.

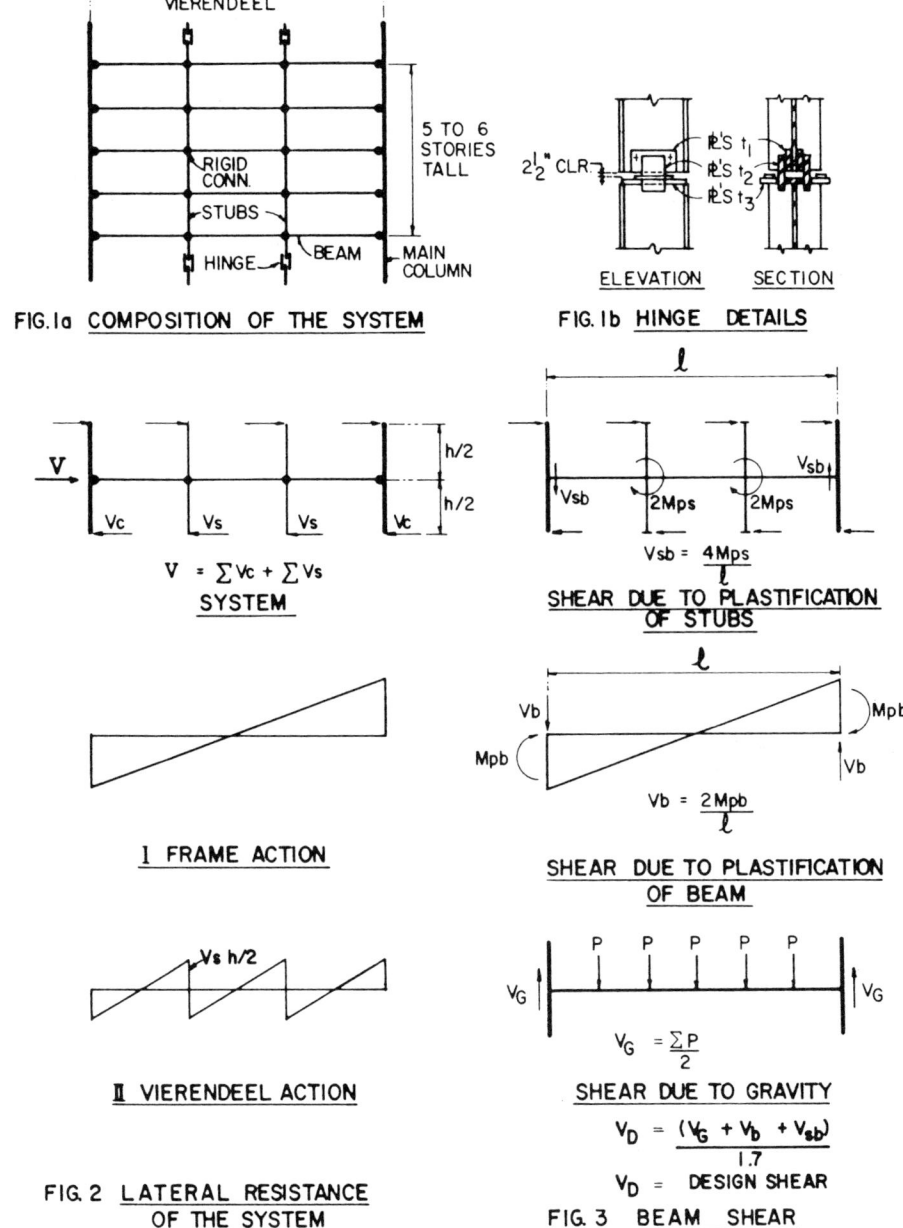

FREE SPANNING
VIERENDEEL

5 TO 6
STORIES
TALL

RIGID CONN.

STUBS

HINGE — BEAM — MAIN COLUMN

FIG.1a COMPOSITION OF THE SYSTEM

FIG.1b HINGE DETAILS

$2\frac{1}{2}''$ CLR

$\cancel{\mathbb{C}}$'S t_1
$\cancel{\mathbb{C}}$'S t_2
$\cancel{\mathbb{C}}$'S t_3

ELEVATION   SECTION

$V = \sum V_c + \sum V_s$
SYSTEM

$V_{sb} = \dfrac{4M_{ps}}{l}$

SHEAR DUE TO PLASTIFICATION OF STUBS

I FRAME ACTION

$V_b = \dfrac{2M_{pb}}{l}$

SHEAR DUE TO PLASTIFICATION OF BEAM

II VIERENDEEL ACTION

$V_G = \dfrac{\sum P}{2}$

SHEAR DUE TO GRAVITY

$V_D = \dfrac{(V_G + V_b + V_{sb})}{1.7}$

V_D = DESIGN SHEAR

FIG.2 LATERAL RESISTANCE OF THE SYSTEM

FIG.3 BEAM SHEAR

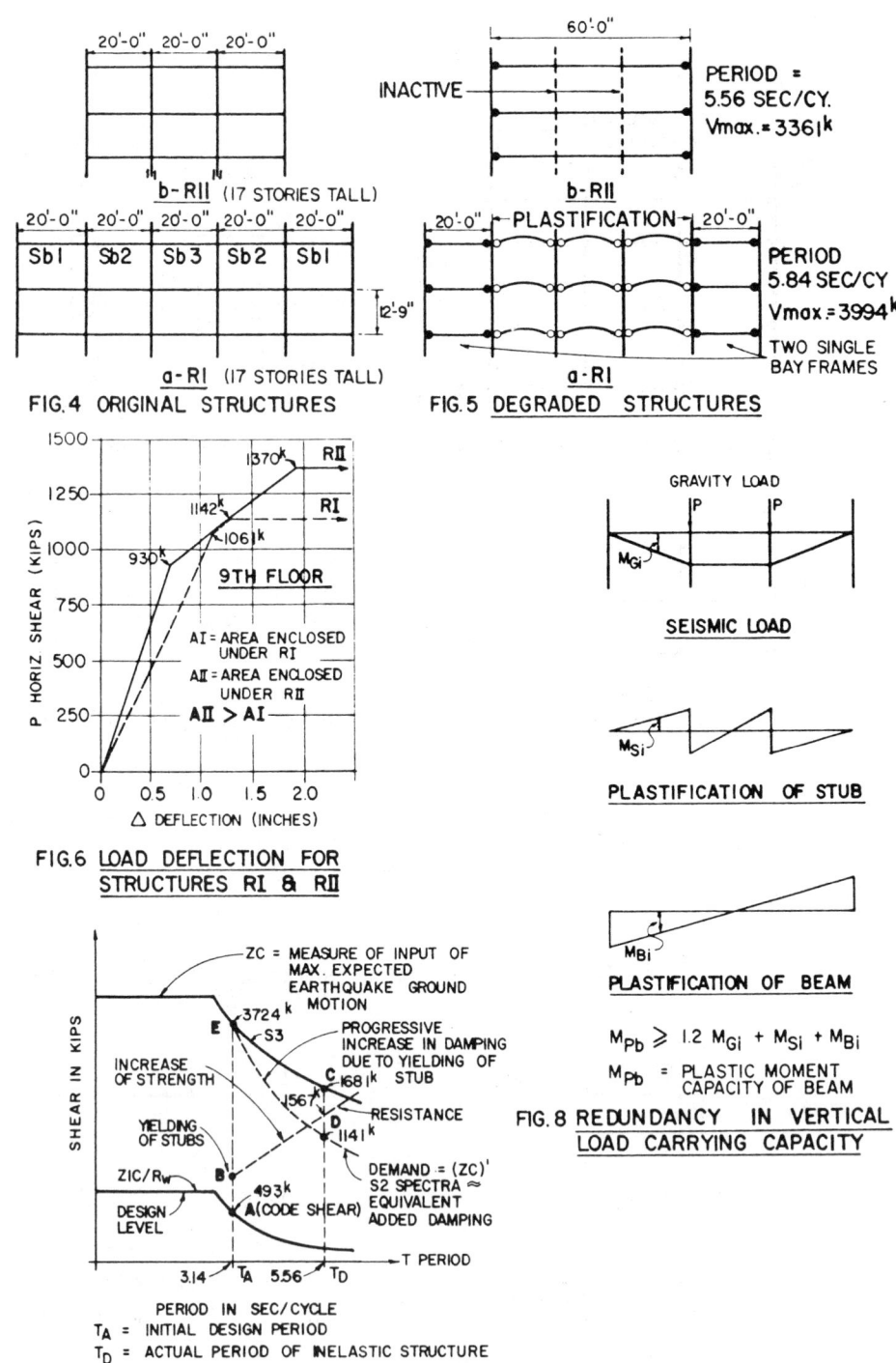

FIG.4 ORIGINAL STRUCTURES

b-RII

PERIOD = 5.56 SEC/CY.
Vmax.= 3361^k

a-RI

PERIOD 5.84 SEC/CY
Vmax = 3994^k

TWO SINGLE BAY FRAMES

FIG.5 DEGRADED STRUCTURES

FIG.6 LOAD DEFLECTION FOR STRUCTURES RI & RII

FIG.7 RESISTANCE VERSUS DEMAND (RII)

T_A = INITIAL DESIGN PERIOD
T_D = ACTUAL PERIOD OF INELASTIC STRUCTURE

$M_{Pb} \geqslant 1.2\, M_{Gi} + M_{Si} + M_{Bi}$

M_{Pb} = PLASTIC MOMENT CAPACITY OF BEAM

FIG.8 REDUNDANCY IN VERTICAL LOAD CARRYING CAPACITY

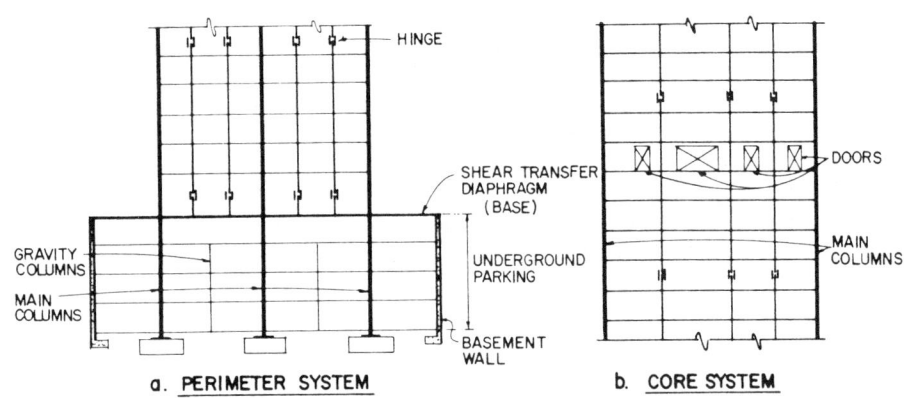

a. PERIMETER SYSTEM   b. CORE SYSTEM

FIG.9 OFFICE TOWER

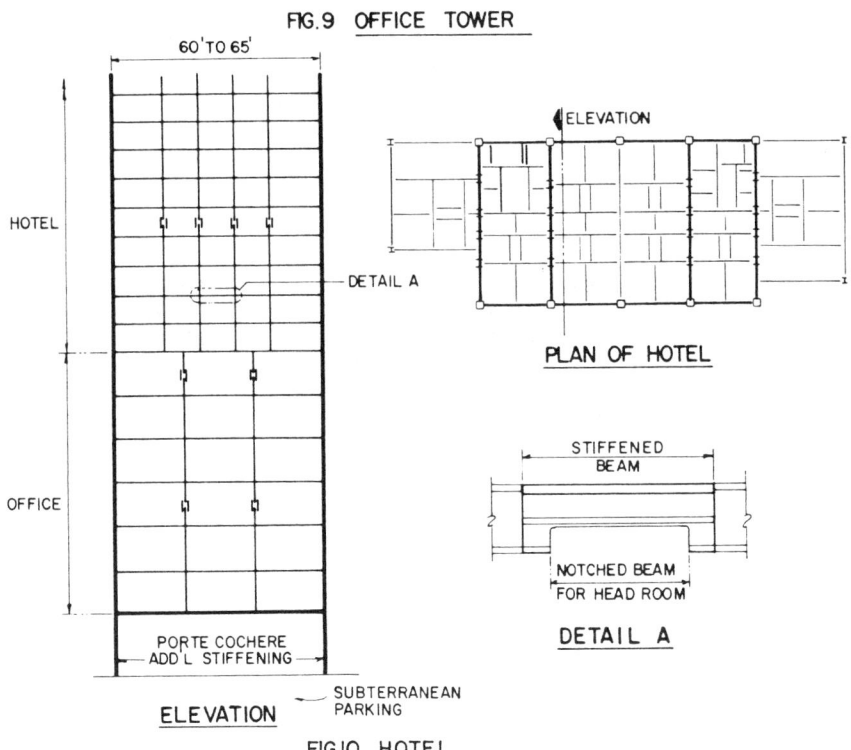

ELEVATION

FIG.10 HOTEL

PLAN OF HOTEL

DETAIL A

TABLE I - COMPARATIVE PROPERTIES OF SYSTEMS RI & RII

	RI	RII
MASS IN KIPS	15321	15321
FUNDAMENTAL PERIOD SEC/CYCLE	3.27	3.14
RESPONSE FOR S3 SPECTRA		
BASE SHEAR IN KIPS	3531	3724
OVERTURNING MOMENT K-FT x 10^6	0.83	0.93
δ AT TOP IN INCHES	36.6	35.4

TABLE II - ULTIMATE SHEAR CAPACITY

FLOOR NO.	RI IN KIPS	RII IN KIPS	RATIO RII/RI
16	727	1059	1.46
13	905	1244	1.37
9	1142	1370	1.2
5	1582	1471	0.93
3	1849	1567	0.85

TABLE III - COMPARISON OF ENERGY AT PLASTIFICATION

FLOOR NO.	ENERGY IN DUCTILE ELEMENTS IN KIP-INCH			TOTAL ENERGIES IN ALL ELEMENTS IN KIP-INCH		
	RI BEAMS	RII BMS+STUBS	RATIO RII/RI	RI	RII	RATIO RII/RI
15	346	703	2.03	470	756	1.61
12	356	784	2.20	493	845	1.71
9	502	846	1.69	670	911	1.36
6	545	857	1.57	748	916	1.23
4	690	846	1.23	951	900	0.95

Proceedings of the Second Conference on Tall Buildings in Seismic Regions
55th Regional Conference
May 16 and 17, 1991, Los Angeles, California

A New Confinement Device
for Tall Ductile Shear Wall Buildings

Hanns U. Baumann, S.E.

A new confinement reinforcement device has been developed and installed in a 17-story concrete shear wall structure in San Francisco, California. Tested initially at the University of California, Irvine (UCI) then at the Earthquake Engineering Research Center (EERC), Richmond, CA. It was also given a "field test" by Loma Prieta Earthquake.

Design, testing, quality control, field construction techniques and performance under actual earthquake conditions are discussed as well as future applications in taller buildings.

A three year long development program has resulted in a new device which promises to improve the ductility of tall concrete and masonry shear wall buildings. Starting at UCI, Civil Engineering testing laboratory in 1987 a full size reinforcement boundary element specimen was subjected to repeated cycles of large tension and compression forces developing stresses in the specimen well beyond the yield strength of the steel.

Based on the positive results of the test at UCI numerous similar specimens at the EERC were cycled well beyond yield point to verify the improved ductility provided by the new device.

After the review of the test results by a peer review board of consulting structural engineers, the device was approved for installations in a 17-story dormitory on the campus of San Francisco State University which is located near the San Andreas Fault.

The proximity of the site to the fault required a novel design to resist the large anticipated site specific dynamic response forces. The actual performance during the Loma Prieta Earthquake is discussed.

Quality control during manufacturing and installation of the proprietary high strength welded wire device is discussed, as well as shop and field assembly techniques, related labor and material savings, and feasibility of its use in ductile shear walls and frames in taller concrete and masonry buildings.

Fig. 1 Residence Apartment Building San Francisco State University

Baumann Research and Development Corporation, Newport Beach, California

INTRODUCTION

Over many years, the structural engineering community has been working to develop the safer use of reinforced concrete structures in regions of high earthquake risk throughout the world. In the forefront of these efforts has been the need to find the right blend of steel reinforcement and concrete such that the structural qualities of each are combined to create strong and secure walls, columns and beams, while at the same time, allowing for a structure that is ductile enough to remain intact under the most extreme destructive seismic forces.

Until recently almost without exception, individual re-bar hooked hoop ties and stirrups have been used to tie the major reinforcing steel elements together to resist the bursting forces that are caused by large seismic forces. These ties have proven to be costly to fabricate, time consuming to install and sometimes inadequate in fulfilling their designated purpose of assuring a more ductile response by the structure.

Recently, a welded wire ladder device, known as Medo-Mesh®, was developed and used in place of the re-bar ties. Although the concept of a welded "wrap around" confinement fabric has been used for many years, it was not until 1987 that an engineering and research firm in Newport Beach, CA invented and developed a new welded confinement reinforcement device to the point of actually having it tested, fabricated, specified and used in concrete structures in seismic zones.

Under the direction of product development engineer and the inventor of Medo-Mesh, Hanns U. Baumann, President of Baumann Research and Development Corporation, the new Medo-Mesh product was approved and used on two projects. The first was a 17 story high rise dormitory on the campus of San Francisco State University (SFSU); the second a 60 unit 4 story apartment complex in Long Beach, CA.

Although the two jobs were quite different in size and scope, the responses to the use of the product were very favorable. On both jobs, Medo-Mesh was found to be faster to install, less costly to work with, created improved ductility, and gave reliable structures including field tested performance under the Loma Prieta earthquake forces at the SFSU site.

Before construction commenced, with the help of many construction industry firms, the new proprietary product was tested first at the University of California, Irvine with Professor Robin Shepherd and then at the Earthquake Engineering Research Center of U.C. Berkeley. The research program there was conducted by Eduardo Miranda and Christopher L. Thompson under the supervision of Professor Vitelmo V. Bertero.

It was found by the above investigators that ". . . Medo-Mesh, if properly designed and installed, can serve to (1) provide stable post-spalling behavior in the ranges of strain that can be induced in rectangular boundary elements of shear walls during severe earthquakes; and (2) substantially decrease the specialized labor requirements for fabrication of the steel reinforcing cages of members of reinforced concrete structures".

It is interesting to note that this new device is a response to the "energy approach" espoused by V. V. Bertero in March of 1988 at the Tall Building and Urban Habitat, Los Angeles, where he stated ". . . There cannot be improvement in earthquake-resistant design if there is no improvement in predicting the stiffness, strength, energy absorption and energy dissipation capacities of real buildings".

62

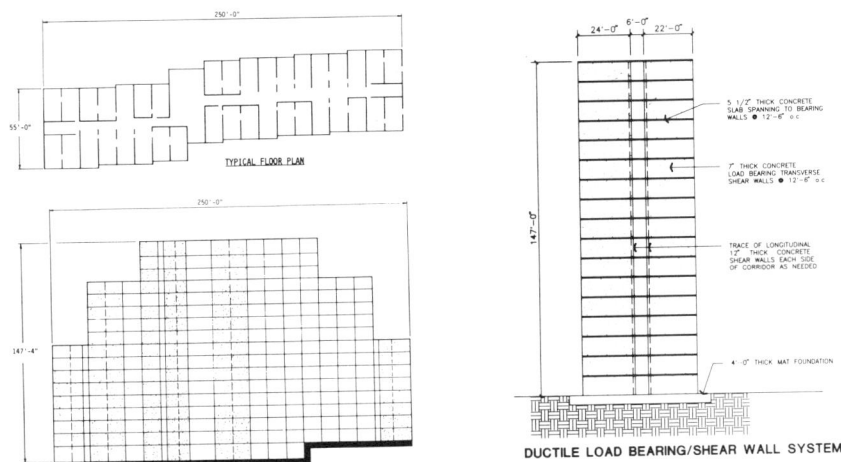

Fig. 2 Structural System

Residence Apartment Building San Francisco State University

In the transverse direction, 7" thick load-bearing ductile shear walls at 12' 6" o. c. support 5 1/2" thick one way slabs. In the long direction are 12" thick ductile shear walls.

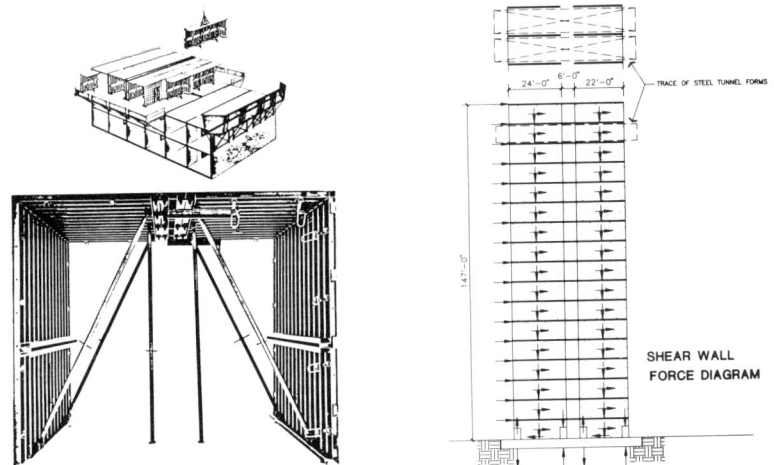

Fig. 3 Tunnel Form System

Tunnel Form Construction Method

The present day Tunnel Form method of construction (Fig. 3) prohibits the use of barbell type shear walls. However its speed and economy still makes it a very attractive way to build shear wall type structures for dormitory, hotel and apartment buildings.

Tunnel Form manufactures are now working on form systems to construct barbell type shear walls for taller buildings.

The problem during the design of the SFSU building was to achieve reliable ductility in 17-story high shear walls just 7" thick.

Building greater ductility in structures is very desirable because they "attract" less earthquake force.

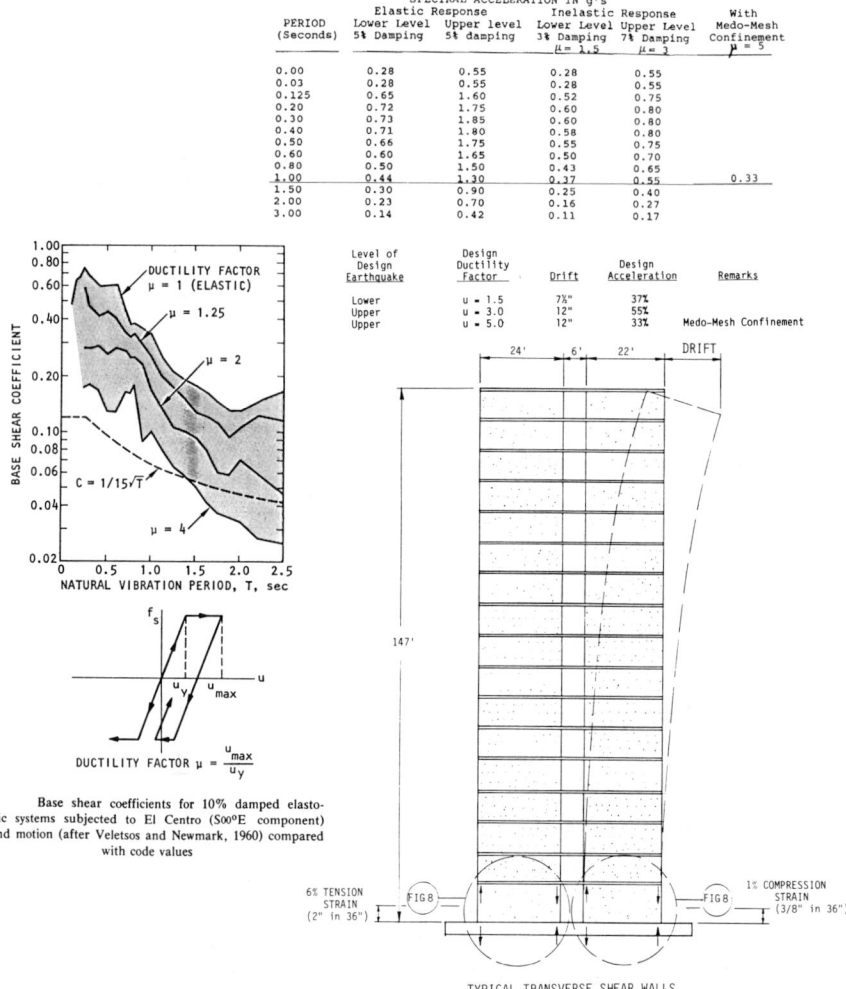

PERIOD (Seconds)	Elastic Response Lower Level 5% Damping	Upper level 5% damping	Inelastic Response Lower Level 3% Damping μ = 1.5	Upper Level 7% Damping μ = 3	With Medo-Mesh Confinement μ = 5
0.00	0.28	0.55	0.28	0.55	
0.03	0.28	0.55	0.28	0.55	
0.125	0.65	1.60	0.52	0.75	
0.20	0.72	1.75	0.60	0.80	
0.30	0.73	1.85	0.60	0.80	
0.40	0.71	1.80	0.58	0.80	
0.50	0.66	1.75	0.55	0.75	
0.60	0.60	1.65	0.50	0.70	
0.80	0.50	1.50	0.43	0.65	
1.00	0.44	1.30	0.37	0.55	0.33
1.50	0.30	0.90	0.25	0.40	
2.00	0.23	0.70	0.16	0.27	
3.00	0.14	0.42	0.11	0.17	

SPECTRAL ACCELERATION IN g's

Level of Design Earthquake	Design Ductility Factor	Drift	Design Acceleration	Remarks
Lower	u = 1.5	7⅜"	37%	
Upper	u = 3.0	12"	55%	
Upper	u = 5.0	12"	33%	Medo-Mesh Confinement

DUCTILITY FACTOR
μ = 1 (ELASTIC)
μ = 1.25
μ = 2
$C = 1/15\sqrt{T}$
μ = 4

DUCTILITY FACTOR $\mu = \dfrac{u_{max}}{u_y}$

Base shear coefficients for 10% damped elasto-plastic systems subjected to El Centro (S00°E component) ground motion (after Veletsos and Newmark, 1960) compared with code values

TYPICAL TRANSVERSE SHEAR WALLS

Fig. 4 Ductility

Ductility

Ductility (μ) is defined as the ratio of ultimate deflection to yield deflection ($\frac{\Delta m}{\Delta y}$). In reinforced concrete and masonry this toughness, or ability to absorb energy without brittle failure, is achieved by resisting large compression bursting pressures after large tension forces are first imposed.

During a violent earthquake, shear walls experience many cycles of large tensile and compression forces at their ends. To resist these large forces, boundary elements composed of reinforcement cages are installed.

It has been shown by many investigators that ductility improves with the improvement of the confinement of the concrete under tension and compression stresses.

The conventional method of confining concrete is depicted in Fig. 5 using re-bar hoops with hooked end that obstruct the passage of wet concrete and the vibrator "stinger".

64

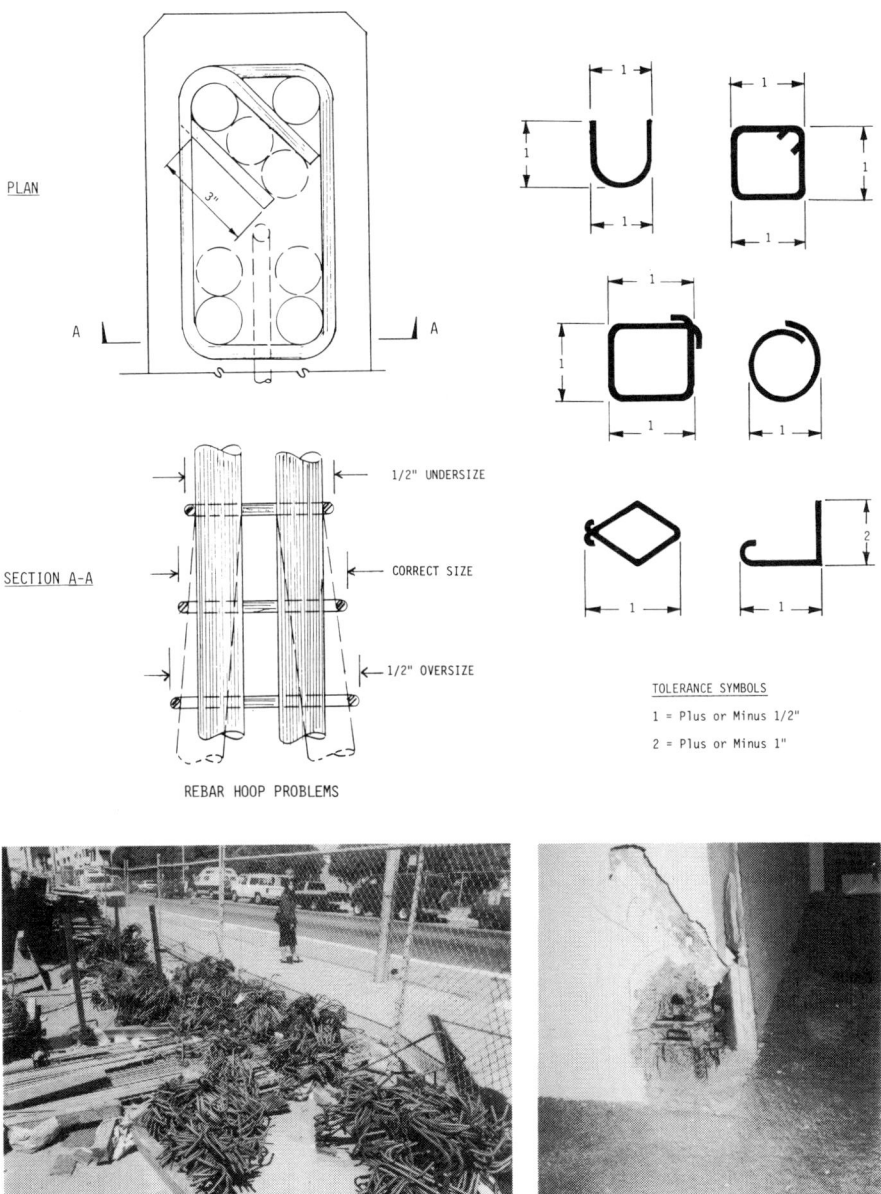

Fig. 5 Confinement Re-bar Problem

Confinement Re-bar Problems

The lack of consistent dimensions in re-bar confinement hoops (Fig.5) as well as the costly labor to install the many individual pieces has prompted a search for a better way to achieve reliable ductility in reinforced concrete and masonry.

65

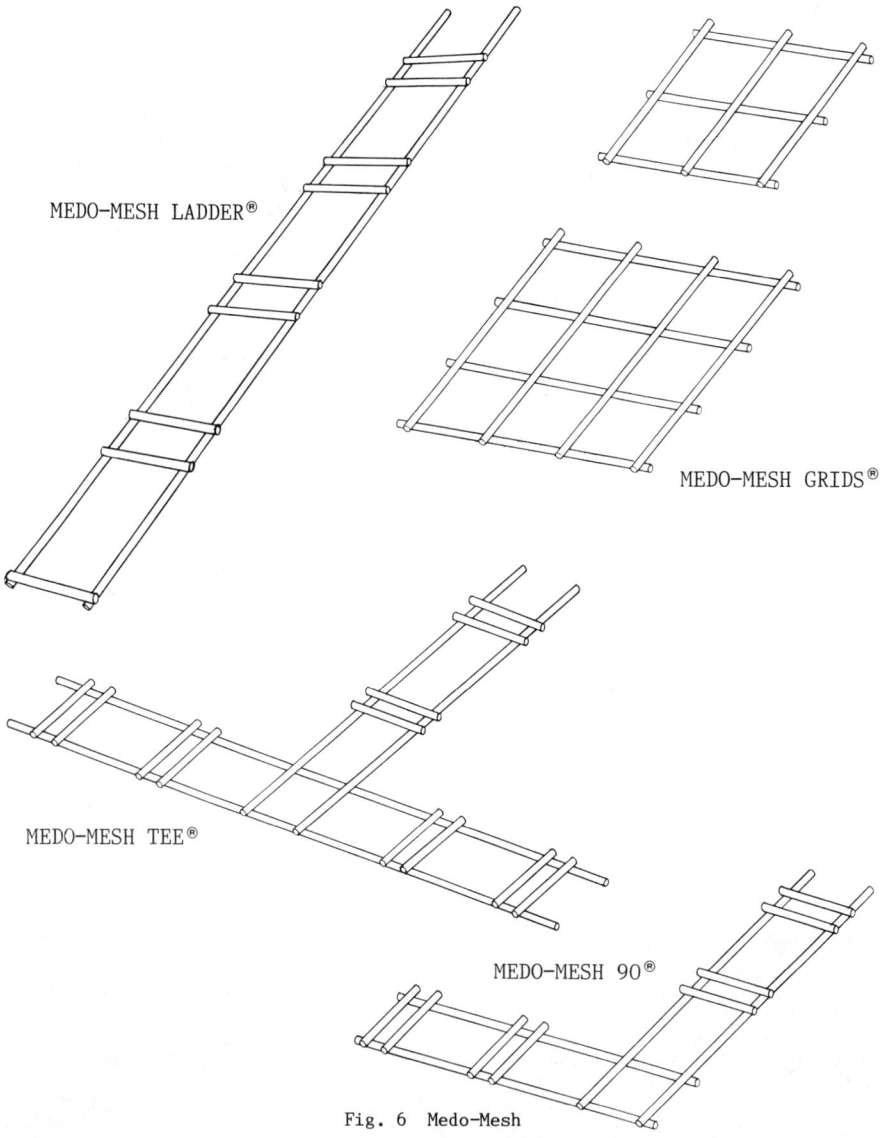

MEDO-MESH LADDER®

MEDO-MESH GRIDS®

MEDO-MESH TEE®

MEDO-MESH 90®

Fig. 6 Medo-Mesh

Medo-Mesh

In place of hooked re-bar hoops, a new proprietary product called Medo-Mesh has recently been introduced. For shear walls the Medo-Mesh has a ladder configuration in strips, tees and L-plan (Fig. 6). Instead of the $\pm$ 1/2" dimensional tolerance of re-bar hoops, MedoMesh is manufactured of high strength wire (72.5ksi) to $\pm$ 1/16" dimensional accuracy.

Cages assembled with large longitudinal re-bar confined by Medo-Mesh (Fig. 8) perform consistently well because of the uniform confinement the full height of the boundary element. During erection the boundary cages stand very erect and do not corkscrew, which saves very significant crane time.

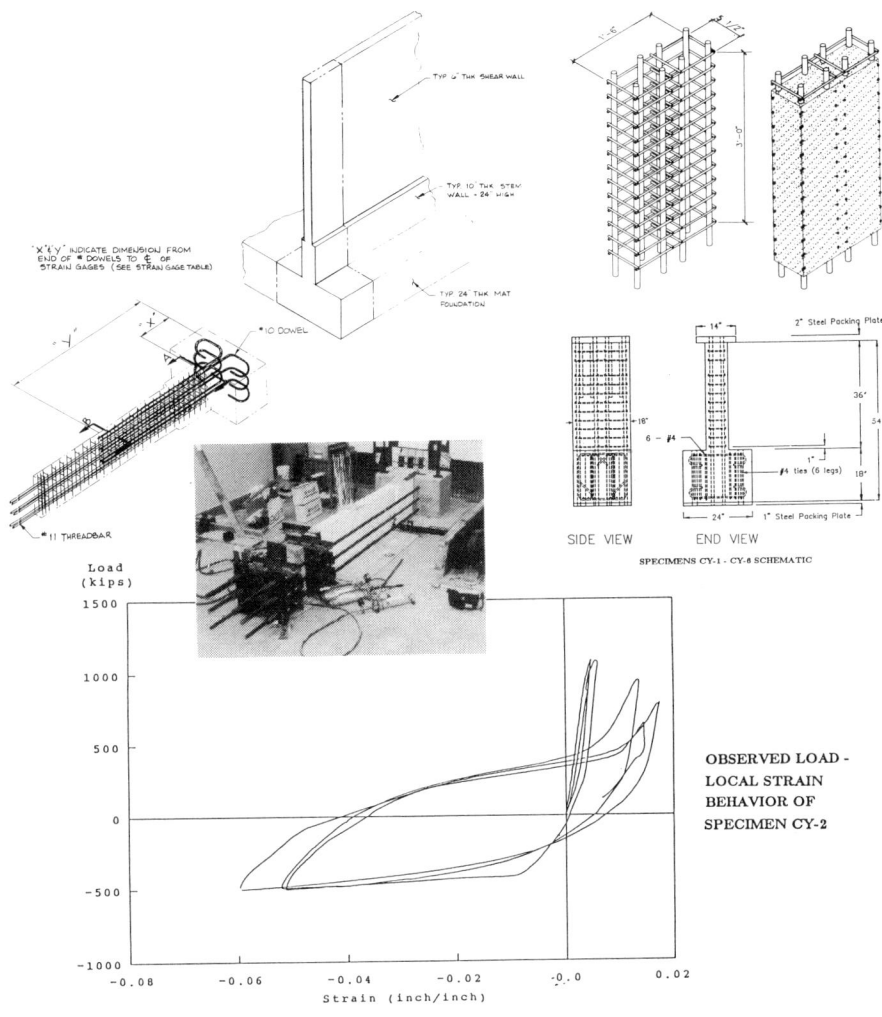

OBSERVED LOAD -
LOCAL STRAIN
BEHAVIOR OF
SPECIMEN CY-2

Fig. 7   Medo-Mesh Testing Program

Tests and Approvals

Full size specimen of shear wall boundary elements were tested in two Civil Engineering laboratories.

A two-foot segment of a 6" thick ductile shear wall was tested first at UCI, in 1988, and at EERC, in 1989, six 5 1/2" wide by 18" long by 36" high specimen were subjected to cyclic loading of up to 520 kips in tension and 1105 kips in compression.

A design ductility factor of $\mu = 5$ was confirmed by achieving a strain of 6% in tension and 2% in compression with a maximum compressive stress of 11,162 psi.

Since the preliminary design used the normal $\mu = 3$, given these higher ductility values, a 40% reduction of the design earthquake forces was permitted by the peer review board.

The final approved design using Medo-Mesh resulted in the thinner 7" thick walls which reduced seismic lateral forces and reduced minimum wall reinforcement further beyond the initial 40% reduction of earthquake resisting concrete and reinforcement steel in the walls and mat foundation.

67

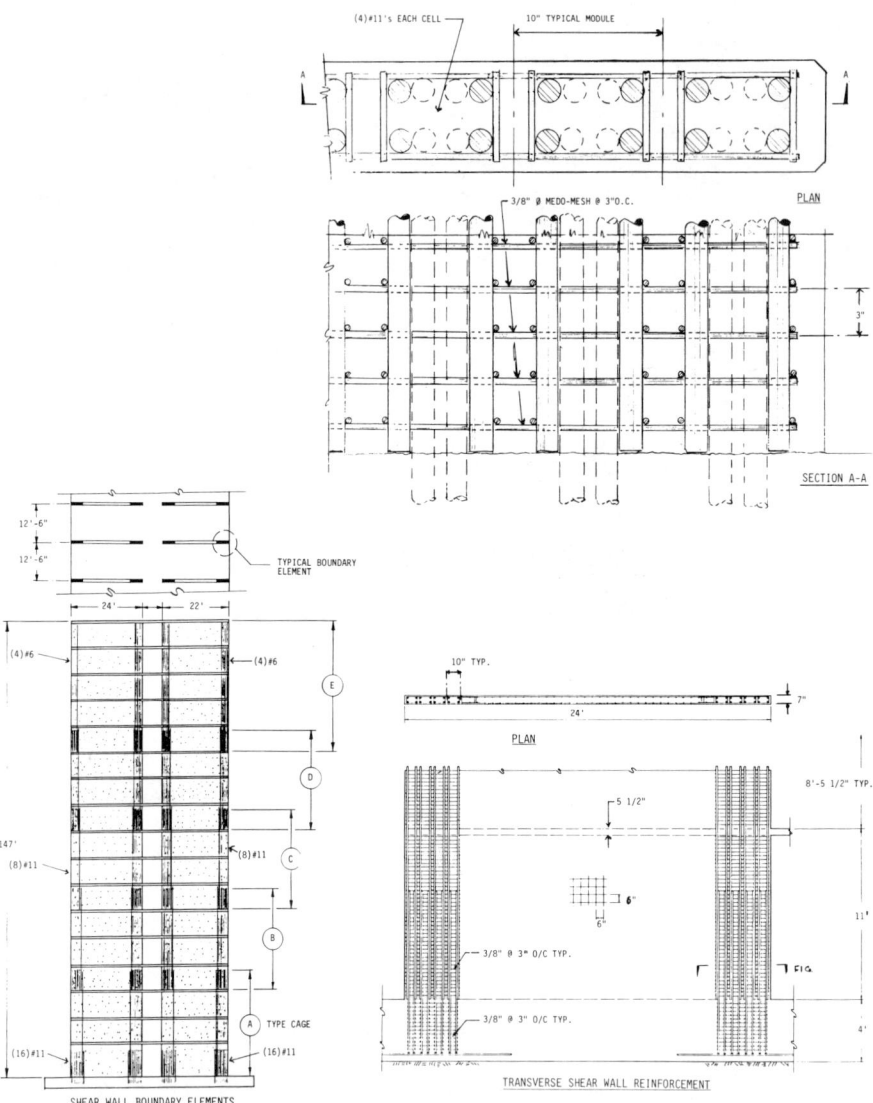

Fig. 8 Final Structural Design

Final Design of 17-story SFSU Structure

Shown in Fig. 8 is a typical 7" thick transverse load bearing ductile shear wall. Important to the economy of the system is locating the resisting reinforcement as far as possible from the center of the wall length. The Medo-Mesh ladders are designed to allow this concentration of re-bar and still leave a vertical passageway for wet concrete and the vibrator each ten inches along the wall's length.

68

Fig. 9 Medo-Mesh Installation Technique

Medo-Mesh Installation

For the construction of the SFSU building, compact bundles of Medo-Mesh ladders (Fig. 9) were delivered to the site. Using conventional horses, the two upper-most re-bar are charged through the bundle. The Medo-Mesh now suspended from these bars is spread out and positioned at the standard 3" spacing. Subsequent longitudinal bars are charged so that each cell holds four #11 re-bar at ten inch (10") on center.

The boundary cages, four floors high, are made and lifted in one piece which saves significant assembly and erection crew labor and crane time.

The inherent rigidity of the many welded joints, and the dimensional accuracy makes possible rapid installation of Medo-Mesh cages in thin wall forms.

Structural Response to Loma Prieta Earthquake

The Loma Prieta earthquake struck the SFSU project when it was at 14th floor, so in essence the concrete structure was 90% completed.

The contruction workers felt very significant movement at the 14th floor where they were pouring concrete.

Upon visual inspection, no evidence of distress was found. In fact the hairline cracks normally observed in reinforced concrete subjected to this level of ground motion were not found.

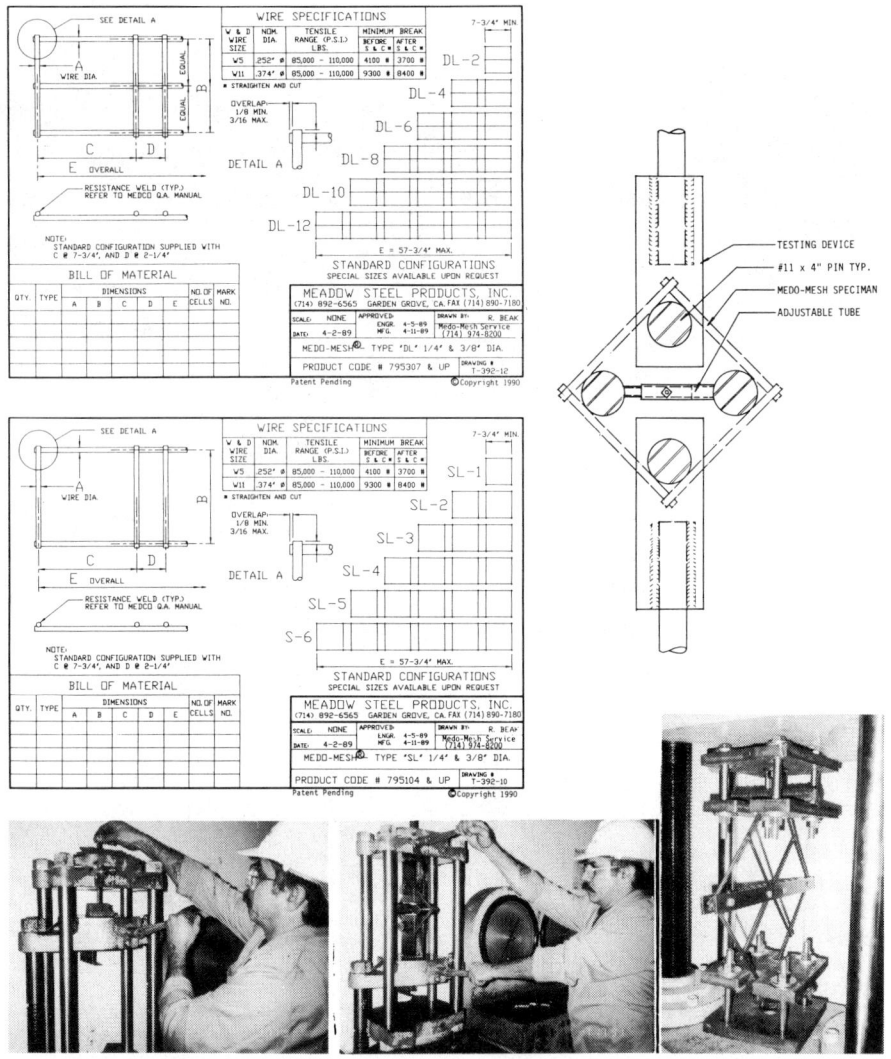

Fig. 10 Quality Assurance Testing

Manufacturing Quality Assurance

Medo-Mesh confinement reinforcement is manufactured under a stringent quality assurance program administered by and independent testing laboratory approved by the International Conference of Building Officials.

The high strength wire for the Medo-Mesh products is first tested in tension for conformance with ASTM and ACI tensile/strain specifications.

The second test is a welded tee-joint test which is used to assure that the wire breaks before the weld joint.

A third test is the "bursting test" which simulates the restraint to the attempted outward movement of the corner re-bar. Again, the weld joint must be stronger than the wire for the part to pass inspection. Two percent (2%) of the units are tested in this way.

70

Proceedings of the Second Conference on Tall Buildings in Seismic Regions
55th Regional Conference
May 16 and 17, 1991, Los Angeles, California

SEISMIC RESISTANCE OF STEEL-COMPOSITE-SECTION (SCS) FRAMES

by Jack G. Bouwkamp (#) and Bernd Schneider (##)

ABSTRACT

The paper presents a composite concrete-steel system with concrete in-
filled steel sections developed originally as a construction system
particularly suited to resist fire exposure. Preliminary findings of an
extensive cooperative experimental research program to develop structural
solutions for the use of this system in earthqake prone regions are
presented. Specifically, the test results of forced-displacement studies
on two full-scale, two-bay long and two-story high moment resistant
frames are presented. The beneficial effect of the concrete is illustrated as well as the excellent ductile behavior of these specific composite
frames designed both with fully welded beam-column connections and with
joints in which the beam endplates are HS bolted to the column flanges.

INTRODUCTION

Considering the characteristic properties of concrete under fire exposure
as compared with steel, it is advantageous to use concrete in a composite
form with standard rolled steel sections. From a strength point of view,
the use of steel sections combined with concrete offers in principle a
reduction of the overall cross-sectional dimensions as compared to pure
reinforced concrete sections.

These considerations have led in the early eighties to the development of
the so-called AF 30/120 System which permits the design of composite
sections for fire resistance times of 30, 60, 90 and 120 minutes. The AF
System was developed by ARBED (1,2) and uses standard rolled steel sections with concrete placed on either side of the web between the flanges
(Fig. 1). In addition, longitudinal reinforcing bars and stirrups are
designed to increase the load resistant capacity and to secure the in-
filled concrete. The stirrups are either placed in halves penetrating the
web through prepunched holes or are bent inward at the web and welded.

A composite steel-concrete composite frame structure designed according
to the AF 30/120 process still has the characteristics of a common steel
structure. The steel beam and column elements are fabricated in the shop
with the reinforcement and concrete being placed by a concrete contractor
either at a yard or at the construction site prior to erection. The
connections are typical to those used in normal steel design. For non-
moment connections standard shear tabs are commonly used. In that case
concrete will be deleted in the joint region to permit the placement of
HS bolts. After erection such regions are either concreted or filled with
fire resistant material to protect the inner steel parts of the sections
and connection from possible future critical fire exposure. In recent
years a beam-seat using a thick steel plate welded to the column flange

(#)Professor, Institute of Steel Construction, Technical University of
Darmstadt, Germany
(##)Research Assistant, same Institution.

has been developed. A single 3/4 in. pin secures the position of the beam horizontally during erection. The overall interconnection between beams on either side of a column is provided by the reinforced concrete slabs which are connected to the beams by means of headed shear studs.

In order to develop a beam-column moment-transfer , beam end-plates are typically used. These plates are welded to the beams and high-strength bolted to the column flanges. In case the design calls for bolts being placed immediately below and above the upper beam flange , the in-filled concrete in the joint region will be deleted to permit erection. These areas are later concreted or filled with specific fire resistant material. With HS bolts located only immediately above and below the upper and lower beam flanges, respectively, in-filled concrete can be placed over the full length of the beam prior to erection.

"SCS" SEISMIC RESISTANT DESIGN CONSIDERATIONS

Although composite moment resistant frames have been designed for horizontal loads, the behavior under simulated earthquake loads had never been studied. In order to assess this behavior and to develop guidelines for the earthquake resistant design of the AF 30/120 System, a cooperative research program was initiated in 1987. Funded by the European Convention for Constructional Steelwork (ECCS) and managed by ARBED-Research, Luxembourg, the experimental research was carried out at the Technical Universities of Darmstadt (Germany), Liege (Belgium), Milan (Italy) and Wuppertal (Germany).

In developing earthquake resistant ductile beam-column moment connections, numerous joint designs for both exterior and interior beam-column connections have been tested. However, in order to assess the potential cyclic moment resistance of "SCS" System beam-column connections in general, it was decided not only to study typical moment connections but also connections designed basically as shear connections. The latter connections, should still exhibit a certain moment capacity due to the internal force couple resulting from the horizontal force resistance of the shear-tab or beam-seat connection and the composite floor slab. Information on the cyclic moment resistant response of these simple connections was deemed important in order to evaluate the effectiveness of such design solutions for structures in areas of low seismicity.

Based on the overall objectives it was decided to study the cyclic behavior of a beam-column joint configuration with a HEB 300 column section and a HEA 260 beam section. The composite beam-floor behavior was to be studied means of a composite concrete slab having a width of 100cm (appr. 40 in) and a thickness of 12cm (appr. 5 in). A HEB 300 section has overall dimensions of 300x300 mm (appr. 12x12 in), a web thickness of 11 mm (0.433 in) and a flange thickness of 19 mm (0.748 in). The HEA 260 beam section has a width of 260 mm (10.24 in) and a height of 250 mm (9.84 in). Web and flange thicknesses are 7.5 mm (0.295 in) and 12.5 mm (0.492 in), respectively. Figure 2 shows the selected beam and column sections with the in-filled reinforced concrete as well as the slab section.

In order to assess the potential earthquake resistance of the different exterior and interior beam-column connections, an extensive experimental research program was performed at the Politechnico di Milano (3). Joints were tested under typical cyclic displacement-controlled load conditions. In all tests the columns had a length of 12 ft (distance between adjacent inflection points). For exterior column joints a 6 ft beam stub was connected to the column at mid-length. For interior column joints beam stubs with a length of 6 ft each were placed on either side of the column section. In developing moment resistant joints both fully welded beam-column joints and connections with beam-end plates, HS-bolted to the column flanges, were studied in detail. In order to evaluate the influence of the concrete on the steel elements, several joints of each basic moment-resistant connection type were studied, both with and without a concrete slab and with and without in-filled concrete.

GENERAL OBSERVATIONS

In developing the joint test program it was recognized that in principle in-filled concrete will increase the local stability of the steel sections; specifically, the beam flanges and the column web in the shear-panel zone. In the latter region concrete also increases the effective shear resistance. Because of the concrete, the width-to-thickness ratio requirement for beam flanges could be less restrictive and the need for doubler plates in the column-web panel zone could be reduced.

Considering the HEA 260 beam sections, the flange (260 x 12.5 mm) has a width-to-thickness ratio b/2t of 10.4. This value is considerably larger than the US SEAOC (4) specified value of 8.0 for a steel with a yield stress of 30 kN/cm^2 or 42.6 ksi. For non-earthquake load conditions the US AISC Code (5) specifies a maximum permissible value of 10.0. European seismic design recommendations (6) specify for a steel with the same yield stress an effective b/2t ratio of 8.85. By the design of the exterior beam-column connections, the opinion prevailed that, with an expected 4000 psi concrete in the column panel zone, the additional compressive concrete strutt action would be sufficient to withstand the panel shear forces. Hence, a doubler plate was not deemed nessesary.

In order to assess the effect of the composite action on the connection behavior it is interesting to consider the test results of three fully welded exterior-column beam connections under cyclic displacement-controlled loads. One of the connections was formed by bare steel column and beam sections. The second joint consisted of a composite column and a bare steel beam. The third joint had fully composite column and beam sections. Typically, all three specimens had continuity plates. Results showed an excellent ductile behavior of all three connections. However, in the first specimen, which exhibited a maximum connection capacity of 400 kNm (288 Kft) with a ductility ratio of 10, the ductile behavior resulted from extensive cyclic shear distortion of the the column shear panel. In the second specimen the concrete in the column prevented virtually any column shear-panel distortion. In fact the ductile behavior was concentrated entirely in the beam. After reaching a connection capacity of 400 kNm at a ductility ratio of about 3, beam-flange buckling reduced the capacity to about 300 kNm. This value remained constant up to a ductility ratio of about 8. The fully concreted third specimen showed an excellent hysteretic behavior with a maximum connection capacity of about 460 kNm at a ductility ratio of about 5. Subsequently, because of concrete shear failure in the panel zone, the connection capacity fell to about 400 kNm and remained constant up to a ductility ratio of about 8. Before reaching the 460 kNm connection capacity the ductile behavior of the beam was about twice as large as in the column shear-panel region. However, following concrete failure in the panel zone the ductile behavior was virtually entirely concentrated in the steel shear panel.

The above described results indicate both the beneficial effect of composite in-filled steel sections but also the potential risk of loss of connection resistance due to concrete failure in the column shear panel zone. In general, the concrete in the beam section significantly increases the beam flange stability at the connection.

Test results of interior beam-column joints (with HEB 300 columns) showed the need for doubler plates if column shear-panel distortion is to be prevented. Although the ductile behavior of the shear distorted steel column web showed extensive strain hardening and an increasing connection capacity with high ductility ratios, the authors consider it desirable to strengthen the column web in the shear panel zone. Only then is it possible to prevent shear panel distortion and to fully develop the potential overall horizontal load resistance of a composite steel frame.

In the second phase of this program both one-story high, single- and double-bay sub-assemblages as well as two-story two-bay frames have been studied. At the Universities of Liege and Wuppertal sub-assemblages were tested with HEB 300 columns (length about 3 m or 10 ft) interconnected at mid-height with about 5 m (16.4 ft) long HEA 260 beams. The full-scale frame tests were carried out at the Technical University of Darmstadt. A description of the test design, set-up and performance, as well as some preliminary results are presented in the following.

FULL-SCALE "SCS" FRAME TESTS

Test Design

The two, two-bay, two-story moment resistant composite frames had overall dimensions of 2 x 5.00 m (2 x 16.4 ft) in length and 2 x 2.75 m (2 x 9.0 ft) in height (Fig. 3). Columns and beams had the same concrete in-filled steel sections as those investigated in all earlier tests. In both frames the steel HEA 260 beams were topped with a concrete composite slab with width-to-thickness dimensions of 100 x 12 cm (39.4 x 5 in) and connected to the beam by headed shear connectors. Both frames were fixed at the base and represented the lower stories of a ten-story frame designed for a base shear coefficient of about 8 percent. A third frame with the same overall dimensions but with only one bay eccentrically braced was also tested. Results of this study will not be subject of this paper.

The two moment resistant frames had different connection designs, namely one with end-plated beams connected with HS bolts to the column flanges and one with fully welded beam-column connections. In the earlier tests, several different end-plated connections had been studied involving different bolt layouts and end-plate thicknesses. Also, the effect of the position of the so-called continuity plates had been studied, considering not only plates typically in-line with the immediately adjacent beam flanges but also plates positioned at the upper and lower ends of the beam end-plates. Based on these earlier tests it was decided to design the first frame with HS bolts located just above the beam upper flange and just below the beam lower flange (Fig. 4). This solution has the practical advantage that the composite beams can be fully prefabricated prior to erection. With the extreme layout and size of the HS bolts (in this case 30 mm or 1 3/16 in) the end-plate thickness will typically be large. Basing the design of the end-plates on a connection-capacity of 1.2 Mp (Mp = plastic moment of composite beam section) and using the yield stress as allowable the end-plate thickness becomes 50 mm or 2 in. Furthermore, based on the large end-plate thickness and associated improved force transfer between beam and column, it was decided to locate the so-called continuity plates directly in line with the upper and lower edges of the end-plates. Realizing that heavy washer plates against the inside of the column flanges would be necessary to accommodate the load transfer of the large bolt forces into the column section, it was decided nevertheless to conform to standard recommended end-plated joint design configurations.

The second test frame was designed with fully welded moment connections (Fig. 5). In this case the continuity plates were placed in line with the upper and lower flanges of the adjacent beams. It should be noted that in both frames doubler plates to increase the column shear-panel resistance had been omitted in order to evaluate the effect of the in-filled concrete in the panel zones of both exterior and interior columns.

Test Performance

In order to provide a reaction to the horizontal forces to be applied at the two floor levels, a special reaction frame had been designed (Fig. 6). The overall system consists of a W14x370 base-beam anchored to a concrete tie-down slab and connected by means of a cross beam at the base to two parallel triangular reaction trusses spaced at 100 cm or about 40 in. Each of those trusses are connected also to two cross beams which in turn are anchored to the tie-down slab. Because of a tight time schedule it was necessary to expedite the testing. Hence, all three frames were erected in an area immediately adjacent to the reaction frame. This arrangement permitted the concreting of the floor slabs in one pour and resulted in an effective sequential testing of the different frames.

The horizontal test loads were introduced by means of double acting actuators with maximum force capacities of about 750 kN in tension and 1000 kN in compression (165 and 220 kips, respectively). The upper and lower actuators had double-amplitude capacities of 130 cm (51.2 in) and 100 cm (39.4 in), respectively. With a hydraulic servo-control system the tests were controlled by defining the displacement history at the upper floor level. The load at the lower floor level was load-controlled at 50% of the load measured at the upper floor level. To prevent direct loading of the left exterior column, special loading jokes (Fig. 7) were used to introduce the horizontal loads at mid-span of the left bay. Vertical loads were not applied. During the tests 24 strain gages, 9 LVDTs and 29 wire extensometers were recorded continuously to capture the response of the test frame under cyclically increasing alternating displacements.

Test Results

The overall test results of the bolted and welded frames are presented in
Fig. 8. This figure presents for both frames the horizontal displacement
at the upper floor level versus the total applied horizontal load (1.5
times the upper floor load). A preliminary observation shows that the
initial stiffness for both frames was virtually the same; a phenomenon
which can be contributed undoubtedly to the composite in-filled concrete
and the very heavy beam endplates. In fact, results seem to indicate that
even after considerable yielding (horizontal top-floor displacement of
-200 mm or -7.9 in) the total horizontal force for both frames is practically the same (-1 MN or -220 kips). This deformation reflects a load
effect whereby the actuators push against the frame. Unfortunately, under
a pull-load the capacity of the actuators is limited to a maximum of just
over + 1MN. This is due to the 1-to-0.5 load ratio set for the test
control (1.5x 750= 1025 kN). Hence, in the first frame test the specified
displacements requiring a load larger than + 1 MN could not be attained
and resulted in an unsymmetrical hysteretic load-displacement response.

Observing the negative (pushing) load ranges in Fig. 8 , it seems that
the frame with the end-plated connections has experienced a larger force
resistance (up to - 1.1 MN) than the frame with welded connections. This
larger resistance is associated with a top floor displacement of about
360 mm (14.2 in), reflecting a total drift of about 6.5%. Although the
composite frame with welded connections reached under a still increasing
load a maximum horizontal displacement of about 230 mm (9 in), fracture
of some welds between the beam flanges and the exterior column caused a
loss of resistance in the third cycle of that displacement. This displacement reflected a total drift of about 4%. In the frame with end-plated
connections failure due to bolt and column flange tear-out failure at
several locations occurred rapidly after the maximum displacement had
been reached.

Other than the above interpretation both frames exhibit serious column
base flange distortions because of a disproportionately high stiffness in
those base regions. Because of the high shear panel forces in interior
beam-column regions both frames experienced high distortions in those
regions and local failure of the in-filled concrete. The exterior columns
in both frames exhibited excellent shear panel behavior without local
concrete failure. However, because of this basically rigid behavior of
the exterior columns, the exterior beam to column connections became
highly strained (following extensive yielding of all column base regions)
and lead ultimately to failures at the connection interface.

CONCLUSIONS

Preliminary results of experimental studies on the earthquake resistance
of a composite steel-concrete system with concrete in-filled steel beam
and column sections illustrated that such a system is well suited to
resist serious earthquake exposure. Other than providing an excellent
fire protection for the steel sections, the concrete improves the local
stability of the steel beam and column sections and increases the shear
resistance of the column shear panel. Results indicate that both welded
beam-column connections and connections with beam-endplates bolted to the

column flanges can be used effectively. Future studies may focuss on further enhancing the column shear-panel resistance and on improving the load transferring behavior of the high strength bolts in beam endplated connections. These considerations would become important by applications in truly high-rise building frames.

REFERENCES

(1) Jungbluth, O., Feyereisen, H. "Verbundkonstruktionen mit erhöhter Feuerwiderstandsdauer", Int. Conf. A Challenge for Steel, Luxembourg, 1980.
(2) Schleich,J.B. et al, "A new technology in fireproof steel construction", Acier-Stahl- Steel, nr.3 1983.
(3) Ballio, G., Plumier, A., Thunus, B.,"Influence of Concrete Behaviourof Composite Connections", IABSE Symposium, Brussels, 1990.
(4) "Recommended Lateral Force Requirements and Tentative Commentary", Seismology Committee, SEAOC, 1988.
(5) "Manual of Steel Construction - 8th Ed" and "Manual of Steel Construction - LRFD - 1st Ed", AISC, New York.
(6) "European Recommendations for Steel Structures in Seismic Zones", ECCS, no.54, Bruxelles.

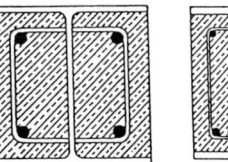

column section   beam section

FIG. 1   STEEL COMPOSITE SECTIONS   (SCS)

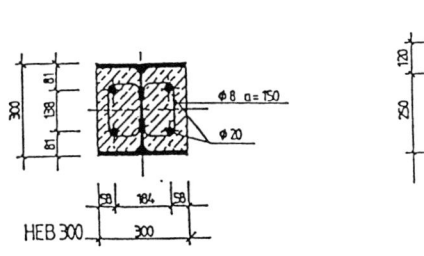

FIG. 2. COMPOSITE TEST SECTIONS
         dimensions in mm (0.04 in)

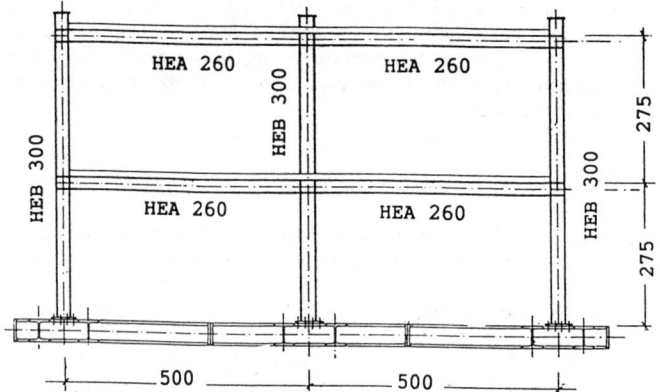

FIG. 3. TEST FRAME - dimensions in cm (0.394 in)

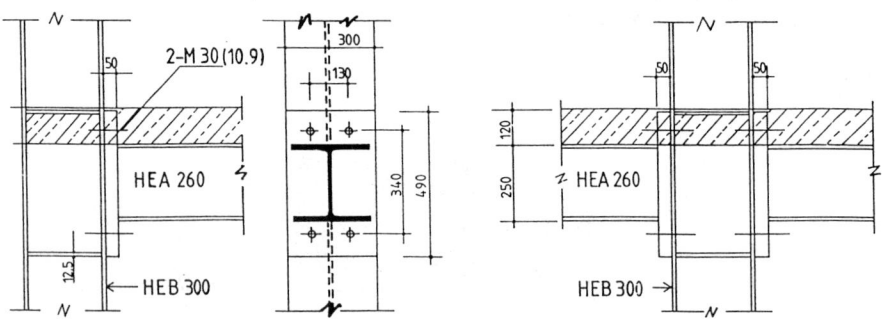

FIG. 4. FRAME 1 - END-PLATED JOINTS dimensions in mm (0.04in)

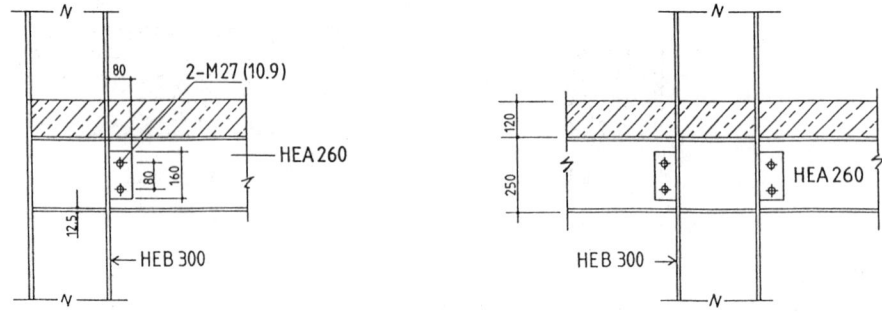

FIG. 5. FRAME 2 - WELDED JOINTS dimensions in mm (0.04 in)

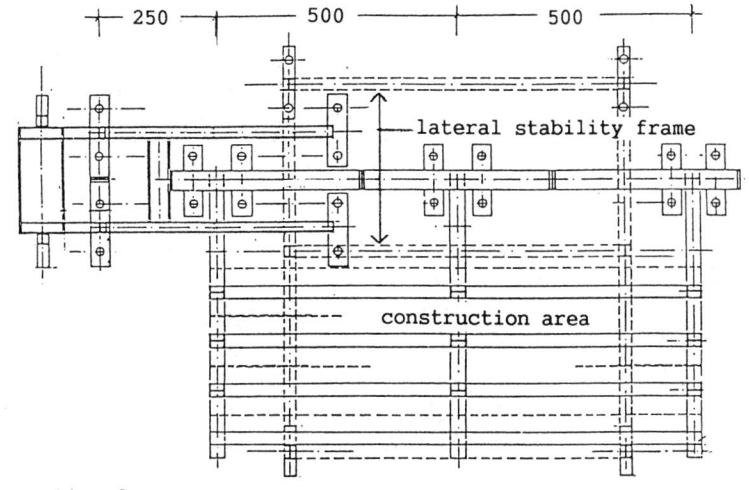

FIG. 6. TEST SET-UP   dimensions in cm   (0.394 in)

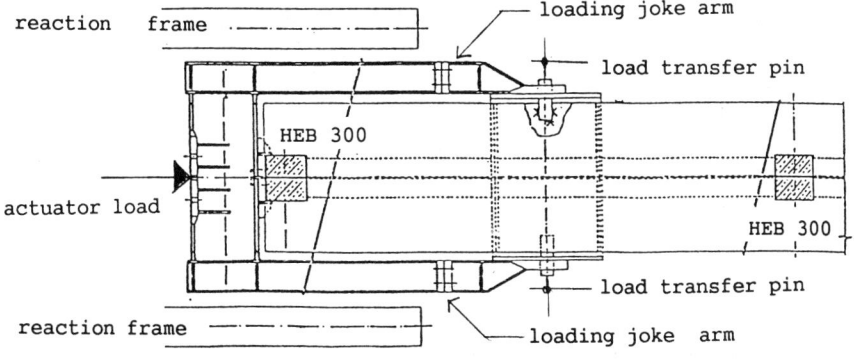

FIG. 7. LOADING JOKE (LEFT BAY)

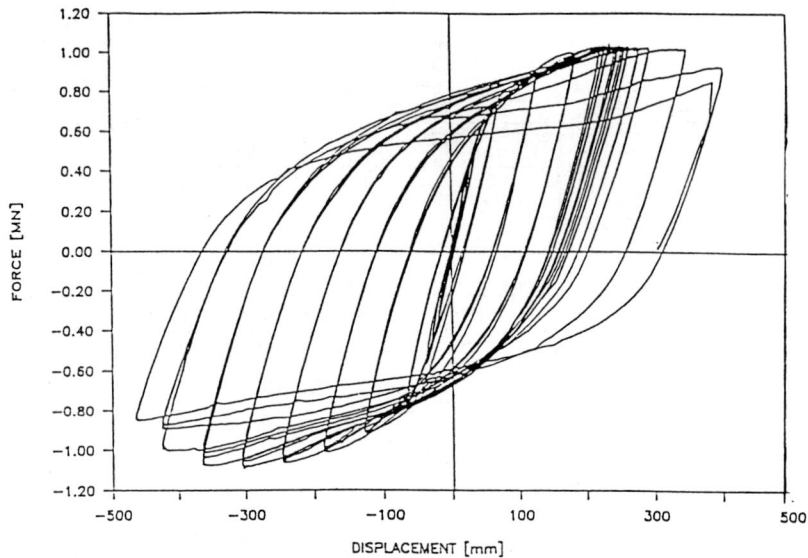

frame with end-plated connections

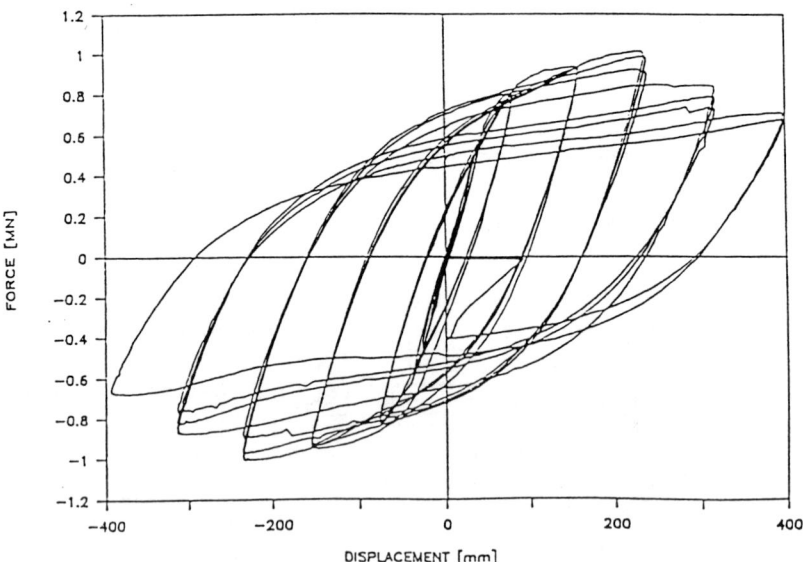

frame with welded connections

FIG. 8.   HYSTERETIC FORCE-DISPLACEMENT BEHAVIOR

80

Proceedings of the Second Conference on Tall Buildings in Seismic Regions
55th Regional Conference
May 16 and 17, 1991, Los Angeles, California

IMPLICATIONS OF THE 1989 NEWCASTLE EARTHQUAKE ON THE DESIGN AND CONSTRUCTION OF TALL BUILDINGS IN AUSTRALIA

by David Brunsdon and James Forbes
Irwin Johnston & Partners
Consulting Engineers, Sydney, Australia

Summary

The December 28, 1989 Newcastle earthquake was the first Australian earthquake to cause loss of life and significant property damage. This Richter magnitude 5.6 earthquake resulted in the loss of thirteen lives and gave rise to an insurance bill which is predicted to total $A 800 million. The majority of this damage in Australia's sixth-largest city was sustained by unreinforced masonry buildings and elements, with well designed and detailed structures generally performing satisfactorily.

While the effects of this earthquake are not major by international standards, it has heralded the realisation that such events could occur in the other major cities of Australia. State capitals such as Sydney, Melbourne, Brisbane, Adelaide and Perth each contain a number of tall buildings in the range 40-60 storeys.

The relatively moderate nature of seismic activity within the Australian continent is such that wind rather than earthquake forces generally govern the strength design of low-rise and high-rise structures. However the Newcastle earthquake has demonstrated that new consideration needs to be given to the seismic detailing of structural and non-structural elements. It is therefore apparent that a probability or risk-based approach which emphasises the importance of laterally restraining building elements adjoining areas of public access is most appropriate.

This paper backgrounds the state of tall building design and construction in Australia with particular emphasis on seismic aspects. The Newcastle earthquake and the resulting damage is summarised and conclusions drawn as to the implications of this earthquake on the future direction of tall building design and construction in Australia.

1. INTRODUCTION

Although the December 28, 1989 Newcastle earthquake was not a major seismic event by international standards, significant damage was sustained by more than 10,000 buildings, many of which were constructed of unreinforced masonry with some nearing the end of their economic life.

Australia has previously experienced earthquakes of greater magnitude and intensity than this event, and the Newcastle region itself recorded magnitude 5.3 earthquakes in both 1868 and 1925. However this was the first such event to significantly affect a highly populated area in Australia, and accordingly neither the people of Newcastle nor the local authorities were prepared for the event and its immediate aftermath. While the damage was similar in extent to that caused in Darwin by Cyclone Tracy in 1974, the nature of the distress sustained by buildings in Newcastle was quite different.

While the Australian Earthquake Code, AS 2121, has been in use for eleven years, the zoning of the continent is such that few structural engineers are familiar with seismic design requirements. In excess of 80% of Australia, including Newcastle, has a Zone 0 designation, as shown in Figure 1. Of the main cities, only Adelaide attracts a rating of Zone 1, at which level seismic considerations begin to influence some aspects of building design.

The philosophies and provisions of AS 2121 are generally based on the SEAOC codes, with zone numbers correlating with the Uniform Building Code zone designations. The minimum force level for buildings in Zones A, 1 and 2 is defined as 0.02 times the building seismic weight.

This code was under review at the time of the earthquake and an updated version is due to be published during 1991.

2. **THE DECEMBER 28, 1989 EARTHQUAKE**

There is a significant lack of seismic instrumentation throughout Australia, and this severely restricted the extent and quality of information recorded from this earthquake. Seismologists estimated the epicentre of the earthquake to be 14 km west-south-west of the centre of Newcastle, with a focal depth of approximately 11 km.

The greatest damage to buildings was recorded in the Central Business District of Newcastle and adjacent inner city suburbs. The intensity of ground motion appears to have peaked in this area at between MMVII and MMVIII on the Modified Mercalli scale. Comparatively little damage to major civil engineering structures and lifeline services was recorded. The earthquake was felt over a radius of approximately 310 km as illustrated in the isoseismal map in Figure 2. There were only two aftershocks recorded, with a maximum magnitude of 2.1.

The variation in depth of alluvial and fill material overlying the bedrock was the predominant geological feature influencing the extent of damage to buildings. There is a strong correlation between the areas where MMVIII intensities were recorded and the zones of deep alluvium and fill as shown in Figure 3. The layers of alluvially deposited sand have total thicknesses generally in the range 10-20m in the inner suburban areas where the worst building damage was recorded. Based on comparisons of MM intensities from adjacent rock and alluvial areas, empirical estimates suggest an average amplification factor of between two and four is attributable to the alluvial material.

Within the greater Newcastle area approximately 25,000 buildings of all categories sustained some form of damage as a result of the earthquake, with 10,000 suffering moderately severe distress. Modern buildings and those framed principally of reinforced concrete and structural steel fared reasonably well. Notable exceptions were The Workers Club (refer Figure 4) and Junction Motel buildings which suffered partial structural collapse and were subsequently demolished.

The majority of the damage was to unreinforced masonry elements, and followed patterns familiar in more earthquake-prone countries. The lack of lateral restraint to masonry elements such as parapets, gable ends and chimneys was a direct cause of most of the failures, including some in relatively modern construction. The failure of a large gable end wall of a public school building is illustrated in Figure 5.

The poor performance of older masonry in general was also caused by the widespread corrosion and failure of brick cavity ties (refer Figure 6) along with degradation of the mortar resulting from an excessive use of lime. This often led to the failure of sections of the outer skins of cavity walls under face loading.

The earthquake itself was of a relatively short duration, and it was apparent that many badly damaged masonry buildings would have collapsed due to a lack of reserve capacity if the earthquake had continued for longer, or if there had been any significant aftershocks.

3. TALL BUILDING CONSTRUCTION IN AUSTRALIA

The history and development of tall buildings in Australia has for the most part parallelled that of other developed western countries. Up until the 1920's most buildings were of load bearing masonry construction with a maximum height of 10 to 12 storeys. Steel and reinforced concrete began to take over in the period between the two world wars with masonry still playing an important role in facades and service cores. Building heights of up to 15 storeys were common.

In the early 1950's fully framed structures began to emerge with lightweight curtain walls, generally of glass, following the then current International Architectural trend. Steel framing was popular and by the early 1960's most major Australian cities could boast a modern building well in excess of 20 storeys.

The late 1960's and the decade of the seventies saw the emergence of more high rise buildings in Australia with both Melbourne and Sydney having a number of buildings well over 40 stories. Reinforced concrete had generally taken over as the preferred structural system for the vertical elements with composite steel and concrete floor systems being common in many buildings.

Facade treatment had swung away from full lightweight curtain walls towards precast concrete with a wide variety of veneer and polished finishes.

The decade of the eighties heralded an unprecedented growth in high-rise building design and construction in Australia. Building heights of 65 storeys were reached, with structural steel and reinforced concrete vying for position as the chosen structural system. Many buildings are of course a composite of both materials. The 65 storey Rialto building in Melbourne (refer Figure 7) is the tallest reinforced concrete building in the country, and the soon to be completed Chifley Tower in Sydney the tallest all steel building at 48 storeys.

The architectural expression of the latest 1990's buildings show a return to the classic era of tall buildings with the widespread use of polished stone in facades, particularly at low levels, and the clear definition of base, middle and top.

Structural design of these buildings has generally followed the "tube in tube" concept with considerable proportions of the lateral load being taken by the structure on the facade. This has been facilitated to some extent by the building code provisions for fire favouring deep spandrel beams and the trend toward smaller window openings for architectural and energy reasons . On a limited number of particularly slender buildings outrigger trusses, usually in steel, have been used to fully mobilise the external facade structure.

Advances in high strength concrete technology and the availability of excellent aggregates has meant that concrete strengths of 80-90 MPa (11600 - 13000 psi) are common in columns and shear walls, particularly in Melbourne.

As mentioned previously wind load is the primary and often the only lateral load required to be considered in the design of tall buildings in Australia. Wind tunnel testing is commonly used to determine wind parameters for use in design. As buildings become taller, parallel and cross wind accelerations are becoming the major design criteria. The passive damping characteristics of concrete frames and the present heavy facades mean that the use of active and tuned mass damping is rare in commercial towers in Australia at the present time.

4. IMPLICATIONS OF THE NEWCASTLE EARTHQUAKE ON TALL BUILDINGS IN AUSTRALIA

As noted above, well designed, detailed and constructed buildings in Newcastle generally performed satisfactorily during the earthquake. However this earthquake highlighted the general lack of awareness on the part of Australian designers and builders of the need to provide lateral restraint for masonry and other elements for all types of lateral load. This observation, coupled with the common use of brittle materials such as masonry in non-structural elements, makes it clear that greater attention must be given to elements where the consequence of failure would be catastrophic in nature.

The following implications are considered to be those most relevant to tall building design and construction in Australia:

4.1 Design of New Buildings

(i) Structural Elements

The consensus after the Newcastle earthquake is that the strength provisions of the existing Australian Earthquake Code are considered reasonable in terms of the assessed level of seismic risk. Furthermore, the historically short duration of Australian earthquakes suggests that it is highly unlikely that members in a tall building would be subjected to inelastic demands of a cyclical nature.

The Newcastle earthquake has highlighted the uncertainty surrounding seismic loadings in Australia and the paucity of recorded data. It is therefore important that key framing members be capable of performing in a ductile manner in the post-elastic range.

While members detailed in accordance with the current Australian steel and concrete codes possess nominal ductility, consideration should be given to enhancing the ductility of critical members such as lower level columns, and link beams between wall sections. This can be achieved at negligible extra cost in concrete sections typically by the addition of transverse reinforcement to increase both the shear capacity and the degree of confinement.

Australian designers and concrete technologists are continuing in their quest for higher and higher concrete strengths in columns to minimise element size. Such columns with code minimum transverse reinforcement can be expected to fail in a very brittle manner in seismic events. The Newcastle experience makes it imperative that a closer understanding by Australian engineers of the nature of even minor earthquake loading on tall buildings goes hand in hand with the development of these high strength concretes.

(ii) Non-Structural Elements

There are three categories of non-structural elements relating to tall buildings that should now be given greater consideration in seismic detailing, namely:

- Heavy exterior cladding (masonry or precast concrete)

- Interior masonry partitions

- Suspended ceiling systems

Heavy exterior cladding panels pose a considerable risk to both passers-by and people escaping from the building itself in the event of connection or restraint failure. As a consequence of wind load considerations traditionally dominating the design of such elements in Australia, most connections are detailed to behave elastically over a single cycle of loading only. Such connections have often also been rationalised to the minimum to suit erection requirements, leaving little margin for loads greater than anticipated by the designer.

Inadequately detailed interior masonry panels pose less of a risk to occupants, although many partition walls are constructed without any top restraint. However the cost of repairing or replacing damaged panels following even a minor earthquake can be appreciable. Of the relatively few modern medium-rise buildings in Newcastle, a significant number sustained cracking to masonry panels and finishes. A predominant cause of much of this damage was the common construction practice of mortaring the ends of masonry walls rigidly against structural columns and walls. The earthquake sway then induces inplane forces in the unreinforced masonry in excess of its capacity.

Similarly the loss of plaster tiles from the suspended ceiling system was reported in a number of Newcastle CBD buildings. In Australia, suspended ceiling systems are generally constructed without any form of restraint for the tiles in the event of differential sway between structural elements.

4.2 Construction of New Buildings

In a number of newer Newcastle buildings the earthquake damage also revealed various sins of omission and/or commission on the part of the original builder. The majority of these related to the masonry trade, involving brick ties being omitted or bent up without being engaged. The destructive forces of the earthquake have thus highlighted the problem with work that is not subject to regular engineering inspection and can be quickly covered up.

The Newcastle earthquake has however underscored yet again the shortcomings associated with the extensive use of subcontractors with a minimum of head contractor input and engineering supervision. In Australia however the principles of Quality Assurance are more rigorously applied in high-rise construction than for smaller scale buildings.

4.3 Existing Tall Buildings

The tallest buildings in Newcastle are only in the range 10 to 15 storeys, and as a group they sustained little damage as a consequence of the earthquake apart from the distress to non-structural items as outlined above.

It is clear that while new tall buildings in Australia can and should have their earthquake safety and integrity enhanced for negligible additional cost, there is not the same economic justification for upgrading the seismic performance of existing tall buildings given the perceived level of seismic risk. The addition of structural strengthening elements after a building has been completed and is in service requires significant expenditure in order to achieve any realistic reduction in risk.

However regular and detailed maintenance inspections have been shown to play a major role in reducing the seismic risks posed by older tall buildings, particularly with respect to facade elements.

5. CONCLUSIONS

The Newcastle earthquake was a relatively minor seismic event internationally in terms of magnitude and intensity, with the nature of damage to buildings being predictable given the predominant type and age of construction.

The significant effect of this earthquake is that it brings a realisation to Australian engineers that such seismic events can strike major cities many of which have a large number of tall buildings. Whilst this realisation will not revolutionise design principles for tall buildings, it is clear that greater attention must be paid to the detailing of structural and non-structural elements using measures common in more earthquake-prone countries. The following items highlight the main areas where this attention should be directed:

(i)   Additional shear and confinement reinforcement should be placed in critical concrete framing members such as wall link beams and lower levels of columns in order to enhance member ductility, particularly where concrete strengths in excess of 60MPa are utilised.

(ii)   Code seismic coefficients for non-structural wall and facade elements should be reformulated in terms of a probabilistic or risk-based approach which would generate much greater design loads for heavy exterior cladding panels and their connections in comparison to interior partition walls.

(iii)   Greater emphasis should be placed on the lateral restraint and seismic separation of non-structural elements, including the need to detail suspended ceilings to accommodate building drifts.

It is considered that the above detailing aspects would significantly reduce both the degree of building damage and risk to individuals in the event of a major earthquake affecting one of Australia's main cities.

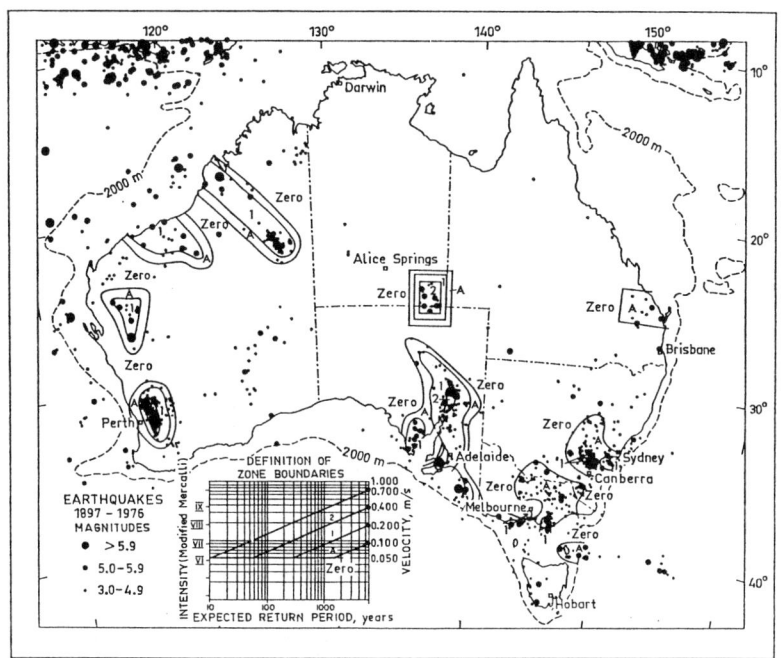

Figure 1:   **Seismic Zone Map of Australia**

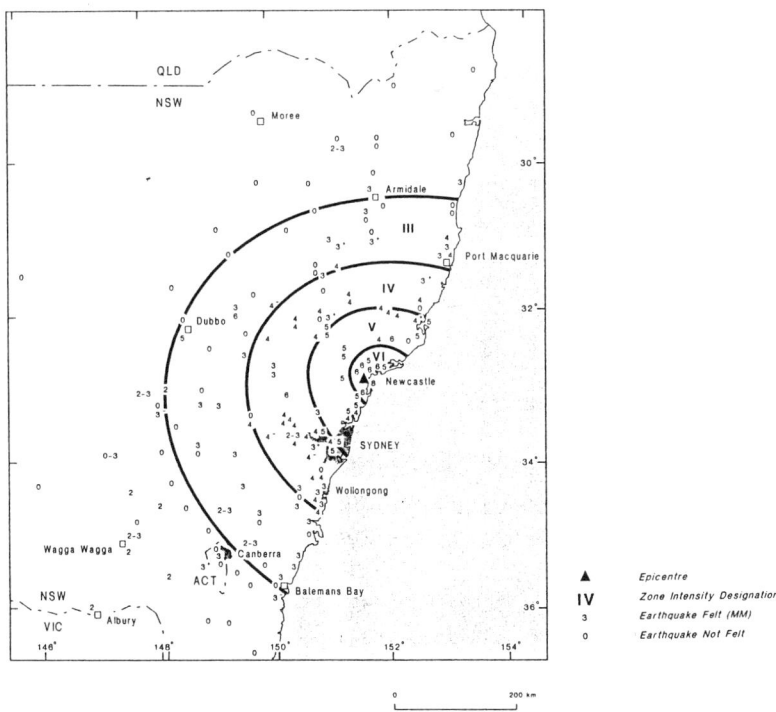

Figure 2:   **Isoseismal Map of the 1989 Newcastle Earthquake**

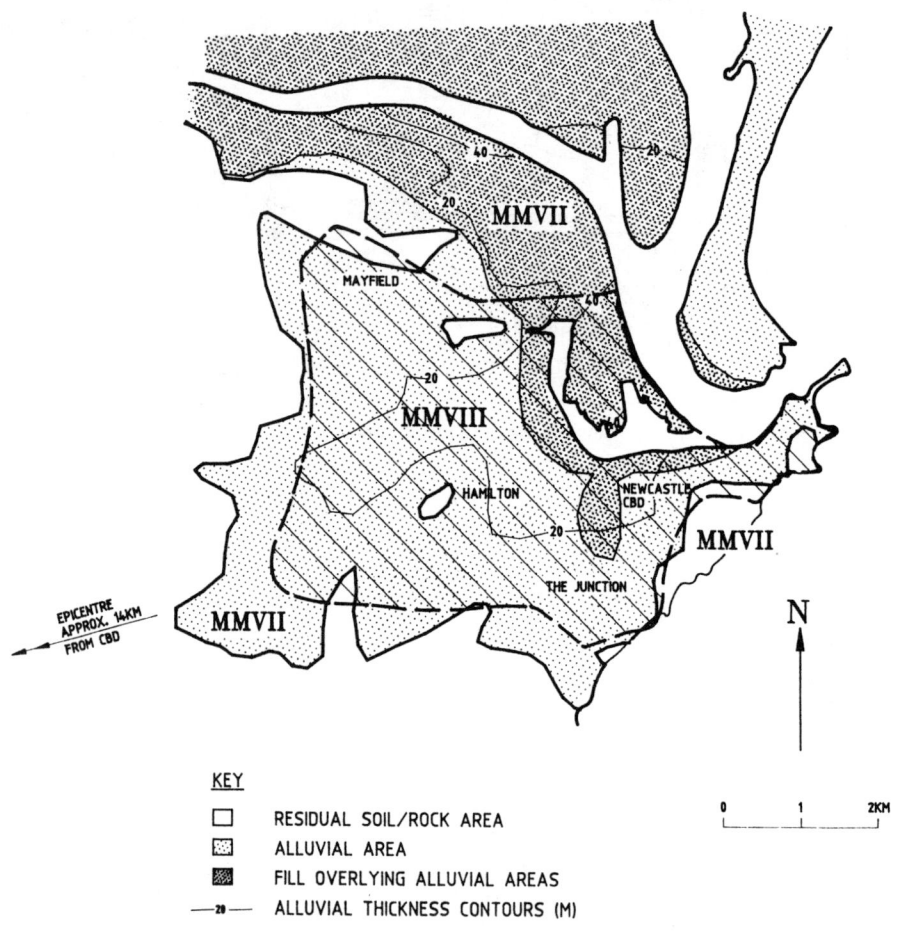

Figure 3:   **Relationship Between Area of Greatest Damage and Alluvial Region**

Figure 4: **The Newcastle Workers' Club**

Figure 5: **Gable End Failure - The Junction School**

Figure 6:   Corrosion of Brick Cavity Ties

Figure 7:   Rialto Towers, Melbourne

Proceedings of the Second Conference on Tall Buildings in Seismic Regions
55th Regional Conference
May 16 and 17, 1991, Los Angeles, California

The Use of Displacement Participation Factors
in the Optimization of
Drift Controlled Buildings

by

Finley A. Charney, Ph.D., P.E.*

Introduction

For many buildings, demands on lateral stiffness control the design of the superstructure. These stiffness demands may be based on seismic or wind drift criteria, perception of motion limits, or stability under gravity loads.

Once a structural scheme is selected, it is the engineer's responsibility to proportion the superstructure such that the stiffness demands are met using a minimum volume of material. Usually, however, the actual optimization process is sidestepped, and instead, the task becomes one of simply meeting the stiffness criterion. The stiffness design is accomplished by carrying out a preliminary design and analysis, followed by one or more cycles of redesign and reanalysis. In carrying out this process, the engineer is faced with several difficulties:

1) judgement and/or experience is not always a sufficient guide for reproportioning members

2) each cycle of redesign requires a time-consuming re-analysis

3) since optimization has not been explicitly performed, the final design is not effectively utilizing materials

One way to avoid these problems is to utilize the method of virtual work in the design process. This method, first described by Velivasakis and DeScenza [1], has been used for years by several tall building design firms. Recent papers by Baker [2], Forrest-Brown and Samali [3], Charney [4], and Henige [5] provide additional details.

Basics of the Virtual Work Technique

Consider the simple statically determinate frame of Figure 1, for which it is desired to compute the horizontal displacement, r, at the top right joint. To compute the displacement, the method of virtual forces is employed:

* Associate Principal, J.R. Harris & Co., and President, Advanced Structural Concepts, Denver, Colorado

$$1*r = \sum_{i=1}^{m} \int_0^L [M(x)m(x)/EI + V(x)v(x)/GS + P(x)p(x)/EA] dx \qquad [1]$$

where   M(x) and m(x) are the real and virtual forces in the member,
        V(x) and v(x) are the real and virtual shears in the member,
        P(x) and p(x) are the real and virtual axial forces in the member,
        E is the modulus of elasticity for the member,
        G is the shear modulus for the member,
        I, S, and A are the section properties,
        L is the member length, and
        m is the number of members in the frame

Usually, a separate stress analysis is required to provide the virtual member forces. Here, however, the virtual forces are proportional to the real forces.

The results from the analysis are shown in Table 1. The entries under the columns FLEXURAL, SHEAR, and AXIAL represent the importance of the various sources of deformation in each member, and the values shown under the entry TOTAL represent each member's total contribution to the displacement. When the individual member contributions, called *displacement participation factors*, are added for all members, the desired displacement, r=1.899 inches, is obtained.

In the planar frame of Figure 1 each member's displacement participation factor has three components. For three dimensional structures, there will be six components: major and minor axis flexure, major and minor axis shear, axial, and torsion. If desired, flexibility of beam column joints may be considered as an additional component for both planar and three dimensional systems [4].

Since displacements for most structures are available from the results of the computer analysis used to develop the member forces, the computations summarized in Table 1 are not traditionally carried out. However, the information provided in the table is of tremendous value when reproportioning the structure to satisfy stiffness demands. By studying the entries in the table, the guesswork is removed from the iterative redesign/reanalysis phase of design; the engineer knows which member to change, and more specifically, which member property to change. For example, from Table 1 it may be seen that the beams (members 2 and 3) cause most of the displacement, and that virtually all of this displacement arises from flexural deformation.

Theoretical Basis for Virtual Work Optimization

While the information in Table 1 is valuable, another measure of member performance is required for optimization; displacement participation factor per member volume.

Consider again the frame of Figure 1. Assume a given volume of material is available, and it is desired to proportion members such that the structure has a minimum top level displacement. Also, there is a requirement that depth be maintained at 20 inches for all members. Based on a non-optimum preliminary design, the top level displacement has been computed using the principal of virtual work (Table 1). Recall that the vast majority of the displacement is attributed to flexural deformation in the individual members.

Starting with the initial design, the engineer has two choices when reproportioning the individual members to meet the optimization criterion:

1) Increase member volume (width), thereby decrease total displacement

2) Decrease member volume (width), thereby increase total displacement

Based on Table 1, the starting displacement participation factors and the factors divided by member volume are shown for each member in Table 2. Three successive tries were made to optimize the frame: giving 4800 cubic inches from each column to the girders, and repeating for a transferral of 9600 cubic inches and 12000 cubic inches. As can be observed from Table 2, the total displacement reduces as the member participation per volume ratios converge.

When the first action is taken for an individual member (girder 2 in Table 2), the total volume increases by the amount δ_v and the total displacement decreases by an amount δ_{d1}. In order to achieve the objective, a second member (column 1) must be decreased in volume by an amount δ_v such that the magnitude of the associated displacement, δ_{d2}, is less than the magnitude of δ_{d1}. If no pair of members can be found such that the displacement reduces by moving volume from one member to another, the structure is optimum. For the structure of Table 2, this occurs when about 13200 cubic inches of material is transferred from each of the columns to each of the girders.

Stated in a more general context, the optimization condition will occur when, for a small change in member volume, the ratio of change in global displacement to change in member volume, δ_d / δ_v, is the same for all members of the structure. This is equivalent to the often stated optimization requirement that there be a constant ratio of member global displacement participation to member volume [2,3]. A rigorous proof of this is given in the paper by Baker [2].

Two additional observations are made about the preceding optimization procedure:

1) There gets to be a point in the process where the controlling displacement becomes very insensitive to transferral of material from one member to another. In this sense, it is not required that all members have the same participation per unit volume— only that these ratios are nearly equal. Experience in virtual work optimization of frames using the program DISPAR [6] has shown that a near minimum volume structure will be achieved once the participation factors per member volume are within a range from about 2/3 to 1-1/2 times the average participation per volume for all members.

2) In three dimensional structures, the engineer must be careful not to transfer material away from members which are not participating in the stiffness of the structure in the direction of loading considered.

Approximate Reanalysis Using Stored Work Quantities

For most building structures, the forces developed in individual members depend on external loading, structural topology, and individual member properties. Since external loading and topology is typically invariant during the redesign process, changes in member forces from one design to another can result only from changes in member properties. Typically the percent change in member force which results from the change of a member property is much smaller than the percent change in member property itself. This is due to the fact that member forces in individual members are dominated by equilibrium requirements associated with the external loading and the structural topology. Many design postprocessors use this fact to allow for stress checks to be made on redesigns without requiring a reanalysis to be carried out.

When changes in member force are relatively insensitive to changes in member properties, overall displacements at selected degrees of freedom may be determined by simply recomputing and summing the participations of those members which have had their properties changed. After changing the properties of one or more members, the new drift at the point of the virtual force may be computed as follows:

$$\delta_{NEW} = \delta_{OLD} - \sum_{i=1}^{n} DPF_{i,OLD} + \sum_{i=1}^{n} DPF_{i,NEW} \qquad [2]$$

where   n is the number of members with changed properties
      δ_{NEW} is the new displacement at the desired degree of freedom
      δ_{OLD} is the old displacement at the desired degree of freedom
      $DPF_{i,OLD}$ is the old displacement participation factor for member i
      $DPF_{i,NEW}$ is the new displacement participation factor for member i

If more than about 50 percent of the members have property changes, it may be more efficient to simply recompute and sum the displacement participation factors for all members in the structure.

In Table 3, the accuracy and speed of the reanalysis procedure is given for a 20-story three dimensional braced frame. In each reanalysis 50 percent of the members had their properties increased by about 20 percent.

The first two analyses, using SAP90 [7] and the virtual work method, respectively, indicate nearly identical displacements, but dramatic time savings using the virtual work analysis. As the structure is changed, the virtual work procedure (based on original member forces and new member properties) is used to recompute the displacement. After the fifth virtual work analysis the final displacement is 4.333 inches. At this point, a final SAP90 analysis is carried out for the structure using the latest member properties, and the displacement is computed as 4.341 inches, which is nearly identical to the virtual work displacement.

The reader should note that certain structures do not lend themselves to this type of reanalysis. Consider, for example, a frame-wall system. Here, a relatively small change in the stiffness of the wall could have a major effect on the forces developed in the frame, and hence, reanalysis based on resummation of DPF's would not be applicable.

Limitations of the Method and Related Research Needs

While the virtual work techniques described herein are useful in the context of "human-loop optimization" they should not yet be used as the basis for developing a "black box" optimization program. There are several reasons for making this statement:

1)   An optimization program useful for three dimensional systems requires that an algorithm be developed for recognizing those members which are meant to contribute to the response in question, and for recognizing those members which are not meant to contribute to response.

2)   Different stiffness requirements may apply to different portions of the structure. For example, optimization for interstory drift will produce a different distribution of material than will optimization for total drift or period. Hence, what is optimum for one criterion is necessarily not optimum for another.

3) For steel structures fabricated from available wide flange sections, a change in member volume can have a significantly different effect, depending on the relative magnitude of deformation sources acting within the member. Consider, for example, the change from a W27X84 girder to a W27X129 girder. This change produces an increase in volume (cross sectional area) of 52 percent, an increase in moment of inertia of 67 percent, and an increase in shear area of only 32 percent. If the displacement participation factor for the member is dominated by flexure, the change in member volume decreases the member contribution to drift by a factor of 1/1.67. If the displacement participation is dominated by shear, the same change in member volume reduces the member contribution by a factor of 1/1.33. Hence, the same change in member volume is more effective for flexure than shear.

In the context of a member dominated by shear deformation, additional material placed in the flanges is completely wasted, while for a member dominated by flexure, material placed in the web is essentially wasted. For axially dominated members, no material is wasted. What needs to be developed, therefore, is an index for the "effective" volume of a member in axial, flexural, and shear deformation modes. Once a mathematical model for effective volume is developed, it may be incorporated into virtual work optimization routines.

Conclusions

The virtual work or displacement participation technique provides valuable insight into structural behavior, and allows the structural engineer to quickly design optimum structures. Currently, however, it is recommended that the procedure be used only as part of a human-loop iterative design, and not as part of a automated optimization environment.

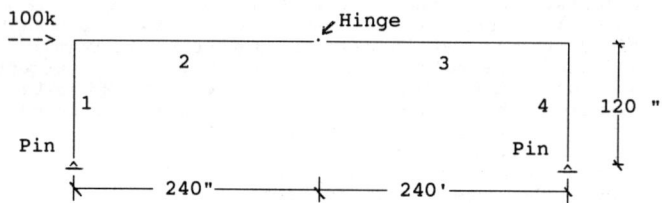

All members: E=5000 ksi
            G=2000 ksi

Column intital size   b=20" d=20"
Girder initial size   b=12" d=20"

FIGURE 1

EXAMPLE FRAME FOR VIRTUAL WORK ANALYSIS

TABLE 1

RESULTS OF VIRTUAL WORK ANALYSIS OF
STRUCTURE OF FIGURE 1

| MEMBER | DISPLACEMENT COMPONENT | | | |
	FLEXURE	SHEAR	AXIAL	TOTAL
1	0.21600	0.00450	0.00038	0.22088
2	0.72000	0.00375	0.00500	0.72875
3	0.72000	0.00375	0.00500	0.72875
4	0.21600	0.00450	0.00038	0.22088
TOTAL	1.87200	0.01650	0.01076	1.89926

NOTE: Values are displacement contributions in
       inches. Displacement at node 4 is 1.8993
       inches.

TABLE 2

OPTIMIZATION OF FRAME OF FIGURE 1 BY SUCCESSIVE
TRANSFERRAL OF MATERIAL

	START		TRY 1		TRY 2		TRY 3	
Member	DPF	DPF/VOL	DPF	DPF/VOL	DPF	DPF/VOL	DPF	DPF/VOL
1	.221	4.60	.245	5.67	.276	7.19	.295	8.19
2	.729	12.65	.673	10.78	.625	9.30	.603	8.74
3	.729	12.65	.673	10.78	.625	9.30	.603	8.74
4	.221	4.60	.245	5.67	.276	7.19	.295	8.19
Tot	1.900		1.836		1.802		1.796	

Note: DPF/VOL values shown are 1000000 times the actual values.

TABLE 3

SPEED AND ACCURACY OF VIRTUAL WORK REANALYSIS

ANALYSIS METHOD	TIME REQUIRED (MINUTES)	COMPUTED DISPLACEMENT (INCHES)
SAP90	11.05	5.458
VIRTUAL WORK 1	0.43	5.433
VIRTUAL WORK 2	0.43	5.220
VIRTUAL WORK 3	0.43	4.896
VIRTUAL WORK 4	0.43	4.683
VIRTUAL WORK 5	0.43	4.333
SAP90	11.08	4.341

References

[1] Velivaskis, E. E., and DeScenza, R., "Design Optimization of Lateral Load Resisting Frameworks", Proceedings of the Eighth Conference on Electronic Computation, ASCE, New York, N.Y., 1983.

[2] Baker, W.F., "Sizing Techniques for Lateral Systems in Multi-Story Steel Buildings", Proceedings from the Conference: Tall Buildings, 2000 and Beyond, Tall Building Council, Lehigh, PA, 1990.

[3] Forrest-Brown, G., and Samali, B., "Practical Optimization of Framed Structures Using Virtual Work Principals", Proceedings from the Conference: Tall Buildings, 2000 and Beyond, Tall Building Council, Lehigh, PA, 1990.

[4] Charney, F.A., "Sources of Elastic Deformations in Laterally Loaded Steel Frame and Tube Structures", Proceedings from the Conference: Tall Buildings, 2000 and Beyond, Tall Building Council, Lehigh, PA, 1990.

[5] Henige, R., "Structural Optimization to Limit Natural Periods", Proceedings of the Tenth Conference on Electronic Computation, ASCE, New York, N.Y., 1991.

[6] Charney, F.A., "DISPAR for SAP90" and "DISPAR for ETABS", Advanced Structural Concepts Division, J.R. Harris & Company, Denver, Colorado, 1991.

[7] Wilson, E.L., and Habibullah, A., "SAP90 Finite Element Analysis Program", Computers and Structures Incorporated, Berkeley, California, 1988.

ANALYTICAL AND EXPERIMENTAL EVALUATION
OF ADVANCED CLADDING SYSTEMS FOR BUILDINGS

James I. Craig[†], *Barry J. Goodno*[‡], *Michael W. Wolz*[*], and *Jean-Paul Pinelli*[*]
Georgia Institute of Technology
Atlanta, GA 30332

Abstract

The results of an analytical study of a building with heavy precast cladding that is integrated together with the primary structural system are described. The analysis was carried out using DRAIN-2D to model a six story 3/4 scale steel frame building previously tested on the shake table at NCEER. Hypothetical cladding panels with "conventional" connection designs are shown to reduce building drift by augmenting the linear stiffness, and the results are shown to be consistent with previous observations by the authors. Significant reductions in seismic response are also shown to be achievable by allowing the cladding panels to remain rigid while two of the four connections are assumed to behave in a ductile inelastic manner. Both conventional bilinear and more complex multi-parameter Takeda models used for the inelastic behavior are developed on the basis of experimental tests in the laboratory of representative precast connection designs.

The studies indicate that response attenuation levels of as much as 70% are possible for representative earthquakes. More specifically, it is demonstrated that these attenuation levels are comparable to those achieved for the study building when either an active mass damping system or an active tension bracing system is used for the same earthquake records. The results strongly suggest that provision for ductile inelastic action in the connections for heavy concrete cladding panels can provide levels of seismic response attenuation that are comparable to those achieved by other means. The paper concludes with a discussion of how these results can be interpreted for the design of practical connections and how the study will be extended to consider a wider variety of connection elements.

Introduction

Attenuation and control of the seismic response of tall buildings is frequently achieved through ductile inelastic action of the structural system. This has traditionally been achieved by deliberate design of structural details, such as eccentric bracing, that will develop well-behaved inelastic action without critical loss of general structural integrity or stability. While this approach is generally used to handle the most severe loading conditions for the primary structural system, it can also be applied to other building subsystems, such as cladding, not traditionally considered to play a structural role. The result can be an overall design capable of achieving significant response reductions for moderate seismic loads.

Studies of building structural and cladding systems [1,2] coupled with experimental observations of the actual dynamic response of these buildings [3], have lead to a recognition that, whether by deliberate design or not, building cladding systems (particularly heavy precast systems) can measurably affect the structural stiffness and therefore the dynamic response. These studies documented essentially linear stiffness augmentation that was confirmed by response measurements at very low (ambient) levels. At the same time other studies have pointed to the potential benefits that could be derived from dissipative or inelastic stiffening action in certain building structural elements. Kelly, et. al. [4] proposed the use of novel ductile connection elements to achieve inelastic interaction between structural members under a variety of conditions. Palsson [5] carried out an early study of cladding interaction with the supporting building structure and proposed the use of "brake pad" material in the panel connections to enhance structural ductility through the addition of coulomb damping. Pall [6] has developed a number of

---

[†] Professor, School of Aerospace Engineering, (404) 894-3042, Internet: jcraig@gtri01.gatech.edu
[‡] Professor, School of Civil Engineering, (404) 894-2227, Internet: bg6@prism.gatech.edu
[*] Graduate Research Assistant, School of Civil Engineering

commercial structural elements that rely on coulomb type damping to augment structural ductility, and applications to building cladding were studied.

The present study is part of a continuing effort by the authors, first to document the actual interaction of cladding with the primary structure, and then to explore the role that this process, if fully understood and effectively utilized, could play in meeting ductility demands under severe loading conditions. These efforts have included both analysis and experiments, but to date the work has examined mostly global structural behavior. The present study provides the first results from a more detailed examination of the cladding connections themselves and their incorporation in total structural models. In addition, these results also pointedly illustrate the importance of including all major building components in computer analyses aimed at the realistic portrayal of overall building seismic response.

Experimental Studies of Connection Properties

Connections for precast cladding systems can be decomposed into three basic parts:
(a)   a steel insert imbedded in the precast concrete panel,
(b)   an anchor in the building structure (a second insert for a concrete structure or an attachment to a steel member), and
(c)   a connector element (often a steel angle) attached between (a) and (b).

Initially, tests were carried out to evaluate the performance of typical precast cladding connection inserts. The inserts tested were weld-plate steel inserts that are typically cast into the rear surfaces of cladding panels. Seven inserts were tested and a detailed description of the test program and test results is given in Ref. [7]. One of these cases (insert 12) is discussed in this paper. It consists of a 10cmx10cm steel plate with two rebars extending 30 cm each on each side of the plate. The insert was located at the middle edge of a concrete specimen, and it was subjected to an applied cyclic moment acting about an axis parallel to the edge. The insert was subjected to repeated cycles of increasing load , and for each cycle, the deformation across the insert (translation, rotation, twist) at the insert-connection interface was recorded.

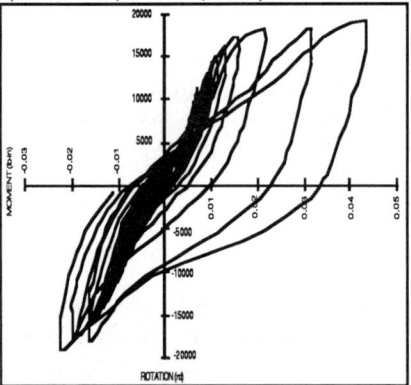

As shown in Figure 1, insert 12 displayed very regular hysteretic behavior with increasing energy dissipation and decreasing stiffness as the moment magnitude was increased in successive cycles. In this case the ductile behavior was attributed to a progressive degradation of the concrete surrounding the insert. Figure 1 shows a smooth and regular increase of the hysteresis loop areas with the loading without any significant stiffness deterioration and strength degradation. The ultimate failure resulted from a failure of the weld between the steel plate and rebars.

Figure 1: Experimental Hysteresis Cycles - Insert 12 (1lb.in=0.113N.m)

Analytical Modeling

Based on the experimental results outlined above, analytical flexural models of the connection inserts which closely reproduced the observed nonlinear hysteretic characteristics were developed. Both empirical and a mechanical modeling philosophies were pursued. The empirical approach based the hysteretic model on purely geometric (graphic) rules that mimic the observed behavior. The mechanical approach was based on use of an idealized mechanical model composed of springs, dampers and other discrete components, and tried to reproduce the observed behavior by various combinations of "atomic" elements.

Empirical Approach

Empirical models developed in these studies were based on a three-parameter model [8] whose parameters were related to particular characteristics of the hysteretic behavior. One parameter defined the stiffness degradation, another defined the strength degradation, and a third parameter controlled pinching of the loops. In order to develop a three-parameter model, the basic geometric properties of the hysteresis loops were identified. Specifically, the monotonic loading curve was identified and was approximated by a trilinear curve, or skeleton curve.

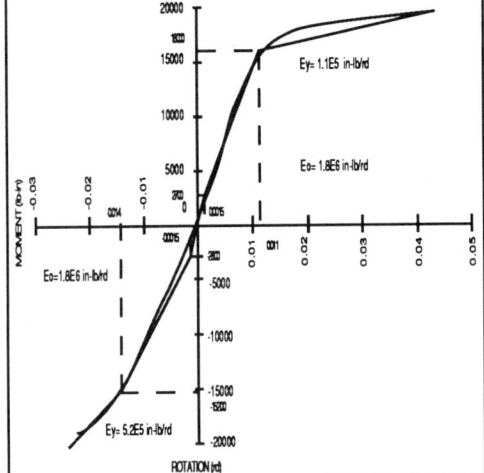

Figure 2: Envelope & Skeleton Curve - Insert 12 (1lb.in=0.113N.m)

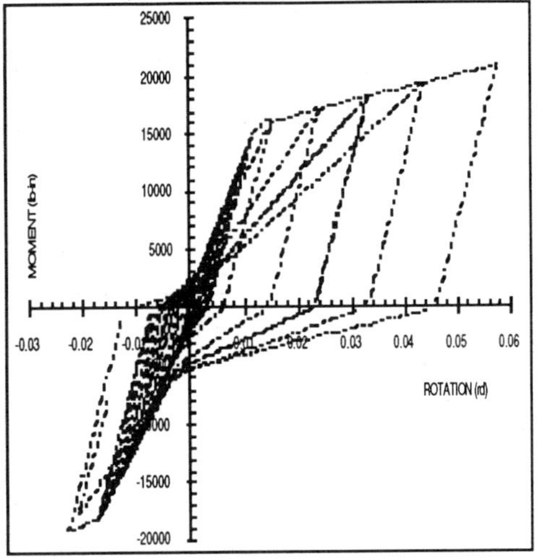

Figure 3: Empirical Hysteresis Loops - Insert 12 (1lb.in=0.113N.m)

For the modeling of insert 12 , it was assumed that the envelope curve of the hysteresis loops is a good approximation of the monotonic loading curve. Then, the envelope curve was linearized into three segments defined by the cracking load, the yield load, and the elastic moduli before cracking and after yielding. The result is shown in Figure 2 . Finally, the remaining parameters were defined through trial and error to yield hysteretic cycles sufficiently close to the experimental results. The resulting analytical hysteretic cycles are displayed in Figure 3 (which can be compared to Fig. 1).

Mechanical Approach
      The empirical approach can yield satisfactory results, especially when applied to regular and continuously varying situations. The case presented above is characterized by such regular behavior and is relatively easy to model in this way. In the case of discontinuous hysteretic behavior, a three-parameter approach cannot be applied directly since the parameters would not remain constant over the entire range of loading. Another serious drawback to the three-parameter approach is the lack of direct physical meaning for the parameters. Clearly, an approach based on the mechanical characteristics of the problem will be more meaningful. The distributed deteriorating element model developed by Iwan et al. [9] in a series of papers over a number of years is one such approach and was adapted in this research. In the mechanical approach a physical system is represented by a parallel-series model made of a combination of linear springs and Coulomb slip elements whose properties (stiffness, friction load, etc) are based on the observed properties of the system.

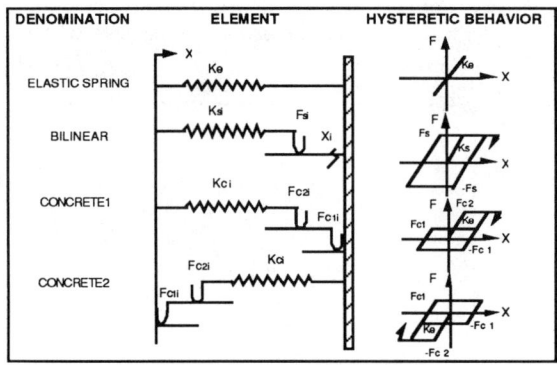

DENOMINATION	ELEMENT	HYSTERETIC BEHAVIOR

Figure 4. Schematic of the Deteriorating Degrading Element Model

Figure 4 shows the basic four elements used in the mechanical models for the present study together with a representation of their hysteretic behavior:

(1) a linear elastic element defined by its stiffness;

(2) a bilinear element defined by its stiffness and yield load, and identified with the steel behavior;

(3) a "pinching" element defined by its stiffness, cracking load, and yield load, and identified with the concrete behavior;

(4) a similar concrete element but with the signs inverted to allow for load reversal.

In addition, the bilinear and concrete elements are assumed to break at a specified value of displacement, and when the fracture point is reached for an element, its contribution to the stiffness and strength of the overall model must be permanently removed.

Table 1. Parameter Values of the Mechanical Elements for Insert 12

Element Type	Quantity	Stiffness (N.m/rd)	CrackingLoad (N.m)	Yield Load (N.m)	Break Point (rd)
Elastic	1	$0.23 \ 10^4$			
Bilinear	1	$0.57 \ 10^5$		1356	10
Concrete 1	2	$0.34 \ 10^5$ $0.45 \ 10^5$	11.3 56.5	226 678	0.02 10
Concrete 2	2	$0.34 \ 10^5$ $0.34 \ 10^5$	11.3 33.9	170 565	10 10

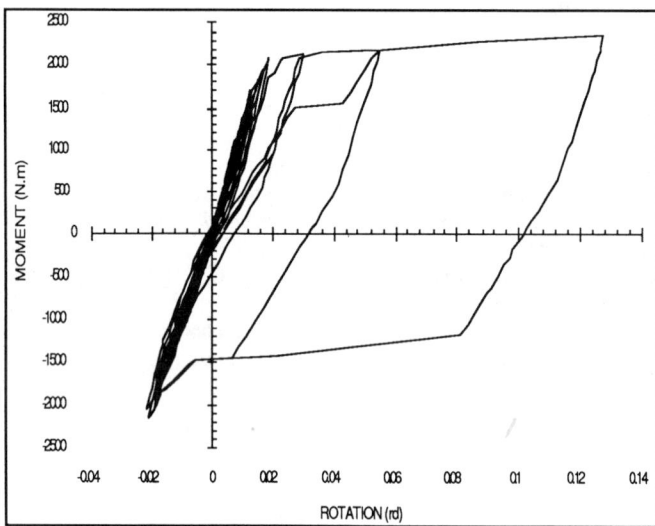

Figure 5: Mechanical Hysteresis Cycles - Insert 12

Insert 12 was also modeled using the mechanical approach, and Table 1 lists the elements used in the model together with the values of the corresponding parameters. These parameters were chosen based on observation and understanding of the experimental results and through trial and error until an acceptable match between experimental and analytical results was obtained. Figures 5 shows the resulting hysteretic cycles for insert 12 and should be compared against the experimental results appearing in Figure 1 and the three-parameter results of Figure 3.

Since the evaluation and control of damping is one of the main concerns of this research, the accuracy with which the proposed models reproduce the observed behavior was evaluated through a comparison of the areas of the corresponding measured and modelled hysteresis loops. These areas are directly proportional to the damping or energy dissipated. In the case of insert 12, shown in Figure 6, the three-parameter model was in very good agreement with the experimental results, while the mechanical model under-estimated the damping. For other inserts the mechanical model was in close agreement while the three-parameter model over-estimated the damping.

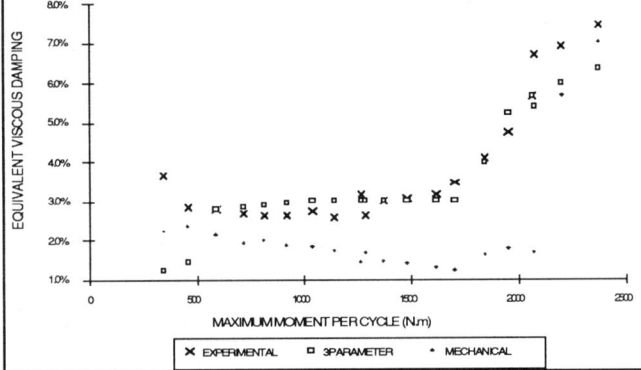

Figure 6. Analytical & Experimental Damping - Insert 12

Each approach has clear and distinct advantages. The empirical approach has two main advantages   First is its simplicity.   Only three parameters have to be defined (although the basic constitutive properties defining the skeleton curve must also be deduced in the form of envelope curves, as shown above), and they remain constant. Second, the approach can be used with either force or displacement as the input without change in the procedure.   There are two principal disadvantages as well. First, a physical interpretation of the parameters may be meaningless, and second, the procedure cannot model discontinuous behavior patterns.

The mechanical approach can be based on parallel-series as well as series-parallel models [10], however, for the present application, only the parallel-series model provided a direct physical meaning for the parameters and allowed modeling the discontinuous states of behavior. It must be noted that in the seismic applications under consideration in the present research program, the independent (input) variable is a displacement (e.g. the interstory drift to which the cladding must respond), and the resulting forces are computed. In this case a parallel-series model provides the most direct means to compute an output force given an input displacement. When the independent (input) variable is force (as is often the case in a lab test), an iterating procedure must be followed, and if the number of elements is large, the computational effort becomes important.

Ideally, for a parallel-series mechanical model, it might appear that any type of behavior could be satisfactorily approximated by a sufficient number of elements with suitable properties, but therein lies a drawback. With no limit to the number of parameters, it can easily become difficult to determine the correct (optimal) values of the parameters (including fracture points), and the computational effort may rapidly increase. This is especially true in the case where force rather than displacement is the independent (input) variable.

Building Model
To study cladding-structure interaction, a modified version of the program DRAIN-2D [11] was used to model the seismic response of a two dimensional structure in which nonlinear behavior of both cladding-structure connections and structural members was considered. Inelastic hysteretic action of cladding connections was tracked by the addition of a new connection element to DRAIN-2D. The goal was to determine both the frame response alterations due to cladding as well as cladding connection force and ductility demands.

A 1/4 scale six story 1 by 3 bay steel space frame, used in recent analytical and experimental studies at the National Center for Earthquake Engineering Research (NCEER) [12], was chosen to investigate the role of nonlinear cladding- structure connections in the seismic response of the frame. Due to the planar behavior of the actual frame observed during previous experiments, plane frame analytical models were developed in the 3 bay direction using DRAIN-2D. For both the clad and unclad models, frame members were modelled as beam-column elements using bilinear hysteretic behavior to define flexural and axial stiffness. In the unclad frame, which served as a reference case, tributary translational mass, due to both the cladding and structural framing, was distributed to the panel nodes. The clad frame model included two rigid panels (represented by truss bar assemblies) per bay, (Fig. 7). No panels were assumed to exist at the first floor level, and no panel-to-panel contact was assumed to occur at any time during the analyses (Fig. 8). Consistent with current design recommendations, the lower panel connections which provide primary panel support and load transfer were assumed to be rigid, while the upper connections were assumed to be flexible. Lateral deformation, due to interstory drift, was assumed to be the dominant form of connection distortion. The flexible connections were therefore composed of a nonlinear horizontal translational spring and linear vertical translation and in-plane rotational springs.

The force-deformation behavior for the nonlinear spring, was represented using both bilinear and degrading stiffness hysteresis models. The form for these models was based on the results from the laboratory studies of representative connection components described earlier. In the clad frame model, the flexible connections, positioned between the panel and structure frame nodes, were represented by a new DRAIN-2D connection element developed for this study. The connection properties were scaled so that the elements would initiate varying levels of cladding participation in frame response. When the flexible connections were assumed to have zero stiffness, no cladding participation was added to the system stiffness because the panels were only attached at the rigid lower connections. In this case, only the additional cladding mass affected the frame response.

<u>Software</u>

DRAIN-2D is a well-known general purpose computer program for dynamic analysis of inelastic planar structures and permits new elements to be added with relative ease [13]. The program performs a dynamic analysis of plane frames of arbitrary configuration under seismic loading with lumped masses at the structure nodes and proportional or structural damping. Time history analysis involves direct integration of the equations of motion at discrete time intervals, assuming constant acceleration and linear structural behavior within each time step. The force level in each element is checked at the end of each step, and if element yielding or unloading occurs, the global stiffness matrix is updated. The program was modified to include the new cladding connection elements used to represent the flexible panel connections. All analyses were performed on a SUN 4/60 workstation.

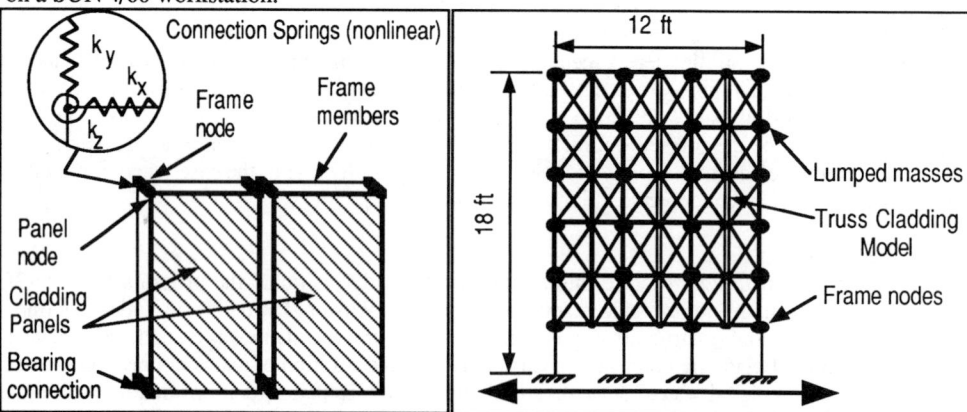

Figure 7. Typical Bay of Truss Cladding   Figure 8. 2-D Frame with Truss Cladding

<u>Results</u>
    Dynamic analyses were performed to determine the variation in structural response due to changes in connection stiffness and yield deformation, and to calculate the lateral force and displacement ductility demands on the connections. The stiffness values were adjusted to correspond to the 1/4 scale frame model and to produce nonlinear behavior of the connections at reasonable yield force levels. A degrading stiffness model was used for the x-translational connection in all of the analyses. First the unclad frame model, which acted as the reference case, was subjected to the first 15 seconds of 25% of 1940 El Centro NS excitation and compared to the response of the clad frame with flexible connection stiffnesses of 100 to 600 kip/inch (typical of relatively stiff conventional connections [3]). Due to low frame distortions for this loading, yielding of connections did not occur. Nonlinear behavior of the connections was then induced by specifying low connection yield distortions (0.002 inches). Other connection properties included a rotational stiffness of 1.0 in-kip/rad, a strain hardening ratio of 0.15, alpha and beta values (which describe the unloading and reloading characteristics of the degrading stiffness model) of 0.2 and 0.4.
    Results of the clad frame analyses demonstrated significant top floor displacement reductions compared to the reference case (Fig. 9). While the differences in top floor displacement reduction for both cases are small for a given connection stiffness, the maximum lateral connection force was reduced by more than 61% for the nonlinear case.
    Response of the reference case when subjected to the first 15 seconds of 100% of 1940 El Centro NS excitation was compared to the response of the clad frame with flexible connection stiffnesses from 5 to 30 kip/inch. In this case, the flexible connections were assumed to yield at a connection distortion of 0.1 inch, and the strain hardening ratio was reduced to 0.05. Again, significant reduction of top floor displacements was achieved (Figures 10 and 11), and connection forces were lower for the nonlinear case. In all of the cases, the maximum lateral force and the maximum displacement ductility occurred in flexible connections at the second floor level.
    Finally, energy dissipated by the nonlinear behavior of the connections during the time history analyses was calculated from the force-displacement hysteresis loops for each connection (Fig. 12). The overall effect of the hysteretic behavior of the flexible connections was also expressed as an equivalent viscous damping ratio. Here, assumed damping ratios for the first two modes were used to determine the proportional damping values necessary to produce approximately equal maximum top floor displacements in both the reference and nonlinear connection cases.

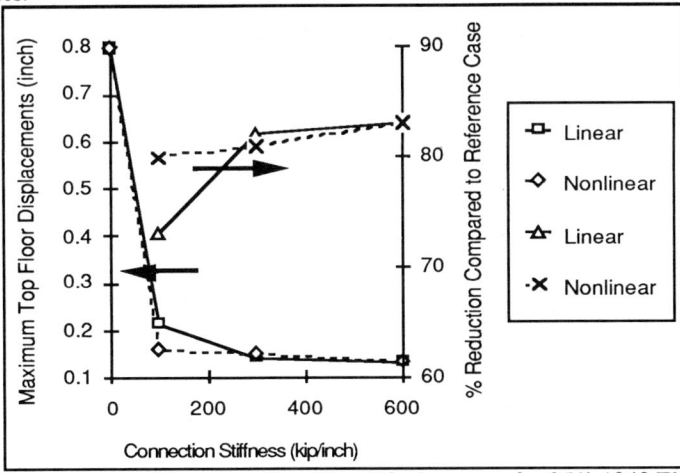

Figure 9. Comparison of Maximum Top Floor Displacement for 25% 1940 El Centro NS Excitation (1 in=25.4 mm, 1 kip=4.45kN)

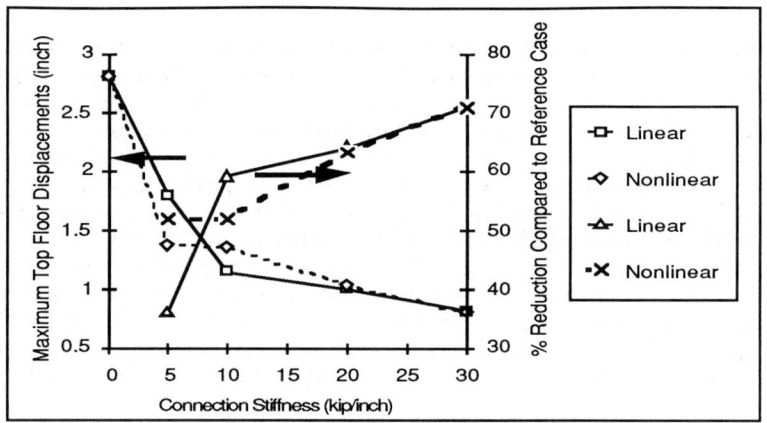

Figure 10 Comparison of Maximum Top Floor Displacement for 100% 1940 El Centro NS Excitation (1 in=25.4 mm, 1 kip=4.45kN)

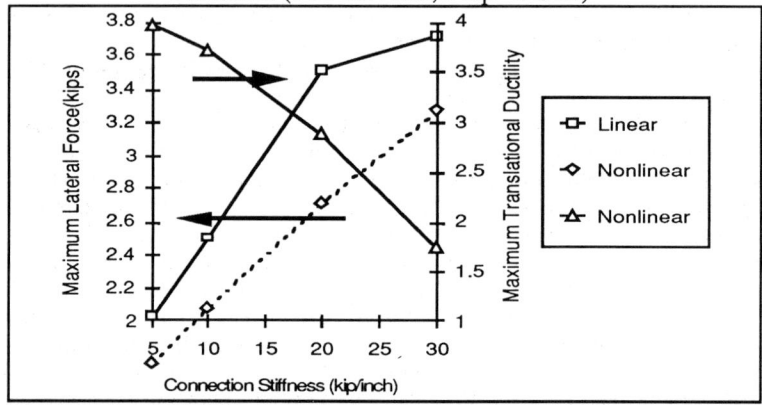

Figure 11. Comparison of Connection Demands for 100% 1940 El Centro NS Excitation (1 in=25.4 mm, 1 kip=4.448kN)

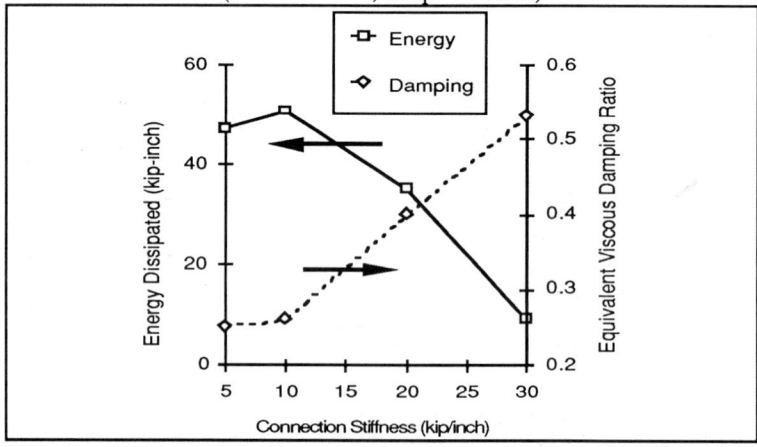

Figure 12.   Connection Energy Dissipation through Cyclic Nonlinear Behavior for 100% El Centro NS Excitation (Yielding deformation = 0.10 inches)

Conclusions

In this study of the effect of cladding interaction, the response of a frame under seismic loading showed that cladding stiffness contributions coupled with inelastic behavior of the cladding connections resulted in significant reductions in the top floor frame displacement. The amount of hysteretic damping provided by the connections in the analysis was dependent on the translational stiffness, yield forces and strain hardening rations chosen for each case. Results demonstrated that significant amounts of kinetic energy could be dissipated by inelastic connection behavior without exceeding realistic connection force and ductility demands.

The empirical and mechanical connection component modeling approaches used to represent the measured connection behavior were shown to provide realistic and accurate models for use in the analytical studies and will be pursued in the future. The mechanical approach is favoured by the authors based on the physical meaning behind the parameters and its ability to model the observed discontinuous behavior. But it is recognized that the simplicity of the three-parameter empirical approach makes it also very attractive. It is felt that the mechanical approach can be efficiently implemented only if coupled with an optimization technique for identifying the model parameters.

In continuing work, advanced connection designs will be tested and analyzed. The tests will isolate the behavior of the connector element from the influence of the insert and the building attachment elements. From these tests results, nonlinear analytical models of the shear and flexural behavior of the connector and the insert-connector-attachment system will be developed using the approaches outlined above. Once nonlinear hysteretic models have been defined for each of the three connection system components, the translational and rotational elements will be combined to simulate a complete connection system, and the models will be incorporated into nonlinear computational models of complete buildings. Analysis of these building models will determine potential cladding performance under strong ground motion conditions and will assess potential cladding influence on damping and on torsional and lateral response of buildings.

Acknowledgement

This study is part of a broader program sponsored by the National Science Foundation that includes identifying, testing, and modeling advanced connection designs. The support of NSF through Grant BCS-8906508 is gratefully acknowledged. However, the findings and opinions expressed here are those of the authors and do not represent the position of NSF.

References

1. Goodno, B. J., Will, K. M., and Craig, J. I., "Analysis of Cladding on Tall Buildings," ASCE National Convention, Developments in Methods of Structural Analysis, Atlanta, GA, October 1979, preprint 3744.
2. Goodno, B.J., Palsson, H.P., and Pless, D.G., "Localized Cladding Response and Implications for Seismic Design," Proceedings, 8th World Conference on Earthquake Engineering, San Francisco, CA, July 1984, Vol. V, pp. 1143-1150.
3. Meyyappa, M and Craig, J. I., "Highrise Building Identification Using Transient Testing," Proceedings, 8th World Conference on Earthquake Engineering, San Francisco, CA, July 1984, Vol. VI, pp. 79-86.
4. Kelly, J. M., Skinner, R. I., and Heine, A. J., "Mechanisms of Energy Absorption in Special Devices for Use in Earthquake Resistant Structures," Bulletin of N.Z. Society for Earthquake Engineering, Vol. 5, No. 3, September 1972, pp. 63-89.
5. Palsson, H., "Influence of Nonstructural Cladding on Dynamic Properties and Performance of Highrise Buildings," Ph.D. Thesis, Georgia Institute of Technology, December 1982.
6. Pall, Avtar S., "Friction Damped Connections for Precast Concrete Cladding," Proceedings,Architectural Precast Concrete Cladding-Its Contribution to Lateral Resistance of Buildings, Precast/Prestressed Concrete Institute, Chicago, IL, November 8-9, 1989, pp.300-309.

7. Pinelli, J.-P., Craig J.I., and Goodno, B.J., "Development and Calibration of Selected Dynamic Models for Precast Cladding Connections," *Proceedings,*.4th U.S. National Conference on Earthquake Engineering, Palm Spring, California, May 20-24, 1990.

8. Park, Y.J., Reinhorn, A.M., and Kunnath, S.K., "IDARC: Inelastic Damage Analysis of Reinforced Concrete Frame-Shear Wall Structures," Technical Report NCEER-87-0008, State University of New York at Buffalo, July 1987.

9. Iwan,W.,and Cifuentes,A.,"A Model for System Identification of Degrading Structures," *Earthquake Engineering & Structural Dynamics Journal,* V.14, 1986

10. Iwan, W., " On a Class of Models for the Yielding Behavior of Continuous and Composite Systems," *Journal of Applied Mechanics*, ASME, Vol 34, pp. 612-617, 1967.

11. Wolz, M., Hsu, C., El-Gazairly, L.F., Goodno, B.J., and Craig, J.I.,"Nonlinear Dynamic Analysis of Buildings with Advanced Cladding Systems", to appear in *Proceedings*, Seventh Conference in Computing in Civil Engineering, Washington D.C., 1991.

12. Reinhorn, A. M., Soong, T. T., Lin, R. C., and Wang, Y. P. (1989), "1:4 Scale Model Studies of Active Tendon Systems and Active Mass Dampers for Aseismic Protection," Report NCEER-89-0026, National Center for Earthquake Engineering Research, SUNY, Buffalo, NY.

13. Powell, G. H. (1973), "DRAIN-2D User's Guide," Report Number EERC 73-22, University of California, Berkeley.

Proceedings of the Second Conference on Tall Buildings in Seismic Regions
55th Regional Conference
May 16 and 17, 1991, Los Angeles, California

STRUCTURAL ANALYSIS AND DESIGN OF THE LOS ANGELES CENTER TOWERS

Roger M. DiJulio, Jr., Robert M. Barker, and Farzad Naeim

John A. Martin and Associates, Inc., Los Angeles, California

ABSTRACT

Structural analysis and design of the lateral load resisting systems in the Los Angeles Center towers, one 40 stories and the other 41 stories is presented. Special attention is given to the complex architectural features which give rise to unusual structural design issues.

INTRODUCTION

The Los Angles Center project consists of two nearly identical towers, one 40 stories and one 41 stories above the shear base. The two towers sit on common lobby and retail levels and four levels of below grade parking structure, which are braced by shear walls (see Figure 1).

The project site is located in Downtown Los Angeles near the Pacific Stock Exchange. The site area is approximately 172,400 square feet. The plaza level of approximately 78,000 square feet includes about 50,000 square feet of retail area. Typical office floors range from approximately 23,500 to 16,500 square feet gross. The total office space in the two towers is approximately 1,800,000 square feet.

The towers have maximum longitudinal plan dimensions of 210 feet and maximum transverse dimensions of 117 feet. The taller tower rises 562 feet above the shear base. The critical aspect ratio is therefore 4.8:1. The typical story height is 13 feet 2 inches.

The buildings are architecturally complex. This leads to an unusually high number of 28 different floor plans in forty stories and a relatively complicated lateral framing systems with a great deal of variation (Figures 2 and 3).

CRITERIA FOR STRUCTURAL ANALYSIS AND DESIGN

Design vertical loads were typical for office buildings.

Seismic loads for the tower were specified by a Maximum Probable design spectra supplied by the Geotechnical Engineer (LeRoy Crandall and Associates). The peak ground acceleration was .33g (Figure 4). Five percent critical damping was assumed and the minimum spectral acceleration was held to .03g.

A Maximum Credible spectra was supplied but did not govern the design of the towers.

The ANSI 70 mile per hour, Exposure B wind loads were used for design of the towers. Structural loads were also supplied by a wind tunnel study performed by Applied Engineering Services, Inc.

Seismic drifts were limited to .0075 times the story height for the Maximum Probable earthquake. Seismic stresses were checked in combination with vertical loads, against AISC allowable stresses with the elastic factors of safety removed. For all intents and purposes this is equivalent to checking against yield stresses.

Wind drifts were limited to .0025 times the story height in the longitudinal direction and .0040 times the story height in the transverse direction. In the transverse direction this corresponds to a 55 mile per hour wind. Wind stresses were checked, for the 70 mile per hour wind, in combination with vertical loads, against AISC allowable stresses including a one third increase.

PRELIMINARY DESIGN

The initial concept for a lateral system consisted of a perimeter frame utilizing WTM24 columns and wide flange beams as deep as 36 inches.

A preliminary model was developed to study this concept. A relatively simple three dimensional model of the frame was made using the Building Design Language (BDL) and analyzed by the CSI-ETABS+ computer program. All special architectural features such as setbacks, transfers and sloped columns were neglected and a single typical vertical load was used throughout.

At this stage it was found that the perimeter frame was not adequate to satisfy wind drift criteria in the transverse direction. The model demonstrated that the single bays (Lines Ve and Ae) at the ends were working alone, with very little help from the diagonal frames. Two interior frames were, therefore added on Lines Pe and Ge.

At this stage a variety of sizes of box columns, ranging from 18" x 16" to 36" x 18", were found to be required. The timely identification of the need for interior frames and a variety of column dimensions facilitated the architectural design process.

LATERAL LOAD RESISTING SYSTEM

The final structural system is a special moment resisting space frame consisting of a perimeter frame and two transverse interior frames. Typical frame beams are grade 36 steel ranging from W24'S to WTM36'S (39 inches deep). There are a variety of transfer girders ranging is depth from 36" to 9'6".

Columns are typically grade 50 steel with the exception of boxes with plates over 4" thick, which are grade 42. A variety of column sizes are utilized. Non-frame columns are W14'S. Many of these "non-frame" columns carry transfer girders, which in turn carry frame columns. Rolled members in the frame are WTM24'S. A large percentage of the columns are boxes. In the lower floors and for tall stories the WTM24'S change to 30" X 16" boxes (Figure 5).

The four columns which are part of both the perimeter and transverse frames (ex.,Pe-17e) are square (24" X 24" below and 20" X 20" above). The four interior frame columns (ex.,Pe-12e) are 36" X 18" boxes; which are necessary to carry large vertical loads and to minimize wind drift. Columns with significant weak axis bending (ex.,Ve-12e, Re-17e) are 30" X 18" boxes reducing to smaller sizes in the upper stories.

FINAL DESIGN AND ANALYSIS PROCEDURE

For final design a complete three dimensional model of each tower, including set backs and transfers, was constructed using the Building Design Language (BDL) modeling environment and analyzed by CSI-ETABS+ computer program. Because of the complexity of the framing system virtually every column line in the building, including non-frame columns was modeled. This resulted in atypical tower model with 50 levels, two frames, 127 column lines and 229 bays (Figure 6).

Accurate dead and live loads were input. Live loads were reduced as permitted by the UBC code. ANSI 70 mph wind loads were input as static loads; and the Maximum Probable spectra with 5% critical damping was used in a dynamic analysis which included 40 modes. These modes accounted for 98% of the modal mass in translation but only 90% in rotation.

The fundamental periods in the final design were 6.62 seconds (.15N) in the transverse direction and 6.33 seconds (.14N) in the longitudinal directions. There is very little coupling between the modes in the principal directions and with the torsional modes (Figure 7). Because of the complex geometry of the buildings there was significant potential for coupling and torsion. Considerable design effort went in to reducing torsion.

The governing load conditions were 55 mph wind drift in the transverse direction and stresses due to vertical plus seismic load in the longitudinal direction. Final stress ratios are generally higher in the beams than the columns and the strong column weak beam provision of the 1988 UBC is satisfied except for transfer girders.

UNCOMMON DESIGN ISSUES

The unusual geometry of this framing system raised a number of design issues, which although they may occur in a regular frame, are not nearly as prominent or significant.

Differential shortening between columns due to vertical load was significant. In a typical building the shortening is between the non-frame core columns and the perimeter frame columns. Generally, the core columns shorten more under vertical loads due to larger tributary areas, and often to their smaller cross-sectional area; because they do not carry lateral load. This problem is handled by adjusting the lengths of the columns during fabrication. It does not present a problem in analysis because the core columns are not generally included in the model.

In this building some core columns are also frame columns and others support transfer girders in the frame. Modeling results under full vertical load indicated differential shortening of up to four inches between columns. This results in significant frame action and high stresses under vertical loads particularly in beams and columns sitting on transfers. Since the building is not actually constructed in this manner the fabrication process must be accounted for in the model. This was partially accomplished by scaling the vertical loads on the non-frame core columns such that their shortening equaled the average for a typical floor. These columns were checked for stress by hand.

This partially mitigated the problem but stresses under vertical load were still high, in some cases unreasonably high. This was due to the fact that the core columns on frame Lines Pe and Ge were still shortening significantly; and partially, it was believed, due to the fact that dead loads are applied, in the model, all at once and not story by story as built. To verify this hypothesis the model was loaded two stories at a time to simulate construction (using an in-house program). Our hypothesis was verified, although columns stresses were typically quite accurate, many of the high stresses in beams and transfer columns disappeared. The results of this study were incorporated into the final design by allowing the stresses in selected members to appear higher than typical.

Another design issue was the use of the weak axis effective length factor (Ky) for columns with beams framing at shallow angles. It is well known that for shallow angles the AISC procedure for calculating the effective length factor results in unrealistically high effective lengths.

In this job, establishing a definition of "shallow" was particularly critical because beams frame into columns from the full range of angles. In the preliminary design phase the calculated values of Ky were used for all but the shallowest angles; to check stresses and to insure that the UBC provision that $KL/r < Cc$ was satisfied. This resulted in a very large number of box columns and a stiff building in the longitudinal directional. The short period resulted in high seismic forces which further increased sizes.

The result was a high cost for strictly adhering to this provision. In the final design Ky was set at 1.2 for columns with beams framing at angles of less than 22 degrees. This resulted in a significant reduction in cost.

Another interesting problem occurred at the base of frame columns which transfer on to floor beams, particularly near the beams support points. Very high weak axis bending stresses were introduced at the base of the columns, due to the curvature of the beam. In most cases this could not be economically mitigated by increasing either the column or beam sizes. The stresses were relieved by detailing the column bases so that they could be modeled as pins (Le-19e on Figure 3). Unfortunately this compromises the frame and in cases where this was not desirable extremely heavy floor beams were required to provide a level platform (Ce-13e on Figure 2)

CONCLUSIONS

Complex architectural features in the Los Angeles Center towers give rise to a great deal of variation in the demand on the components of the structural system. This resulted in considerable variation in member shapes and sizes and presented unusual analysis and design issues.

REFERENCES

LeRoy Crandall and Associates, *Report of the Crandall Investigation for the Proposed Los Angeles Center Phase1,* September 1989.

American Institute of Steel Construction, *Specifications for the Design, Fabrication and Erection of Structural Steel for Buildings*, 8th Edition, 1978.

American National Standards Institute, *Minimum Design Loads for Buildings and Other Structures*, ANSI A58.1-1982.

Trade Arbed, *Arbed-Rolled Wide Flange Beams -- 40" & 44" Standard and Tailor-Made Series*, 1989.

Naeim, Farzad and Dehghanyar, T.J., *Building Design Language -- User Reference*, John A. Martin and Associates, Inc., Los Angeles, 1987.

Habibullah, Ashraf, *ETABS -- Three Dimensional Analysis of Building Systems, User Manual*, Computers and Structures, Inc., Berkeley, 1989.

(a) small scale model of the project

(b) composite photo depicting the completed project in the Los Angeles skyline

FIGURE 1. THE LOS ANGELES CENTER PROJECT

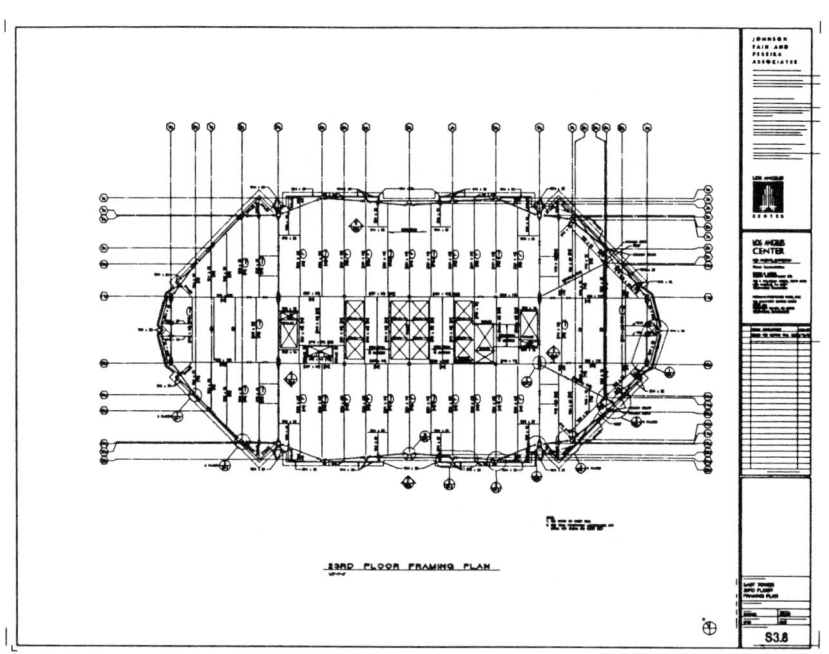

FIGURE 2. The 23RD FLOOR FRAMING PLAN

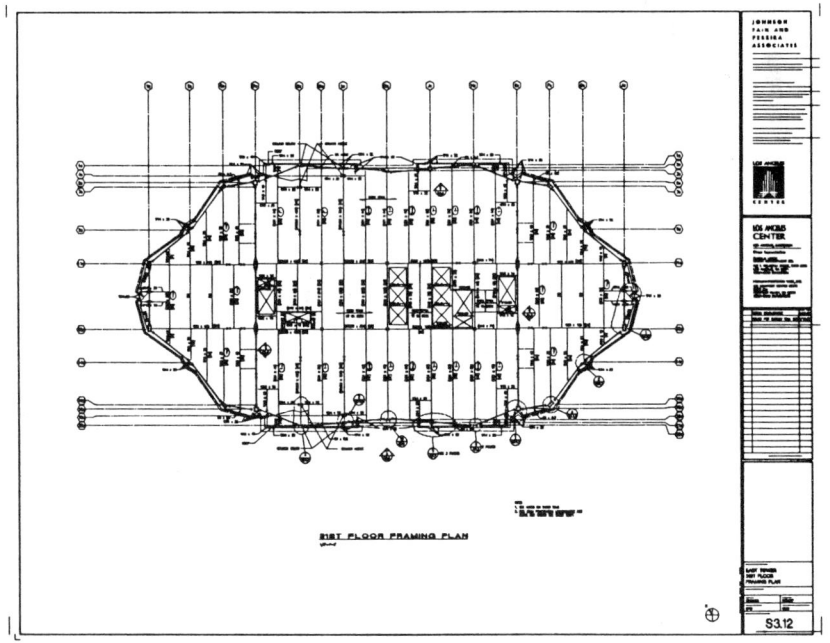

FIGURE 3. The 31ST FLOOR FRAMING PLAN

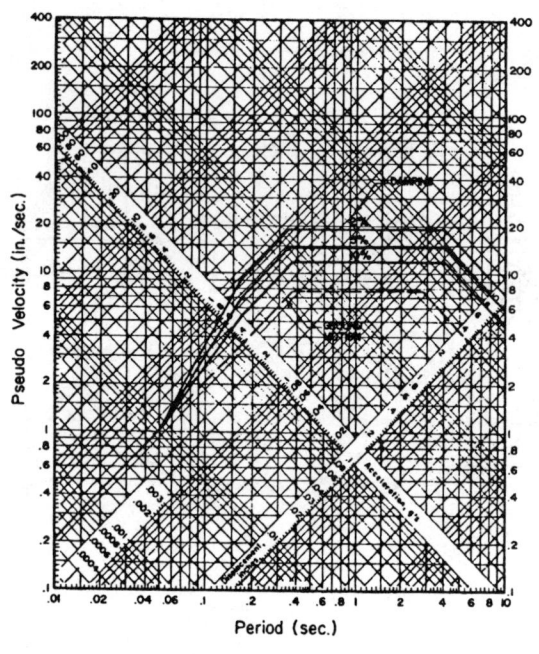

RESPONSE SPECTRA

MAXIMUM PROBABLE EARTHQUAKE "C"
Coyote Pass, or Santa Monica-Hollywood, or Newport-Inglewood Faults:
Mag. = 6.5; Dist. = 3.5 to 5.9 Miles
(50 year recurrence interval)

FIGURE 4. MAXIMUM PROBABLE DESIGN SPECTRA

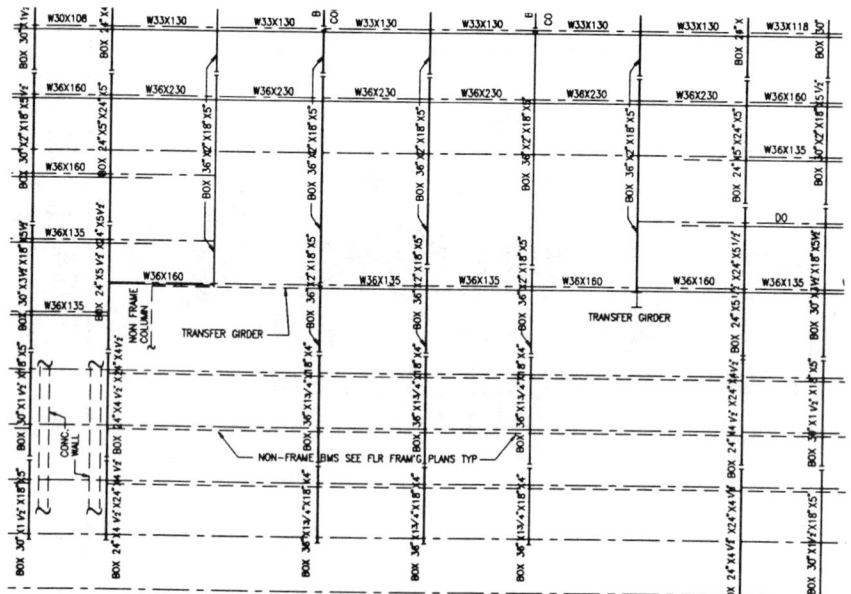

FIGURE 5. PARTIAL FRAMING ELEVATION AT THE VICINITY OF SOME TRANSFER GIRDERS

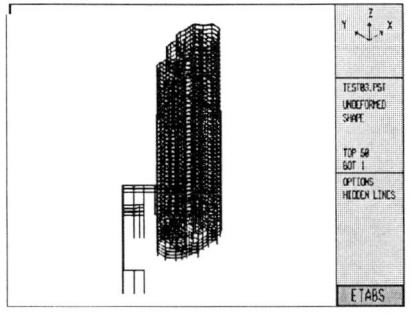

(a) model of one tower and the base

(b) model view from the top

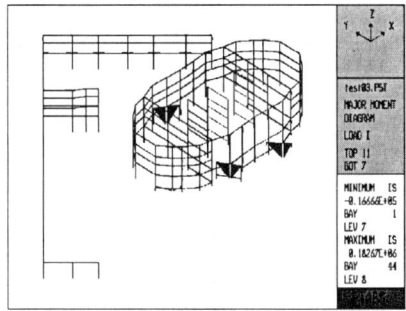

(c) partial view of the model showing the relative magnitude of transfer girder moments under gravity loads

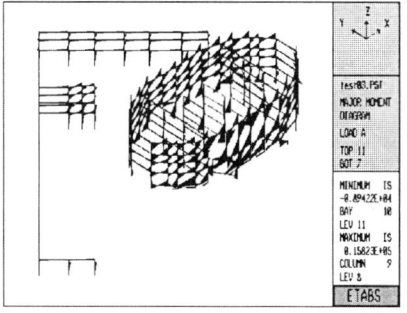

(d) partial view of the model showing the relative magnitute of major moments under lateral loads

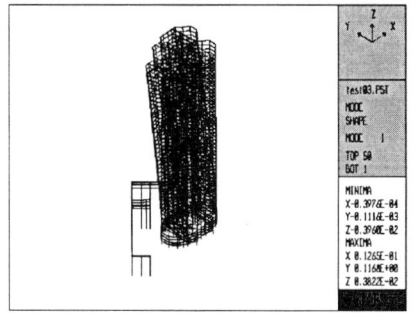

(e) dynamic mode shape 1

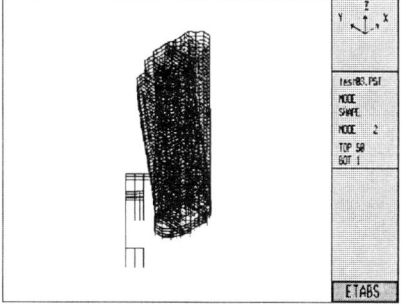

(f) dynamic mode shape 2

FIGURE 6. COMPUTER MODEL OF LOS ANGELES CENTER TOWER

STRUCTURAL TIME PERIODS AND FREQUENCIES

MODE NUMBER	PERIOD (TIME)	FREQUENCY (CYCLES/UNIT TIME)	CIRCULAR/FREQ (RADIANS/UNIT TIME)
1	6.62384	0.15097	0.94857
2	6.33788	0.15778	0.99137
3	3.50743	0.28511	1.79139
4	2.45231	0.40778	2.56215
5	2.29489	0.43575	2.73790
6	1.52448	0.65596	4.12154
7	1.38050	0.72438	4.55140
8	1.34748	0.74212	4.66290
9	1.07011	0.93449	5.87155
10	0.99680	1.00321	6.30333
11	0.84632	1.18158	7.42410
12	0.81841	1.22189	7.67734
13	0.78171	1.27925	8.03775
14	0.66612	1.50124	9.43254
15	0.64946	1.53975	9.67454
16	0.62857	1.59092	9.99605
17	0.54969	1.81921	11.43043
18	0.54681	1.82878	11.49056
19	0.51820	1.92977	12.12512
20	0.46907	2.13189	13.39504
21	0.46116	2.16844	13.62468
22	0.44305	2.25706	14.18155
23	0.40876	2.44645	15.37148
24	0.39169	2.55307	16.04140
25	0.37815	2.64442	16.61538
26	0.35932	2.78304	17.48634
27	0.34451	2.90268	18.23807
28	0.32682	3.05978	19.22515
29	0.31574	3.16715	19.89981
30	0.30279	3.30267	20.75130
31	0.28669	3.48806	21.91615
32	0.28071	3.56240	22.38319
33	0.26424	3.78448	23.77859
34	0.25678	3.89432	24.46874
35	0.25242	3.96159	24.89143
36	0.23400	4.27353	26.85139
37	0.23054	4.33760	27.25394
38	0.22809	4.38427	27.54716
39	0.21092	4.74111	29.78930
40	0.20900	4.78480	30.06377

EFFECTIVE MASS FACTORS

MODE NUMBER	/--X TRANSLATION--/ %-MASS	%-SUM	/--Y TRANSLATION--/ %-MASS	%-SUM	----Z ROTATION---- %-MASS	%-SUM
1	0.57	0.6	62.23	62.2	0.05	0.1
2	65.34	65.9	0.53	62.8	0.03	0.1
3	0.03	66.0	0.81	63.6	46.01	46.1
4	0.03	66.0	12.50	76.1	2.81	48.9
5	11.74	77.7	0.05	76.1	0.01	48.9
6	0.01	77.7	4.35	80.5	4.54	53.4
7	4.61	82.3	0.07	80.5	0.47	53.9
8	0.52	82.8	1.07	81.6	6.65	60.6
9	0.02	82.9	3.34	85.0	0.20	60.8
10	3.06	85.9	00.00	85.0	00.00	60.8
11	0.13	86.0	0.78	85.7	4.22	65.0
12	0.17	86.2	2.02	87.8	2.83	67.8
13	2.48	88.7	0.01	87.8	0.03	67.8
14	0.06	88.8	2.53	90.3	00.00	67.8
15	2.48	91.2	00.00	90.3	2.73	70.6
16	0.28	91.5	0.16	90.5	3.43	74.0
17	1.12	92.7	0.82	91.3	0.36	74.4
18	0.87	93.5	1.12	92.4	0.75	75.1
19	0.12	93.6	0.12	92.5	3.11	78.2
20	0.74	94.4	0.70	93.2	0.29	78.5
21	0.30	94.7	0.80	94.0	0.36	78.9
22	0.22	94.9	0.16	94.2	1.49	80.4
23	0.35	95.3	0.68	94.9	0.29	80.7
24	0.48	95.7	0.18	95.1	1.85	82.5
25	0.01	95.7	0.51	95.6	0.02	82.5
26	0.25	96.0	0.30	95.9	0.06	82.6
27	0.36	96.3	0.08	95.9	1.90	84.5
28	0.00	96.3	0.65	96.6	0.20	84.7
29	0.40	96.8	0.01	96.6	0.29	85.0
30	0.10	96.9	0.09	96.7	1.21	86.2
31	0.01	96.9	0.44	97.1	0.15	86.3
32	0.44	97.3	0.06	97.2	1.09	87.4
33	0.01	97.3	0.00	97.2	0.30	87.7
34	0.02	97.3	0.31	97.5	0.09	87.8
35	0.36	97.6	0.08	97.6	1.04	88.8
36	0.00	97.7	0.04	97.6	0.01	88.9
37	0.16	97.9	0.06	97.7	0.04	88.9
38	0.15	98.0	0.21	97.9	1.05	89.9
39	0.00	98.0	0.06	98.0	0.04	90.0
40	0.14	98.1	0.04	98.0	0.06	90.1

FIGURE 7. SIGNIFICANT DYNAMIC PROPERTIES OF ONE OF THE TWO LOS ANGELES CENTER TOWERS

Proceedings of the Second Conference on Tall Buildings in Seismic Regions
55th Regional Conference
May 16 and 17, 1991, Los Angeles, California

APPLICATION OF HIGH-STRENGTH CONCRETE
IN REGIONS OF HIGH SEISMICITY

S. K. Ghosh
Portland Cement Association
Skokie, Illinois

The application of high-strength concrete in highly seismic regions has lagged behind its application in regions of low seismicity. One of the primary reasons has been a concern with the inelastic deformability of high-strength concrete structural members under reversed cyclic loading of the type induced by earthquake excitation.

This paper discusses the current state of application of high-strength concrete (with specified compression strength in excess of 6000 psi) in buildings across the United States, including major west coast cities. It then discusses the properties of high-strength concrete that are relevant to structural applications in high seismic regions. Particular attention is paid to inelastic deformability of reinforced concrete structural members under reversed cyclic loading.

HIGH-STRENGTH CONCRETE IN BUILDINGS

As late as the early 1950's, the tallest concrete buildings were in the 20-story height range. By 1975, the 74-story high Water Tower Place, till recently the tallest concrete building in the world, had already been constructed. This virtual revolution within a very short time span was made possible by a number of factors, the most important amongst which were the availability of: new, improved construction methods; bigger cranes; high-strength materials; innovative structural systems; and high-storage, high-speed computer hardware and the corresponding software that gave the structural engineer unprecedented analytical capabilities. It is futile to speculate which of the factors was more or less important than the others; all of them contributed to the dramatic growth in height of concrete buildings. This paper will focus in particular on high-strength concrete.

Figure 1 shows a series of nine concrete buildings, each of which, with the exception of Two Prudential Plaza, was the tallest concrete building in the world at the time of its completion. It is clear that the growth in the height of concrete buildings has gone hand-in-hand with the availability of higher- and higher-strength concrete.

Almost incredibly, seven of the nine record-setting buildings are located in Chicago, a city that in many ways has pioneered the evolution of high-strength concrete technology. However, very recently, there has been an impressive spread in the availability of ultra-high-strength concrete (with specified compression strength in excess of 10,000 psi). Figure 2 shows that 12,000 psi or higher-strength concrete has been used in the last three or four years in Atlanta, Cleveland, Minneapolis, New York, and most significantly, Seattle which is in Uniform Building Code Seismic Zone 3.

In fact, the highest concrete strength ever used in a building has been 19,000 psi in the composite columns of Seattle's 62-story, 759-ft high Two Union Square (Skilling Ward Magnusson Barkshire Inc., Structural Engineers). The strength was obtained by use of : what may be a record low water to cementitious ratio of 0.22 (this is the single most important factor in increasing strength and reducing shrinkage and creep); the strongest of available cements; a superplasticizer which reduces the need for water, and provides the necessary workability; a very high cement content; a very strong, small (3/8 in.), round glacial aggregate available locally; silica fume (increasing strength by about 25%); a design strength obtained at 56 rather than the usual 28 days; and an extraordinarily thorough quality assurance program. The 19,000 psi strength was the byproduct of the design requirement for an extremely high modulus of elasticity of 7.2 million psi. The stiffness was desired in order to meet the occupant-comfort criterion for the completed building. The same concrete strength was later used by Skilling Ward in the composite columns of the shorter 44-story Pacific First Centre.

9500 psi concrete has been used by Skilling Ward at 600 California in San Francisco, and at 1300 Clay in Oakland, both composite buildings. The Watry Design Group has used 8000 psi concrete in several all-concrete Bay Area buildings, including the 19-story Fillmore Building.

The spread in the use of high-strength concrete in Southern California has been hampered by the City of Los Angeles Code provision restricting the strength of concrete to a maximum of 6000 psi. Even then, concrete strength in excess of 6000 psi has been used in the Great American Plaza office-hotel-garage complex in San Diego, a 14-story residential building at 5th and Ash in San Diego, and in the 22-story Pacific Regent (senior citizen housing) at LaJolla.

ADVANTAGES AND PROPERTIES OF HIGH-STRENGTH CONCRETE

The three biggest advantages of high-strength concrete that make its use attractive in high-rise buildings are that it provides more
- strength/unit cost
- strength/unit weight
- stiffness/unit cost

than most other building materials including normal weight concrete.

Commercially available 14,000 psi concrete costs significantly less in dollars per cu. yd. than 3 1/2 times the price of 4000 psi concrete. In fact, the unit price goes up relatively little as concrete strength increases from 4000 to 10,000 psi. Thus, high-strength concrete gives the user more strength per dollar.

The unit weight of concrete goes up only insignificantly as concrete strength increases from moderate to very high levels. Thus, more strength per unit weight is obtained, which can be a significant advantage for construction in high seismic regions, since earthquake induced forces are directly proportional to mass.

The modulus of elasticity of concrete remains proportional to the square root of the compressive strength of concrete at the age of loading even for high-strength concrete.[1] The user thus obtains a higher stiffness per unit weight and unit cost. Indeed, it is quite common for a structural engineer to consider and specify high-strength concrete for its stiffness, rather than for its strength.

It is important to mention that the specific creep (ultimate creep strain per unit of sustained stress) of concrete decreases significantly as the concrete strength increases. The most recent verification of this is available in Ref. 2. This is indeed a fortunate coincidence without which the application of high strength concrete in highrise buildings would have been seriously hampered. Because of the lower specific creep, high-strength concrete columns with their high stress levels suffer no more total shortening than normal-strength concrete columns with their lower strength levels. Otherwise, the problem of differential shortening of vertical elements within highrise buildings would have been aggravated by the use of high-strength concrete in columns.

INELASTIC DEFORMABILITY UNDER REVERSED CYCLIC LOADING

The inelastic deformability of high-strength concrete structural members under reversed cyclic loading of the type induced by earthquake excitation has been a concern that has slowed the application of high-strength concrete in high seismic regions.

The inelastic deformability of ultra-high-strength concrete members (compressive strength ranging up to 15,600 psi) under monotonic as well as reversed cyclic loading has been investigated through two series of tests.[3] The specimens tested were columns under zero axial load. The principal test variables were the compressive strength of concrete, the percentage of longitudinal reinforcement and the spacing of confinement reinforcement. The ratio ρ/ρ_b turned out to be the most dominant factor influencing the magnitude of inelastic deformability, with the latter decreasing for increasing values of ρ/ρ_b (ρ = tension reinforcement ratio, ρ_b = reinforcement ratio producing balanced strain conditions). However, even at very large ρ/ρ_b ratios, substantial amounts of inelastic deformability were found to be available. Such deformability was generally found to increase with increasing concrete strengths. It was concluded that in the absence of axial loads acting simultaneously with flexure, high-strength reinforced concrete members subjected to reversed cyclic loading possess as much inelastic deformability as is likely to be required of them in practical situations.

In a very significant recent study, Muguruma and Watanabe[4] have investigated the possibility of improving the ductility of high-strength concrete columns through the use of lateral reinforcement. Eight column specimens confined by lateral reinforcement having 47.6 ksi and 115.0 ksi yield strength were tested under reversed cyclic lateral loads with constant axial compressive load levels from 0.254 to 0.629 times the axial load carrying capacity. The concrete compressive strengths were 12.4 and 16.8 ksi. Volumetric ratio of lateral reinforcement was 1.6% in all specimens. Test results indicated that very large ductilities could be achieved by using high yield strength lateral reinforcement even for such high-strength concrete columns.

CONCLUSIONS

The use of high-strength concrete (with specified compression strength in excess of 6000 psi) in multistory buildings is spreading across the country, and is now not uncommon in major cities in Uniform Building Code Seismic Zones 3 and 4. The big advantages which make such application attractive are: 1) more strength per unit cost, 2) more strength per unit weight, 3) more stiffness per unit cost and wight, and 4) lower specific creep for high-strength concretes. One remaining concern that has somewhat slowed the spread of high-strength concrete in regions of high seismicity has been about the inelastic deformability of high-strength concrete structural members under reversed cyclic loading of the type caused by earthquakes. However, reassuring test results have begun to come in, indicating that even high-strength concrete columns under high levels of axial loading can be made ductile under reversed lateral loading, with proper confinement of the concrete.

REFERENCES

1.   Russell, H. G., "Shortening of High-strength Concrete Members," High Strength Concrete - Second International Symposium, Publication SP-121, American Concrete Institute, Detroit, Michigan, 1990, pp. 1-20.

2.   Bjerkeli, L., Tomaszewicz, A., and Jensen, J. J., "Deformation Properties and Ductility of High-Strength Concrete," High-Strength Concrete - Second International Symposium, Publication SP-121, American Concrete Institute, Detroit, Michigan, 1990, pp. 215-238.

3.   Shin, S.-W., Kamara, M., and Ghosh, S. K., "Flexural Ductility, Strength Prediction , and Hysteretic Behavior of Ultra-High-Strength Concrete Members," High-Strength Concrete - Second International Symposium, Publication SP-121, American Concrete Institute, Detroit, Michigan, 1990, pp. 239-264.

4.   Muguruma, H., and Watanabe, F., "Ductility Improvement of High-Strength Concrete Columns with Lateral Confinement," High-Strength Concrete - Second International Symposium, Publication SP-121, American Concrete Institute, Detroit, Michigan, 1990, pp. 47-60.

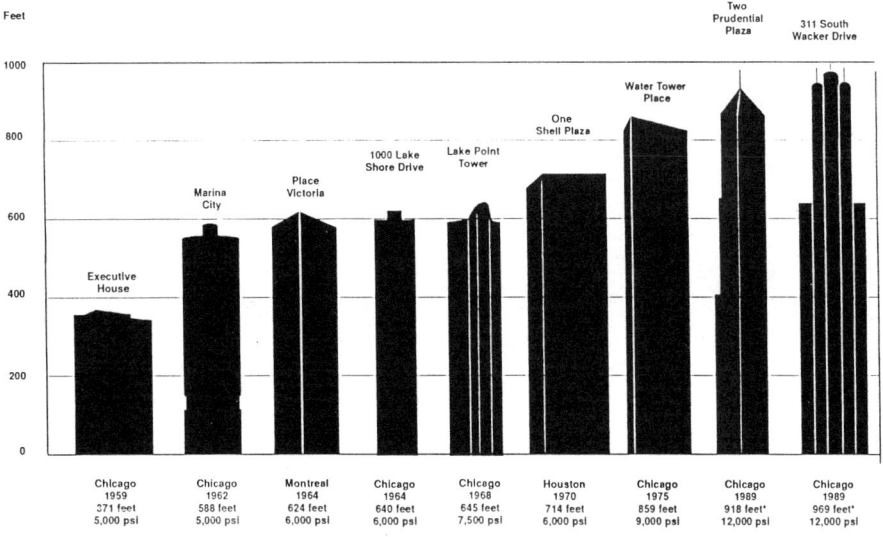

Feet

| Chicago 1959 371 feet 5,000 psi | Chicago 1962 588 feet 5,000 psi | Montreal 1964 624 feet 6,000 psi | Chicago 1964 640 feet 6,000 psi | Chicago 1968 645 feet 7,500 psi | Houston 1970 714 feet 6,000 psi | Chicago 1975 859 feet 9,000 psi | Chicago 1989 918 feet* 12,000 psi | Chicago 1989 969 feet* 12,000 psi |

* Silhouette shown includes steel structure at top of building.

Source: Portland Cement Association

Fig. 1 - High-Strength Concrete in High-Rise Construction

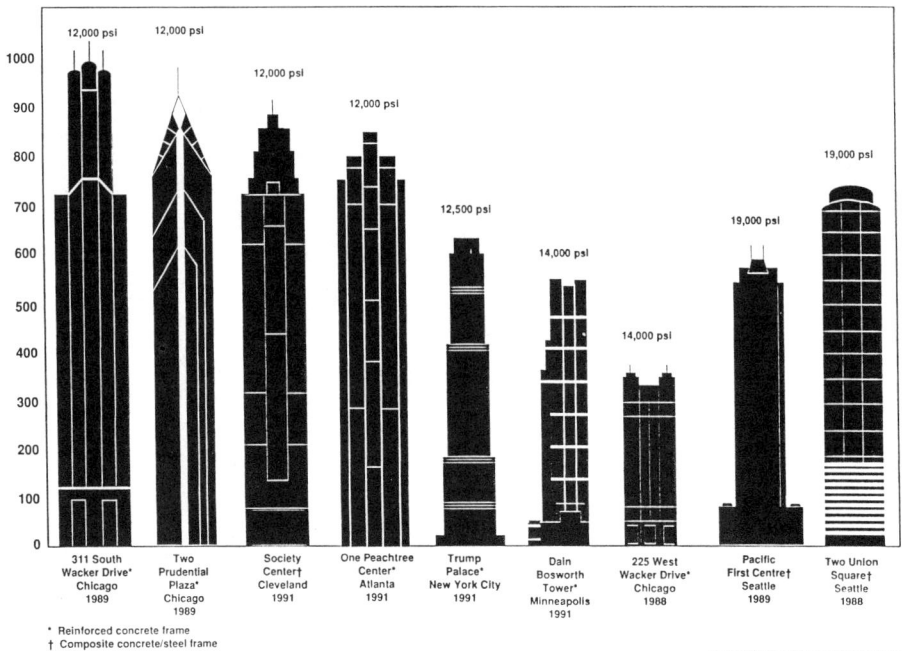

* Reinforced concrete frame
† Composite concrete/steel frame
Center

Source: Portland Cement Association

Fig. 2 - Ultra-High-Strength Concrete Shapes New Skylines

Proceedings of the Second Conference on Tall Buildings in Seismic Regions
55th Regional Conference
May 16 and 17, 1991, Los Angeles, California

Performance of Steel, Concrete and Masonry Structures in recent earthquakes of Armenia and Iran Design and Construction Deficiencies

Zareh Gregorian, Principal, Gregorian Engineers [1]
Garen Gregorian, Project Engineer, Gregorian Engineers [2]

Major earthquakes have occurred during the past two years in the Northern parts of Iran, Armenia, Pakistan and Turkey, killing more than 75,000, causing billions of dollars of damage and destruction, and leaving more than half a million people homeless. These earthquakes are the result of the tectonic movements of the Arabian plate moving with a rate of one inch per year, North, towards the Eurasian plate.
The collision zone of the plates occur on a line stretching form Northern parts of Afghanistan passing through Pakistan, Iran, Armenia and Turkey (Ref.1).
The Armenian earthquake occurred on December 7, 1988 with a magnitude of 6.9 and an aftershock of 5.8 on the Richter scale. The Iranian "Manjil" earthquake occurred on June 21, 1990, with a magnitude of 7.3 and an aftershock of 5.7 on the Richter scale.
The epicenter of the Armenian earthquake was located near the city of Spitak, a city located at about 80 Kilometers North of Yerevan, the capital of Armenia (Ref.2). A forty Kilometer long fault, which originally was thought to be eight Kilometers, was created at the town of spitak extending North West towards the city of Leninakan causing
vertical and horizontal movement of 70" and 20" at the fault.
Due to shallow focal depth of the earthquake, the energy released affected a small radius. The earthquake caused drastic damage and destruction to three major cities of Spitak, Leninakan, and Kirovakan, and a smaller city Akhorian. The town of Spitak was levelled with the ground and only about 25% of buildings survived in Leninakan.
In Spitak most of the buildings were built with hollow core concrete slabs, resting on masonry walls. These buildings were four to five stories high and suffered the most damage due to lack of horizontal or vertical ties.
In Kirovakan the subsoil conditions provided minor amplification of the earthquake forces resulting in less damage to buildings even though they were closer to the epicenter than Leninakan.
In Leninakan, which is located at a flat sedimentary alluvial and volcanic Tuff basin, amplification occurred due to poor soil conditions resulting in failure of most of the buildings. These areas are not suitable for constructing earthquake resistant structures designed for conventional forces and construction systems. Sedimentary subsoil conditions caused similar destructions in down town Mexico City during earthquake of 1985 (Ref.3).
Due to the extreme shortage of housing in the Soviet Union, programs were developed in the past two decades to construct high rise buildings in various densely populated areas within a short period of time. Prefabricated concrete building

---

1) Zareh Gregorian, Principal, Gregorian Engineers
2) Garen Gregorian, Project Engineer, Gregorian Engineers
75 Spring Valley Road
Belmont, Massachusetts

systems were selected for mass production and fast erection purposes. Standard drawings were developed for construction all over the country which could be modified for earthquake prone areas. Typical systems include:

1- Precast Reinforced Concrete Buildings.
Built with precast bearing walls and hollow core slabs, the bearing walls are connected, at their intersection, with vertical bars slipped through U-shape reinforcing bars extended from the walls. The vertical joint is grouted after erection of intersecting walls. Corrugations are provided on the horizontal surfaces of the vertical panels to overcome the shear forces developed from lateral loads. The "Bearing Wall Buildings" performed relatively well during the quake, the reason being the simple erection procedures, and joint connections, which could be performed without much default.

2- Lift Slab Construction.
The other commonly used system is the lift slab system, which in general needs a lot of precision detailing, high quality execution and supervision during construction, lack of which caused the complete destruction of such buildings.
Lift slab buildings have not acted properly during earthquakes. One such example was a six story lift slab building which collapsed during the 1964 Prince William Sound Earthquake, in Anchorage Alaska. Unfortunately there are numerous lift slab buiHxings in other areas, specially in the capital city of Yerevan, which needs immediate attention for retrofitting.

3- Precast Concrete Frame Systems.
The other most commonly used structural system is the precast frame system with either moment frames, or precast panel shear walls in one or two directions.
Seventy two such structure, which were built in Leninakan, totaly collapsed and fifty five were heavily damaged and had to be demolished.
We have acquired an original plan of one of the nine story buildings (Ref.4), which collapsed during the quake, and examined the design and construction deficiencies which contributed to the failure.

The building is nine story precast frame system, with three 6 Meters (20'-) bays in each direction, and two panel shear walls located at end bays of two middle frames. The shear walls are made of precast panels welded to special steel plates preset at the edges of beams and columns. The floor is constructed with 22 centimeter (9") hollow core concrete slabs, with virtually no physical attachment to the load carrying beams and without concrete topping and precast concrete stairs.
The structural system consists of the following:
a) Precast concrete columns two story high, with splice joints created at 1/3 story height by welding the

extended bars from the adjoining columns and placement of concrete in the area.

b) Precast main direction beams with bars extended at ends are welded to bars protruding from precast columns at girder elevations. Concrete is then placed in the connection area.

c) Precast transverse beams with extended end bars are welded to bars inserted through sleeves preset in the columns. Again, the connection is completed by placing concrete in the connection area.

d) Precast hollow core slabs are placed on main girders, in most cases with no concrete topping, which acts as floor diaphragms.

e) Precast shear panels with preset steel plates welded to preset columns and girders.

f) Precast stairs attached to stairwell girders.

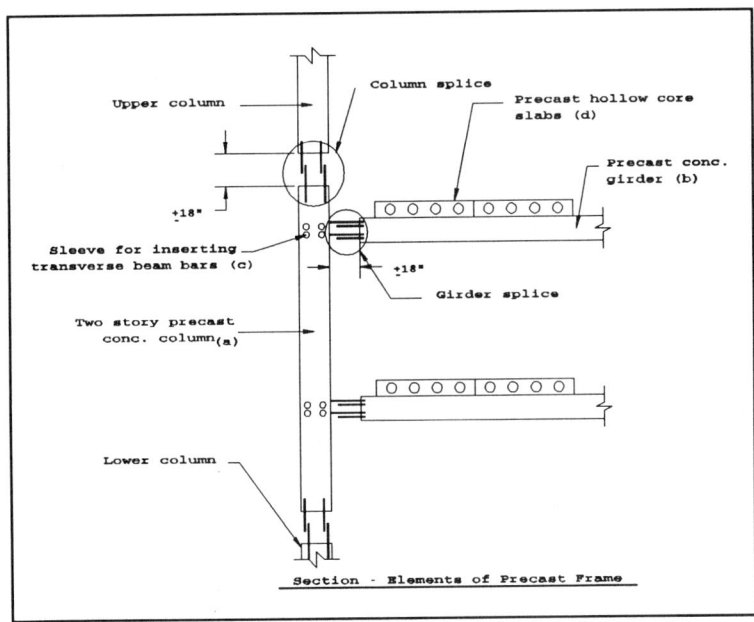

Figure 1

A case study following the UBC code (Ref.4) in zone 4 with poor soil conditions yields the following forces on the typical nine story building constructed in Leninakan. The building has shear walls in one direction and moment frames in the other direction.

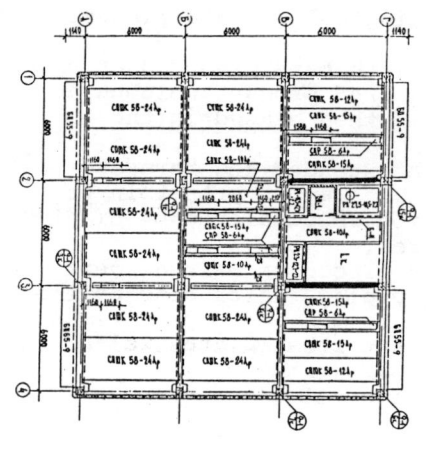
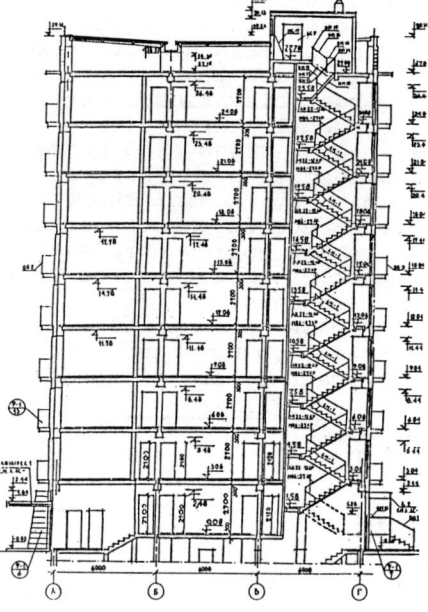

NINE STORY BUILDING

PLAN & SECTION

Very high!

The following assumptions are made.

Roof D.L.= (160) psf. ; Roof L.L.= 40 psf.
Floor D.L.= (180) psf. ; Floor L.L.= 40 psf.

The UBC yields the following base shear values for the direction with moment frames.
Using the equations below:

$$V = \frac{ZICR}{R_W} \qquad C = 1.5 \frac{S}{\sqrt[3]{T}} \qquad T = C_t (h_n)^{3/4}$$

With I=1, Z=0.4, Rw=5 we get the following:
T= 0.9 sec.
C= 2.01
W= 60x60x(160+40+8x180)/1000 = 5904 Kips.
V= 0.4x1x2.01x5904/5 = 948 Kips.

Shear taken by frame B-B at 1st floor = 948/3=316 Kips.
Shear taken by middle column frame B-B at 1st floor = 316/3 = 105 kips.

Moment at splice point= 105(1/2-1/3)h
                      = 105(1/2-1/3)x13= 277 K-ft.
Vertical load at 1st floor= 20x20(200+8x220)/1000= 784 kips.
Shear= 105 kips.

A 20"x20" column with 1% steel was checked (ignoring the effects of shrinkage, and assuming the section is intact), to

128

resist the above forces and the following stresses were obtained: fs=63 ksi, fc=3.75 ksi.
These values are much higher than allowable stresses of conventional materials used for construction of such buildings.

The following design, details, and construction deficiencies are noted:

Design Deficiencies.
1. According to reports prepared by US engineers and scientists visiting Armenia after the quake, the forces acting on structures were four times more than what the building was designed for. The discrepancy in forces could be the consequence of resonance which occurred due to the coinciding magnitudes of natural and forcing frequency(s) of the buildings and the earthquake (Ref.6).
2. Soil amplification factors were not considered properly.
3. The precast frames assumed as moment resisting are not designed with the ductility requirements of such structures, and according to recent codes such frames can not be considered as moment resisting frames (Ref.7, 8).
4. The shear wall panels attached to beam column panels have weak connections. Placing of shear walls at middle bay will minimize the effects of planar torsion.

Detailing Deficiencies.
1. The splice details, as designed and constructed, have the following deficiencies:
In beams and columns: Bars welded by lapping causes unprecedented eccentricity in the steel bars.
In beams and columns: Improper quality, placement, and shrinkage of concrete at the joints resulted in the transfer of loads mostly by steel. Most failures occurred at beam and column splices.
2. Lack of concrete topping on the hollow core slabs diminishes the capability of the floor diaphragm for transfer of lateral loads.

Construction Deficiencies.
1. The erection of columns were performed by lifting precast columns in place, by cranes, and welding the splices while the crane was holding the column. This causes considerable out of plumpness in columns resulting in eccentricity and additional stresses.
2. The quality of concrete was much lower than specified due to lack of materials, specially cement. Savings were made on steel using less bars.
3. Filling the joints with concrete needs special attention to diminish the effect of shrinkage.
4. In general the quality of workmanship was very poor.

The Manjil Earthquake of Iran (Ref.9).

The epicenter of the quake was located at the North of Roudbar province, and an area of approximately 2200 square kilometers was shaken. Three major cities, Manjil, Loshan, Roudbar and more than 87 villages and small towns suffered major damages, and 85 villages were completely destroyed.
Two large cities of Quazvin and Rasht also shook causing a 10 percent damage to buildings and public utilities.
The seismotectonic map of Iran (Ref.10), shows the occupance of earthquakes recorded since the beginning of the 20th century, some of which have a magnitude close to 8 on the Richter scale.
Due to shallow focal depths of the quakes occurring in the area, the released energy affected only a small area with high intensity causing extensive damage and destruction in the immediate areas as follows.

Houses Constructed with Adobe and Brick Walls.

With log joists and woven straw roofing or steel I-beams and jack arch roofing, and brick masonry load bearing walls, the houses collapsed completely due to lack of horizontal tie beams and vertical ties. Buildings with brick bearing walls with vertical and horizontal floor tie beams did not suffer considerable damage during the quake.
In some cases where massive masonry walls existed in the building (tied or untied to other elements), the quake energy was absorbed by the shear failure of the masonry walls, causing large cracks but leaving the structure in tact (Ref.9).

Concrete Structures.

Concrete structures behaved poorly during the quake, except one large gravity concrete dam in the Sefid Rood area, which suffered almost no damage.
An elevated concrete reservoir was completely destroyed in Rasht. The major destruction occurred in a group of four to five story reinforced concrete buildings which were under final stages of construction in the Rasht province.
The buildings were designed with no shear walls, no ductile frame details, and with minimum column sizes. The failures occurred at first floor corner columns, where the columns were in some parts two stories high and were pounded to an S shape, absorbing most of the quake energy, and leaving most of the remaining structural elements in place.

Steel Framed Structures.

Steel structures in the area were severely damaged, specially where liquefaction occurred causing non-uniform settlement and distribution of forces in the structures. An eight story steel structure with no lateral force resisting elements completely collapsed, and a six story structure nearing completion was damaged and tilted 25 inches at the top.

We have examined the eight story building, using UBC (Ref.5) earthquake code zone 4, poor soil conditions, and compared the values obtained by using the "Iranian Code for Seismic Resistance Design of Buildings" (Ref.11).
The building is assumed to have a square plan, four 20' bays on each side, and eight stories high, with 12' height for the first floor and 10' hight for the upper seven floors.

The dead and live loads according to the Iranian code are:

Roof D.L. = 120 psf.   Roof L.L. = 40 psf.
Floor D.L. = 140 psf.   Floor L.L. = 40 psf.

Using the equations given in the Iranian code:

$$V = cw \qquad C = \frac{ABI}{R} \qquad T = 0.08H^{3/4} \qquad T = 0.08H^{3/4} \qquad \text{With A=0.35, To=0.7, I=1,}$$

R=6 and
W=7424 Kips results in:

T=0.89 sec.   B=1.70   C=0.1 and finally V=742.4

The UBC code gives value of V=960 Kips for the base shear, which is 30% more than the Iranian code. Using the Iranian code, the gravity plus earthquake moments acting on first floor girders is calculated to be 382 K-ft.. Two spandrell European 28cm I-beams was assumed to be used as first floor girders, with a cpacity of 110 K-ft. allowable moment (for the girders and connections).
The resisting moment Mr is 3.5 times less than the Me the earthquake moment Mr=110 K-ft. << Me=380 K-ft., which has resulted in the failure of the structure.

Conclusions:
The buildings constructed in Armenia need retrofitting to overcome the design and construction deficiencies discussed in this paper. some refrofitting work has already started as can be seen in the accompanying photographs.
The buildings in Iran need more attention, due to poor design and construction conditions. Retrofitting must be planned and performed as soon as possible, to avoid disasters specially in large cities like Tehran.

Acknowledgements:
The authors would like to thank Professor Mishac Yegian Chairman, Vahe Ghahraman, Graduate Student, of Civil Engineering Department of Northeastern University, Boston, MA, for valuable information provided during preparation of this paper.

References:

1. F. Naeim, The Seismic Design Hand Book, 1989, Van Nostrand Reinhold, NY.
2. Siesmotectonic map of Armenia.
3. H.B. Seed et al. The Mexico Earthquake of 1985 E.Q. Spectra 5, 687-729.
4. Information and photos provided by Prof. M. Yegian and V. Ghahraman.
5. Uniform Building Code, UBC, 1988 Edition.
6. Earthquake Spectra EERI, August 1989.
7. ACI, Building Code Requirements for Reinforced Concrete (ACI-318-90).
8. R.E. Englekird, "Seismic Design Considerations for Precast Multistory Buildings". PCI Journal May/June 1990.
9. Zareh & Garen Gregorian, "June 1990 Earthquake of Iran, A Study". Introductory Workshop, Oct.6, 1990, MIT.
10. Geological Survey of Iran, Seismotectonic Map of Iran.
11. Iranian Code for "Seismic Resistance of Buildings", February 1989.

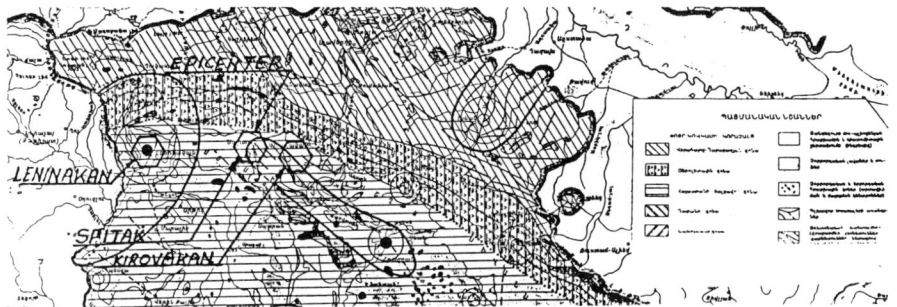

PART–SEISMOTECTONIC MAP OF ARMENIA

TOP–DAMAGED NINE STORY
BUILDING IN LENINAKAN

TOP RIGHT–COLUMN SPLICE FAILURE

RIGHT RETROFITTING DAMAGED
NINE STORY BUILDING

133

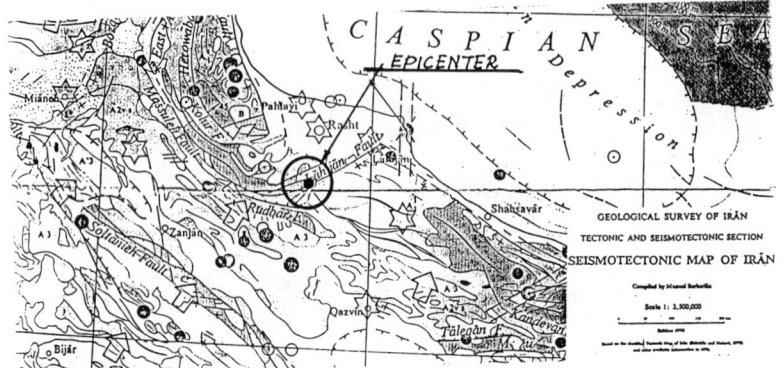

PART–SEISMOTECTONIC MAP OF IRAN

FAILURE OF R/C STRUCTURE

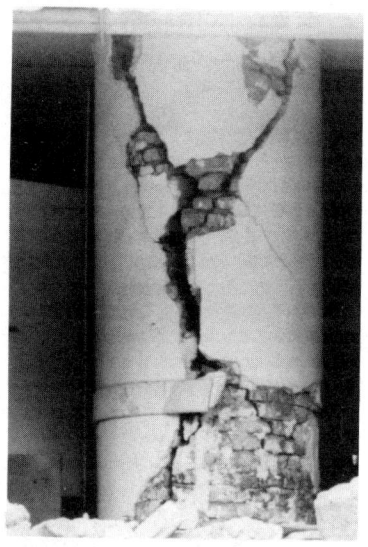

SHEAR FAILURE OF MASONRY PEIRS

COMPLETE COLLAPSE OF 8 STORY STELL STRUCTURE

Proceedings of the Second Conference on Tall Buildings in Seismic Regions
55th Regional Conference
May 16 and 17, 1991, Los Angeles, California

A STUDY OF SEISMIC LOAD BEARING WALL STRUCTURES BUILT FOR THE PUBLIC HOUSING SECTOR OF VENEZUELA

Arch. Luisa Teresa GUEVARA P., Ph.D.[1]

Eng. Mario PAPARONI M.[2]

Introduction

In Venezuela the Public Housing Program was officially started in 1928 with the founding of Banco Obrero (Workers' Bank). In 1975, it became the "Instituto Nacional de la Vivienda-INAVI."

Until the 40's its role was purely financial, then it changed to a source of internally developed projects, the main target being always low-income families. INAVI, in over 62 years of existence, has built a great variety of documented housing projects, with a remarkable degree of coordination and inspection activities, encompassing form small family dwellings to 17-story buildings. The construction activities have been done by private contractors.

At the end of the 60's, in an effort to catch up with demand, it was decided to build multifamily dwellings with industrialized methods, using both cast in place and prefabricated load bearing wall structures. The peak of this wave was reached in the 70's with large-scale multistory developments.. Most of these structures (80%) are located in active seismic areas, where the Venezuelan population is concentrated.

Many problems arose due to the great variety of industrialized systems proposed, their majority coming from non-seismic countries, needing a general evaluation for acceptance, and changes in construction methods, project criteria and inspection procedures, both from the architectural and engineering point of view, with an attempt to quantifying quality criteria and satisfying seismic demands.

The main point extracted from was the vital influence of floor-plan irregularities in the earthquake performance of this type of construction; it was clear that a certain architectural criteria have to be applied in seismic areas to such developments. All attempts to patch up or to adapt non-seismic projects or technologies always produced costly mistakes, being another typical case of wrong technology transfer.

This paper presents some of the most useful results of the study of the application of industrialized formworks to produce multistory crosswall dwellings. Local technologies and simple and quantifiable decision rules were the outcome of this effort in the form of Manuals of Practice for design, construction and inspection, which were planned to make them available to the Construction Industry to be applied in future developments.

This study was up-dated with some guidelines for the architectural design of irregular floor-plan buildings, based on the Uniform Building Code 1988.

1. Chairwoman of the Housing Research Unit, Consejo Nacional de la Vivienda, Venezuela.
2. Dean of Engineering, Facultad de Ingeniería, Universidad Metropolitana, Venezuela.

This paper has been sponsored by the "Fundación de la Vivienda Popular," Caracas, Venezuela.

Methodology of the Study

This study was conducted through the following steps:

1. Problem Identification. A survey of INAVI's crosswall buildings was done, encompassing existing buildings, industrialized formworks and contractors using them. The routine field reports, pertaining to damages observed in these building and inspection reports, were studied in detail. Pertinent publications and interviews to experienced structural engineers were also sources of information.

2. Data Base. Twelve projects were studied in detail. The studied subjects were architectural and structural design; production processes. The evaluated parameters were configuration; type of formworks and characteristics of the main contractors.

a. <u>General Characteristics of the Building System</u>. This system is the natural result of the evolution of traditional formworks in order to achieve the monolithic arrangement of both walls and slabs. The components of the buildings are cast on site and not in a factory. Because of this, they are considered semi-industrialize. As advantages they can claim speed of erection, labor efficiency and acceptable finishing within the process. Reduced waste of materials and the possibility of employing non-highly qualified workers, are also advantages.

b. <u>The Structural System</u>. In the traditional framed systems, the forces are taken by a skeletal structure and the infills are considered passive, but in this case both, walls and slabs, are at the same time fillers and structure. In active seismic zones, these systems need cast-in-situ structural walls arranged in two orthogonal directions. The floor slabs could be cast-in-situ, completely precast slabs, or partially precast. This system differs from the prefabricated large panel system in being monolithic, completely or at least for the majority of its structural elements.

c. <u>Formworks</u>. In Venezuela de most widely used systems are: (i) *Tunnel or semitunnel*, the walls and the floor slabs are poured at the same time; both, complete or split tunnels are used, their width varying between 3 and 4 meters. These systems require heavy formworks, heavy cranes and to be economically justified they require large volumes of work (20,000 square meters of flooring or more.) (ii) *"Flying tables" and wall formworks*, slabs and walls are cast separately; walls require long panelled metallic formworks; the slabs are cast over big "tables" mounted on rolling supports with a height equal to the floor height. The horizontal casting surface is made, generally, of plywood, easily replaced. This system is more flexible, from the user point of view, than the tunnel system and can be applied to smaller projects, but their production times are longer, because of the two-step casting process. (iii) *Modular formworks*, are used in small projects, they have an artisan character because thy must be assembled and disassembled in small modules for each job. They, nevertheless, are very flexible in their configurational possibilities.

d. <u>Enterprises</u>. In Venezuela, the industrialized form works were introduced after 1965 by Venezuelan companies, either by buying the forms or their patent from foreign companies. Later, many of these companies became form manufacturers, some of them developing their own technology. Selling or renting formworks became a good business. Some of the manufacturers started their activity after being repair or maintenance service shops for the forms. The overall result was a technology well suited to Venenzuelan conditions, allowing both big and small contractors.

3. Conclusions.

a. INAVI has applied two technologies for the crosswall monolithic systems: (i) the completely monolithic cast-in-situ systems, being the most widely used; (ii) the mixed systems, with prefabricated slabs and cast-in-situ walls.

b. The early projects were straight adaptations of existing framed, cast-in-situ traditional solutions. This procedure lead to forced solutions, with wedged spaces, cantilevers-on-cantilevers, no modular coordination, and excess of unnecessary or too strong lintels, and long walls mixed with too short ones (wall-columns). Other deficiencies were create by the originally irregular shapes, like moving around the formworks, or the requirement of several cranes, with consequent increases costs. The lack of rationalized spans, with differences of less than one inch between some of them, lead to span multiplicity, imposing the need of too many different types of formworks, with disastrous consequences whenever one of them had to be repaired. Large investments were then wasted in this multiplicity and the methods lost potential efficiency.

c. Long, simple, oblong buildings led to easy construction, but 15-story buildings precluded their vicinity, too much ground was wasted and maintenance expenses became untenable for the low income residents.

d. To reduce cantilevering, some of the solutions used produced passageways narrower than the Venezuelan Fire Code allowed (5 feet)

e. Other series, with asymmetrical lay-outs, caused undesirable seismic behavior.

f. Until 1979, it was allowed to build buildings without mutually orthogonal systems of walls; this situation created severe problems due to the lack of rigidity of this structures in one direction, including partitions cracking.

g. Buildings with large spans; this either delayed formwork removal or imposed the use of temporary supports, leading to increased costs.

h. Inspection manuals; there were no special manuals developed for this type of construction process. The lack of specific procedures and specifications demanded by the different construction methods produced severe and numerous problems. This is a very important issue.

i. No coordination between architects, structural engineers and contractors was established from the beginning of the design project. Projects were not tailored to the new demands of these construction technology. This situation came from the habits established by the traditional design and construction processes.

j. The advantages and disadvantages of the construction methods were not well known by the designers, therefore they could not take advantages of the virtues or eliminate or, at least, attenuate the deficiencies of these methods.

k. The same happened to the on-site inspectors, since the specific points to be looked at were not clear to them, due to lack of information, knowledge, and experience with the new methods.

l. In many cases, incorrect foundations were built, due either to the defective data transmitted from the structural engineer to the soil specialist, or to the unusual values of moments.

m. Low buildings were usually overdesigned, at least in one direction, due to the fear and misunderstood behavior of the new structural systems. Many lintels used on these low rise buildings could have been eliminated because they can withstand the required horizontal forces acting only as cantilevers, if they have adequate foundations.

n. It was found, in many of the studied cases, that commercial computer programs developed for framed buildings which included design routines were improperly applied to crosswall structures; crosswall structures should be capable of being analyzed through simple, almost shorthand procedures to be able to understand its whole general behavior and to take the proper conceptual decisions; they are, really, not as complicated as many people think.

4. Recommendations. Venezuela is a predominantly seismic country where seismic design imposes constraints that are not applicable in other countries where the seismic problem does not exist or where it is only of academic interest. After examining the data already presented in this paper, we could establish the following rules, pertaining to the most frequently observed shortcomings detected in this study, or to the most desirable goals to be attained. Some of these recommendations are also applicable in non-seismic zones. It is important to remember, though, that the effects produced on a building by an earthquake, will depend not only on other aspects such as seismic zoning, site geology and soil characteristics and local building codes and/or regulations.

a. Professional Team

(i) The whole process of Planning, Conception, Architectural Design, Structural Engineering, Soil Studies, Construction Inspection, must be considered as a single system from the beginning. In this type of construction, there is no freedom of decision between stages of design as it happens in the traditional framed buildings. If there in no coordinate plan, any aspect might interfere with any of the others, due to lack of "conceptual" free spaces to absorb slacks or changes. (ii) An especially delicate area is the interface building-foundation. The infrastructure for the crosswall building is completely different from the ones applied to traditional framed structures. (iii) The whole process depends entirely on the following issues which must be known by all the members of the team: (a) Constraints imposed by the construction process, its advantages and peculiarities, and the requirements for effective timing; (b) Type of construction equipment available in the area; (c) Construction processes and, at least, an outline of the construction chronogram, to be known to all and followed through.

b. Building Configuration. There are two main aspects to be considered regarding building configuration, construction and structural and architectural design:

Construction:

(i) Modular coordination is a must, due to the "hard" nature of the objects involved in the construction processes. There is no room for corner cutting. It brings also flexibility in planning and the possibility of using exchangeable hardware. (ii) Simple building configurations (floor-plan and elevation), rectangular and avoiding, as much as possible, reentrant corners simplifies the design and construction of the formworks, allows repetitive uses, and simplifies lifting and handling equipment; the regular cycle forming-pouring-stripping, must be routinely maintained in order to obtain an economically efficient process. (iii) Façades must be simple; it must be remembered that formworks for the construction of façades, have to be supported form the ground, or conversely require special structural detailing. (iv) Stairways must not hinder the mobility of the construction equipment; they must be located either out of the way or at the extreme of the building as subassemblages.

Structural and Architectural Design:

(i) Building configuration must be simple, avoiding asymmetries in plan and elevation. The main reasons are: (a) asymmetric structural arrangements are vulnerable to seismic torsion; (b) structural analysis and design are more complicate and less reliable; (c) seismic codes provides minimum requirements for regular configurations; the structural analysis and design of irregular configurations are the responsibility of the professional(s) in charge. (ii) If it is impossible to avoid irregularities, then this paper presents a series of guidelines (Ref.3) mostly based on the Uniform Building Code (UBC) 1988 which must be considered for the design of "irregular rectangulate" floor-plan shapes, such as "U," "L," "H," and "+" floor plans; It must be clear that these guidelines are not totally inclusive.

(a) *Identification of Irregularities*. These guidelines mainly concern the influence of the geometrical characteristics of floor-plan shapes on the response of buildings to torsional effects, only plan irregularities of Types A and B are analyzed. The reentrant corner problem is one of the most important factors in the evaluation of floor-plan shapes regarding the torsional effects induced in buildings by earthquakes. This design aspect is one of the factors that permits the identification of a shape as regular or irregular for the application of a specific procedure of analysis in earthquake-resistant design. Dimensions of the reentrant area produced by the reentrant corner define the seriousness of the structural irregularity. Irregular shapes determine the distribution of structural components and are the cause of variations in stiffness in the various rectangular blocks or "wings" of the building. and, in turn, influence the diverse seismic performance of these wings. In "irregular rectangulate" floor-plan shapes, the structure is determined and formed by different the wings, if these shapes are not treated appropriately, they can produce stress concentrations in the horizontal diaphragms at the intersections or reentrant corners, because each wing tends to vibrate in a different way under lateral forces. Since a reentrant corner is not always worthy of special treatment because of the dimensions of the reentrant area, the UBC 1988 in Table No. 23-N classifies and defines plan structural irregularities which require special design consideration, as follows:

"Plan configurations of a structure and its lateral force-resisting system contain reentrant corners, where both projections of the structure beyond a reentrant corner are greater than 15 percent of the plan dimension of the structure in the given direction." Fig. 1, presents an example: if A>0.15 B and C>0.15 D, then the floor-plan is irregular.

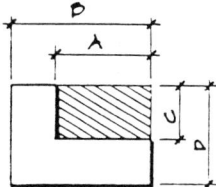

Fig. 1

This means that the structural analysis and the resulting solution will be more complicated and contain more unpredictable factors than if the plan were a regular. The UBC 1988 imposes two penalties for this type of irregular shape: Section 2312(h)2I(v) and Section 2312(h)2I(vi). The number and the location of significant reentrant areas are also important. The larger the number of reentrant areas in a floor plan, the greater the uncertainties in predicting the response of the building. The location of the reentrant area is also a very important factor. The following figure illustrates, in the vertical direction of the matrix, the comparison between three different groups of floor-plan shapes with the same number of reentrant areas, but with different locations for the reentrant areas. It can be observed in Fig.2, that in floor-plan shape "a" of the figure one perimeter reentrant area with both open areas set in the perimeter produces two wings or blocks and one critical corner floor-plan shape "e," with one perimeter reentrant area with only one side set in the perimeter, produces three blocks and two critical corners.

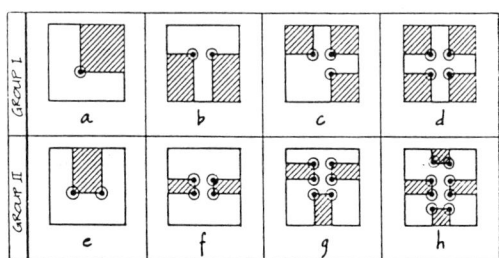

Fig. 2

A reentrant area is significant when the length of the sides of the building meeting in the reentrant corner both exceed fifteen percent of the length of the building in the direction in which the reentrant sides extend. The greater the number of "significant" reentrant areas, the worse the overall structural condition.

(b) *Analyzing Symmetry.* Geometric asymmetries in floor-plan shapes are often the reason for structural asymmetries and lead to undesirable torsional eccentricities in buildings. Geometrically symmetrical floor-plan shapes do not guarantee that the floor plan is structurally symmetrical. There are geometrically symmetrical floor-plans with problematic reentrant corners , such as those illustrated in Fig. 3. The dimension and location of the reentrant area(s), as well as proportion of the wings can produce high torsional flexibility and diaphragm deformation. Also, because of the various blocks that constitute the building, any construction deficiency will cause accidental imbalances in the structure described by these floor-plan shapes resulting in undesirable torsions.

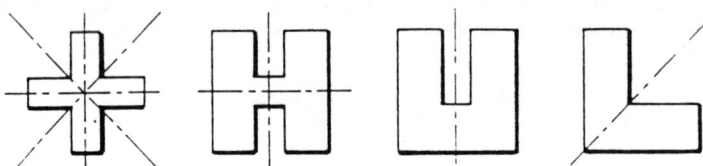

Fig. 3

In some cases the floor-plan shape is considered irregular for the effects of lateral force requirements even though the shape is geometrically symmetrical.

Fig. 4 illustrates examples of floor-plan shapes which present one or more axes of reflective symmetry but are considered structurally irregular because the axes of reflection are neither parallel nor perpendicular to the two main directions in the floor plan. This means that the structural solution would result in structurally irregular floor plan.

ONE AXIS   ONE AXIS   TWO AXES

Fig. 4

To minimize the torsional effects produced by irregularities in plan, it is recommended that: floor plan be symmetric; distribution of mass be symmetric; structural layout be symmetric; and yielding resistance be symmetric.

(c) *Identifying Problematic Proportions of Wings.* Another important factor has to be considered in identifying the level of irregularity of floor plans. This is the **proportion of the blocks or wings** generated by the reentrant area. Long, slender wings, if not treated with adequate structural solutions, can lead to significant horizontal diaphragm deformations. Long diaphragms, with structural elements only at the ends of the diaphragm perpendicular to the longest direction, will be flexible. In order to reduce the risk of torsional effects in a rectangular block due to an accidental eccentricity produced by any unplanned imbalance, it is desirable to restrict the length of a rectangular component, wing or block to three times its width (Ref.8). This proportion, is only allowed when intermediate structural components such as shear walls are included. Otherwise, the recommended proportion is two times width (Ref.2).

(d) *Possible Solutions for Geometrically Problematic Floor Plans.* **Solution 1. Structurally interconnect the different blocks of the building** (Ref. 1). Recommendations for this solution are given in Section 2312(h)2E "Ties and Continuity" in "Structural Framing Systems" of the UBC 1988. Examples of alternative structural solutions are: collectors or drag struts; collector walls; splayed reentrant corner; and coupled walls (Ref. 3) and couple girders (Ref. 1). If the selected solution is either drag struts or collector walls, the UBC 1988 imposes two penalties in Section 2312(h)2I(v), and Section 2312(h)2I(vi). **Solution 2: Structurally separate the building into simple rectangular blocks** (Ref.3). This approach implies the use of seismic joints which might create new design problems because of the allowed size of the seismic joint (UBC 1988, "Building Separations" Section 2312(h)2K.) The resulting rectangular blocks must still obey the rule for proportion between the width and the lenght of the horizontal diaphragm.

c. Slab Spans

(i) It is possible to cast spans of up to 6 meters, but the best results are obtained with medium spans, 4 to 5 meters. (ii) In relation to construction efficiency, the best spans are those between 3 to 3,50 meters, because they permit fast formwork removal and they do not require temporary props; they also dispense the need of additional partition walls; the temptation of building "cubby holes" should, though, be resisted. (iii) The most important consideration is to repeat the spans in the building, as often as possible in order to permit the maximum efficiency by the multiple use of every formwork; this is the most expensive and hard part of the construction system.

d. Wall and Slab Thickness

i) As a general rule, walls should not be thinner than 15 cms. (6"), less than this means difficulty in placing the reinforcements and pouring the concrete; honeycombing is very frequent with smaller thickness. (ii) Aggregate and reinforcement sizes should be commensurate with the structural thickness; large aggregates and large reinforcing bars should be out of them (the practical maximum for bars being N° 7) (iii) Acoustically speaking, a minimum thickness of 14 cms. (5.7") is required; less than that, hinders the privacy of every individual dwelling. (iv) It is recommended to keep wall thickness constant in the two directions, unless shear or reinforcing requirements impose due changes. (v) Slabs should be 1 cm. thinner than the walls, to let them crack before the walls during seismic events, at the critical crossing of walls and slabs; breaking these rules for economies sake brings rather inferior quality building.

e. Crosswall Orientation

(i) At least, two clearly identified structural systems must exist for every building; both systems must include crosswalls, except, perhaps, in the very low buildings (up to two stories); mixed wall and the so-called "frames" formed by the slabs and the walls, simply do not perform in practice, with usual dimensions of walls and slabs. (ii) The two systems must be, whenever possible, equivalent in rigidities and bearing areas; being rigidity critical to control lateral deflections and areas critical to control shear strength, the weakest link in seismic design; this aspect must be fixed at the conceptual stage and no later. (iii) Where more than ten floors are required, it is convenient to associate both the longitudinal and transverse systems, in order to get the necessary increases in rigidity and also to make more efficient use of the reinforcements set into the walls, to get double duty. See Fig. 5.

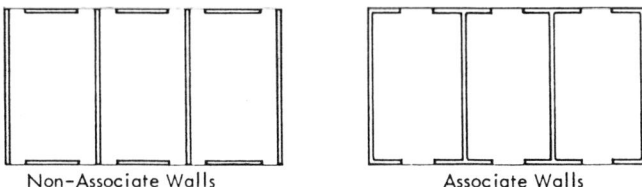

Non-Associate Walls   Associate Walls

Fig. 5

(iv) Settlement sensitivity is dependent on seismic design, the latter favoring the former. (v) The aforementioned rules should not be applied to low-rise buildings (4 to 5-story high) because they bring more problems than benefits; simple, rectangular, crosswall are the best for this category of buildings. (vi) Sometimes, the construction complication derived from the application of these rules may hinder the use of some formwork systems (laterally moving ones); systems that permit removal in close quarters should be preferred; a typical example being those which work with the "open sky" schemes, and with separate elements to cast walls and slabs.

f. Crosswall Shapes

(i) Width to height ratio should be greater than two, per floor. (ii) The widest wall should not be more than 1,5 times longer than the narrowest. See Fig. 6.

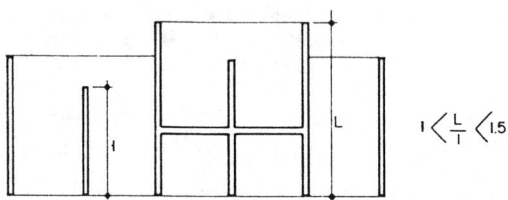

Fig. 6

(iii) It is preferable to keep wall shape constant along the height of the building; any sudden change generates "shear whips" at that level when lateral loads appear. (iv) Whenever changes of thickness in the walls occur, they should happen at the same level and in a proportionate way, generally these changes are not justified because they produce more construction complications than material savings. (v) Excessive rigid walls, in comparison the rest of the walls of the floor, are not desirable; they tend to attract all the seismic shears, with the resulting necessary over-reinforcement and extra strong foundations; it is better to "spread" the lateral forces amongst all walls. (vi) Because of the lapping provisions in the Design Codes, in this case vital for seismic survival of this type of construction, floor-to-floor height of crosswall buildings should be limited to 2.65 meters (104") to be able to match commercial lengths. (vii) Due to right-of-way requirements and sheet steel commercial sizes, most formwork manufacturers limit the floor-to-floor height to 2.45± 0.30 meters (8'±1').

g. Wall Openings. The following rules should be complied:

(i) No more than 25% of the walls should be pierced. (ii) Openings should be spaced away at least one width of the smaller opening. (iii) External solid portions of the walls should be at least 1 meter (3.3') wide. See Fig. 7.

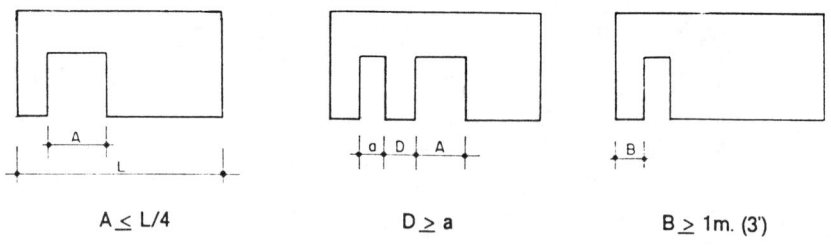

A ≤ L/4   D ≥ a   B ≥ 1m. (3')

Fig. 7

(iv) The vertical arrangement of openings should lead to good lintel (connecting beams) arrangements. If the exist, they should have a clear structural behavior as the one shown in Fig. 8 "c", or little influence in the rigidity and strength, as in Fig. 8 "d". The arrangement shown in Fig. 8 "e" has to be avoided, because it does not allow the placing of the necessary reinforcements along with the door edges, or to get a clear behavior at the "lintels."

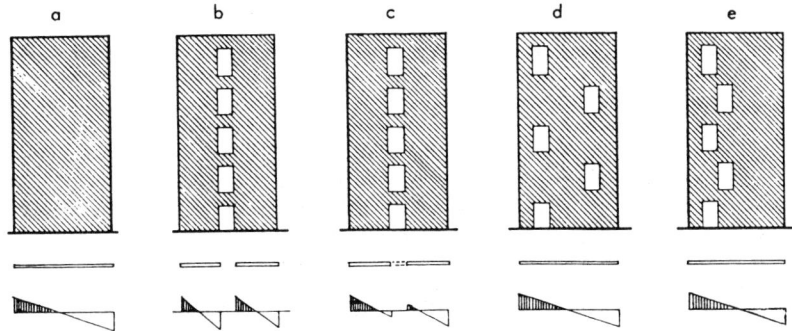

Fig. 8

(v) Careful use of connecting beams, or redesigns, to either avoid them or make them big enough is a good way of getting good results; it must be said that connection beams are desirable in tall structures (more than 10 stories) and, even, undesirable in very low structures (up to 4 or 5 stories.) (vi) One absolute taboo with crosswall structures in seismic areas is the existence of vertical discontinuities in stress flow such as openings in the ground floor, or the so-called "soft-story"; the resulting "legs" are prone to fail due to the concentration of energy on them; if the "soft-story" is unavoidable, the following rules are recommended: (a) increase substantially the thickness of these "legs," so that they become real columns. (b) the thickness increase should go over at least one more floor. (c) "soft-story in the ground floor should not be included in low-income dwellings, they are extremely expensive.

h. Building Height

(i) Each particular layout or building configuration has a height limit; the wall arrangement and the necessary connectivities are a function of building height. (ii) Planners should restrain form the temptation of "stretching" well performing configurations beyond its limits, especially in seismic areas. (iii) The number of necessary connectivities, and its efficiency, generally increase with the building height, the necessity of keeping wall dimensions alike and without sudden changes also increases with height. (iv) Low buildings are more tolerant of this requirements; sometimes these requirements are put upon them by designers which do not understand at all the differences of behavior between buildings with different heights or with different structural systems. (v) In the case of very low buildings (4 stories or less) it pays not to connect, accepting larger wall thickness and larger amounts of reinforcement in them, whose costs will be offset by the constructional advantages. The so-called "cantilever bundle" type of structure is, perhaps the best choice for this case.

i. Horizontal Cantilevers

(i) It is very convenient to avoid external cantilevers in this type of buildings; They are always prone to cracking, excessive deflections and client's complaints. (ii) They can be tolerated only on open, external, narrow, passageways and in those places where the walls do not reach the roof, for instance, semi-open washing areas.

j. Foundations

(i) The foundation design for crosswall buildings is more delicate and laborious than in the case of framed buildings; large local moments and shears are the rule and not the exception; connections amongst vertical elements must also be stronger and, generally, not at all easy to model in its structural behavior; even the architects must have this situation in mind when they design with this type of structural system. (ii) Structural engineers are never recognized for the extra work involved in the design of this type of structural system; soil engineers and piling companies which have handled only traditional framed buildings tend not to believe on the existence of the large moments and horizontal shears generated by the lateral seismic loads; they still believe in the old cliché of " consider vertical forces only, the large lateral loads are just non-existing phenomena, but just nuisance imposed by theoretical code-writers."

k. Final Remark

Crosswall buildings, when well understood, well planned and well built are, intrinsically better structures than frame structures, from the point of view of durability, resistance to abuse and survival under extreme environmental forces; but this type of buildings are much more prone to mistakes in the planning-design-building process. Thy cannot be considered amenable to solutions based upon modifying existing traditional solutions. They are a different kind of "animal" which has to be domesticated with different methods and fresh thinking.

References

1. ARNOLD, Ch., and R. REITHERMAN, "Building Configuration and Seismic Design," A Wiley-Interscience Pub., N.Y., 1982.
2. DECANINI, Luis, "Influencia de la Configuración y Regularidad Estructural sobre el Comportamiento Sísmico de las Construcciones Nº 1 I.S.." Universidad Nacional de Córdoba, Facultad de Ciencias Exactas, Físicas y Naturales. Depart. de Estructuras. Córdoba, Argentina. 1982.
3. GUEVARA, L.Teresa, "Architectural Considerations in the Design of Earthquake-Resistant Buildings: Influence of Floor-Plan Shape on the Response of Medium-Rise Housing to Earthquakes," Ph.D. Dissertation,Department of Architecture, College of Environmental Design, University of California, Berkeley. December, 1989.
4. GUEVARA, L.Teresa, and Mario PAPARONI, et al., "Estudio de Edificaciones de Muros Vaciados en Sitio Realizadas por el B.O.-INAVI de Venezuela." Paper presented to the "III Congreso Venezolano de Sismología e Ingeniería Sísmica," Caracas, Venezuela, March, 1981; and to the "Simposio Latinoamericano de Racionalización de la Construcción," Sao Pavlo, Brasil, Oct. 1981.
5. GUEVARA, L.Teresa, and Mario PAPARONI, "A study of Cast-in-Place Crosswall Structures Built for the 'Instituto Nacional de la Vivienda' (INAVI), Venezuela," International Journal for Housing Science and its Applications, Volume 8, number 1, pp. 089-100. Miami, Florida. 1984.
6. PEÑA, José A. and Nancy DEMBO, "Sistemas Constructivos Industrializados para Edificios de Viviendas." Paper presented to the "Simposio Latinoamericano de Racionalización de la Construcción." Sao Pavlo, Brasil. Oct. 1981.
7. The International Association of Earthquake Engineering, "Guidelines for Earthquake Resistant Non Engineered Construction." Tokyo, Japan. 1986.
8. International Conference of Building Officials, "Uniform Building Code, 1988 Edition." Whittier, California, USA. May, 1988.

Proceedings of the Second Conference on Tall Buildings in Seismic Regions
55th Regional Conference
May 16 and 17, 1991, Los Angeles, California

SEISMIC RESPONSE OF A
TALL MASONRY BUILDING

by
Gary C. Hart
Jing-Wen Jaw

INTRODUCTION

This paper seeks to utilize the experience of the Loma Prieta earthquake to study the performance of a tall reinforced masonry building located in the City of Santa Cruz which is in the epicentral region. The building did not experience any observable damage in spite of a probable maximum ground motion of 30 to 50%g. It is important to explain why.

Figures 1 shows a plan view of the building. The building was designed using the provisions in the 1970 UBC. This paper models the building as a two dimensional system and addresses the response of the building in the east/west direction. The lateral seismic force resisting system in this direction for the top 9 stories is provided by eleven load bearing 8-inch concrete block walls. All story heights are 8 feet 9 inches for a total building height of 87 feet 6 inches. The walls are solid grouted with a design minimum specified compressive strength of masonry of 1,500 psi and typically Grade 40 reinforcing steel bars. The corresponding values of these material parameters used in the following calculations are 1,500 psi and 40 ksi. The floor system is a 5 and 1/2 inch poured in place lightweight concrete slab.

The shear walls in the east/west direction can be divided into three basic types. Two L-shaped shear walls occur at the ends of the building. Four T-shaped and five rectangular shaped shear walls are located inside the building. The 10-inch reinforced concrete walls in the center zone of the rectangular walls are not included in the initial phase of this study and will be discussed in a later section of the paper. These shear walls are all load bearing and thus resist both lateral and vertical loads. The steel detailing of the walls especially the concentration of steel at the ends of the walls is not what is typically recommended based on the latest masonry research. However, it is typical of masonry buildings of this general age. The building is located on drilled caissons and goes to a depth of 10 feet.

(GCH) Professor, Department of Civil Engineering, University of California, Los Angeles and Englekirk, Hart & Sabol, Inc.
(JWJ) Engineer, Englekirk, Hart & Sabol, Inc.

SHEAR WALL COMPONENT LOAD/DEFLECTION CHARACTERIZATION

A good structural engineering perspective of the expected seismic response of the building can be obtained by performing a behavior and limit state analysis. Figure 2 shows a moment curvature plot for the rectangular reinforced masonry walls. The symbol ϕ_y denotes the curvature at first yield of the steel and the symbol ϕ_u corresponds to a masonry compressive strain of 0.003. The factored axial dead and live load on the wall is approximately 0.13 $f'_m A_g$ or alternately approximately 40% of the balanced design axial load. This is actually a reasonably large axial load but even so, it is clear from Figure 2 that these rectangular concrete masonry walls are expected to respond in a ductile manner. If these walls are considered only by themselves then, as noted in Table 1, they have a curvature ductility of 6.1 and a system displacement ductility of 2.9.

TABLE 1
CURVATURE AND DISPLACEMENT
DUCTILITY OF COMPONENT SHEAR WALLS

WALL TYPE	CURVATURE DUCTILITY	DISPLACEMENT DUCTILITY
Rectangular	6.1	2.9
T-Shape	24.0* 2.3**	13.0* 1.7**
L-Shape	19.0* 4.9**	14.2* 3.6**

*   Wall Flange in Compression
**   Wall Flange in Tension

Figure 3 shows a plot of the moment curvature for the T-shaped flange wall. Note that when the flange of the wall is in tension that the capacity of the wall is at its maximum but the curvature ductility of the wall is at its minimum. For the case of the flange in tension the wall has a curvature ductility of 2.3 and if only these type of walls existed and if they were loaded in this manner then as shown in Table 1 the shear wall system would have a displacement ductility of 1.7. To illustrate the strong dependence of the wall performance on the direction of loading notice that when the flange is in compression the moment curvature and displacement ductilities are 24 and 13, respectively.

The structural engineering modeling concepts that Priestley presented and used in our analysis of T-shaped walls can be extended and applied to the L-shaped walls. Figure 4 shows the wall moment curvature plots. Similar to the T-shaped walls the direction of loading is an important consideration as shown in Table 1.

COMPONENT AND SYSTEM BEHAVIOR STATE ANALYSIS

Two behavior and limit states are considered in this paper. The first behavior state reflects the wall performance up to the first limit state which occurs at the start of tension yielding of the extreme reinforcing bar. The second behavior state then occurs and it ends at the second limit state which corresponds to a masonry strain of 0.003 which is a reasonable estimate of the unconfined concrete masonry crushing strain. Table 2 provides a summary of the lateral load and the deflection at the top of the wall at these two limit states for each wall component. The load corresponds to the shear force at the base of the concrete masonry wall. In parenthesis next to the deflection values are the corresponding building system drift ratio values (in percentage) which are defined as the lateral displacement of the top of the wall divided by the height of the wall.

TABLE 2
LATERAL LOAD AND DEFLECTION OF INDIVIDUAL WALL
COMPONENTS AT YIELD AND ULTIMATE LIMIT STATES

WALL TYPE	YIELD		ULTIMATE	
	LOAD (kips)	DEFLECTION (in)	LOAD (kips)	DEFLECTION (IN)
Rectangular	79	2.9 (0.28)	96	8.5 (0.81)
T-Shape	267 349	1.6* (0.15) 2.8**(0.27)	301 360	21* (2.00) 4.6**(0.44)
L-Shape	338 433	1.2* (0.11) 1.6**(0.15)	411 503	17* (1.60) 5.7**(0.54)

*   Wall Flange in Compression (West Load)
** Wall Flange in Tension (East Load)

The building under consideration has a floor system of a poured in place concrete slab. It is reasonable to assume that this floor system can, in the context of this paper, be considered as a rigid floor system. Thus, it can be assumed for the purposes of this paper that at each floor level that each shear wall experiences the same lateral displacement. The high torsional rigidity of this building enables us to neglect the torsional motion of the floor. Therefore, when we combine the load deflection characteristics of all of the walls it is possible to obtain the expected building system behavior and limit state characterization.

The north-south lateral load resistance is provided by the flanged masonry shear walls and 10 inch concrete shear walls. The concrete shear walls contribution to strength and ductility again depends on the direction of the loading. When the loading is such that the concrete shear wall is in compression the addition of the concrete shear wall to the model of the rectangular wall does not significantly influence the moment capacity or the curvature ductility. On the other hand, with the concrete shear wall in tension, the behavior is significantly different from that of the rectangular wall. As indicated in Figure 5, the addition of the concrete wall results in a significant enhancement in moment capacity and decrease in ductility capacity. Since the participation of concrete shear wall has a significant impact on the behavior of masonry rectangular component wall it is included in the characterization of east west building response.

Figure 6 shows the load deflection response of the building system when the seismic loading is in the east direction. Figure 7 shows the corresponding plot when the loading is in the west direction. In obtaining the building system load deflection responses, the shear capacity of walls has been computed and shown to be greater than the flexural strength. For seismic loading in the east direction the first yielding of the tension steel occurs in the L shaped wall at a lateral displacement amplitude of 1.6 inches or 0.15% drift ratio. As the lateral system displacement increases up to 4.6 inches (i.e. a drift ratio of 0.44%) the slope of building systems load deflection curve is always positive. Note that the L-shaped walls yield first, then the T-shaped walls and finally the rectangular walls. At the lateral displacement amplitude of 4.6 inches the ultimate limit state of T-shaped walls is reached and approximately a 40% reduction of the lateral load capacity of the building system occurs with the loss of the load carrying capacity of the T-shaped wall. Then at a displacement of 5.7 inches (i.e. a 0.54% drift ratio) the lateral load capacity of the L-shaped wall is lost and only the rectangular wall provides the buildings lateral load resisting capacity.

TABLE 3
SYSTEM DUCTILITY CHARACTERIZATION

	DISPLACEMENT AT FIRST YIELD	DISPLACEMENT AT MAXIMUM DEFLECTION	SYSTEM DUCTILITY
• West load, T Wall Flange in Compression	4.1	5.0	1.2
• East Load, T Wall Flange in Tension	2.5	4.6	1.8

Figure 7 shows the performance of the building system to west loads. Because the L and T-shaped walls are very ductile when their flanges are in compression the building system load deflection curve increases up to a displacement of 5.0 inches (i.e. a 0.48% drift ratio). The building system ductility characterization in the east/west direction is summarized in Table 3.

Figure 8 shows the composite curves from Figures 6 and 7 over the drift ratio 0 to 1.5%. The building performance for west loading retains a significant load carrying capacity past the 0.5% drift ratio level and at a 1.5% drift ratio level still carries approximately 2,000 kips of load. In contrast to this we see that for east loading we experience approximately an 80 percent reduction in load carrying capacity and then we also have very little additional drift ratio capacity.

148

THE EARTHQUAKE

The Loma Prieta earthquake with a Richter Magnitude of 7.1 occurred at approximately 5 PM on October 17, 1989. The earthquake was the result of a rupture along a 25-mile long segment of the San Andreas fault in the southern Santa Cruz Mountains. The point of initial fault rupture was approximately 12 miles below surface and 10 miles northeast of Santa Cruz. The maximum displacement on the fault was approximately 6 feet horizontal and 4 feet vertically.

Earthquake ground motion records were obtained at stations maintained by the California Strong Motion Instrumentation Program and the U.S. Geological Survey. The Corralitos, Gilroy No. 1 and the Santa Cruz stations are at rock sites close to the source of the earthquake.

BUILDING RESPONSE

The maximum ground acceleration at the building site located in Santa Cruz can realistically be expected to be in the 40 to 45% g range. The level of ground shaking in the period range of this building is significantly less than the 1988 UBC design spectra. It is reasonable to assume and for a reasonable estimate of this building's period that the base shear experienced by this building during the earthquake was approximately 6% of the building weight or a total of 2572 kips. When this base shear is compared with the composite base shear strength as shown in Figure 8 it is clear that we would not expect the building to experience major damage though some local cracks of the walls would be expected. The roof drift ratio of the building is estimated to be 0.25% which is less than the UBC allowable drift of 0.4%.

The maximum compressive strength of the masonry and concrete and the yield stress of the reinforcing steel were minimum specified design values as stated in the general notes of the design. Therefore, it is reasonable to assume that the yield and maximum load carrying capacity of the building exceeds to a small extent the values calculated and plotted in this paper. Therefore, it is reasonable that for the Loma Prieta earthquake we experienced no damage to this building.

ACKNOWLEDGEMENTS

The research presented in this paper was funded by the Concrete Masonry Association of California and Nevada and by the national Science Foundation as part of the TCCMAR research project.

149

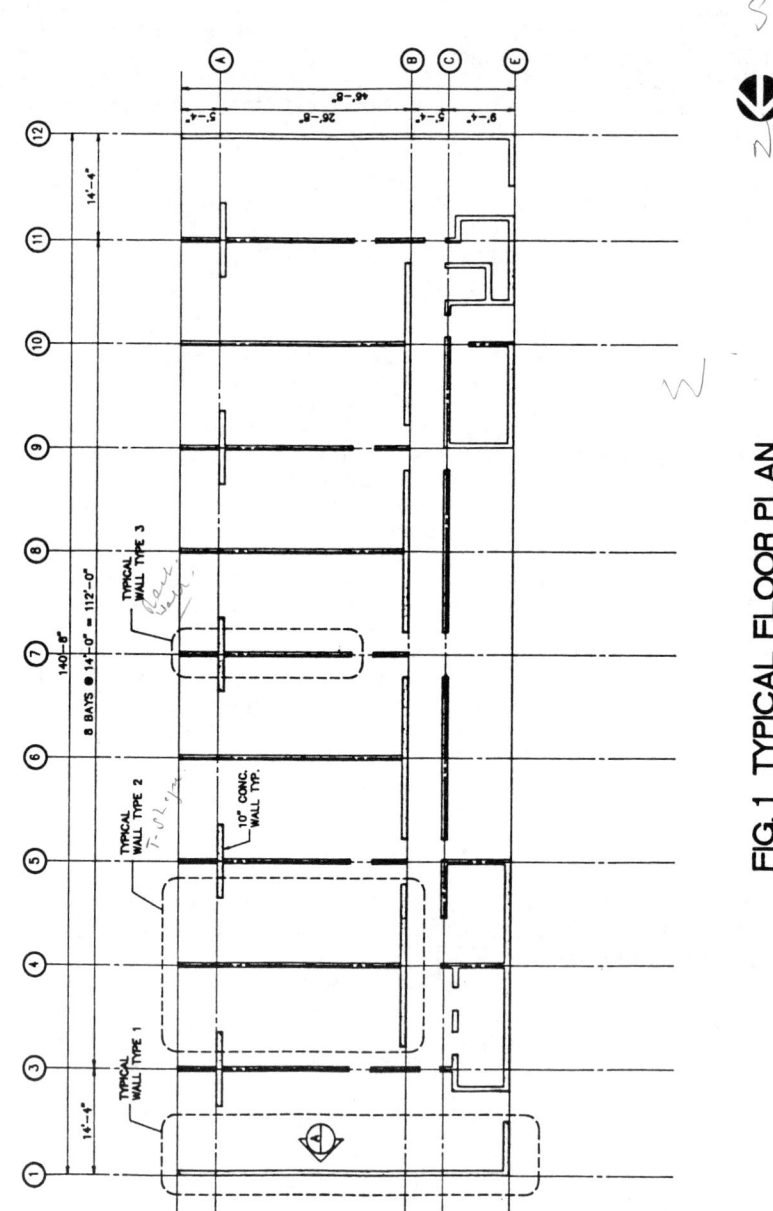

FIG. 1 TYPICAL FLOOR PLAN

150

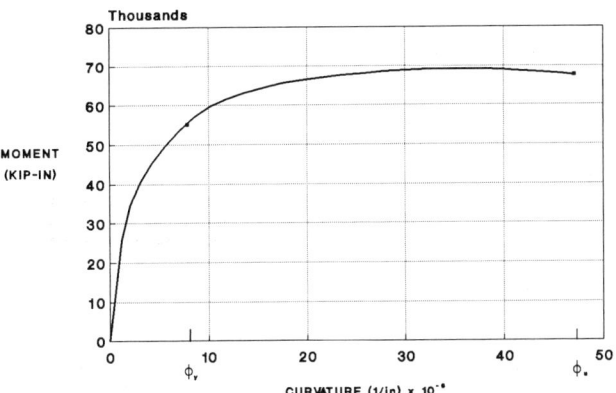

Fig. 2 Moment Curvature Plot for Rectangular Wall

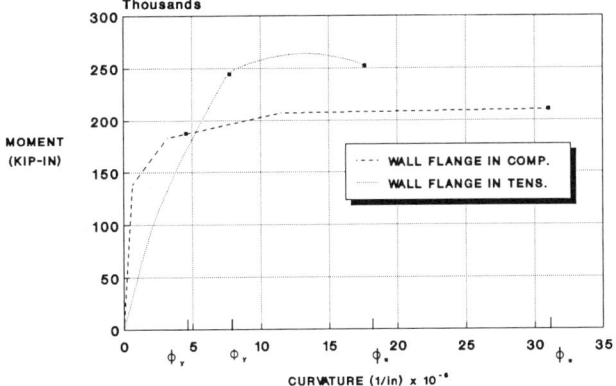

Fig. 3 Moment Curvature Plot for T-shaped Shear Wall

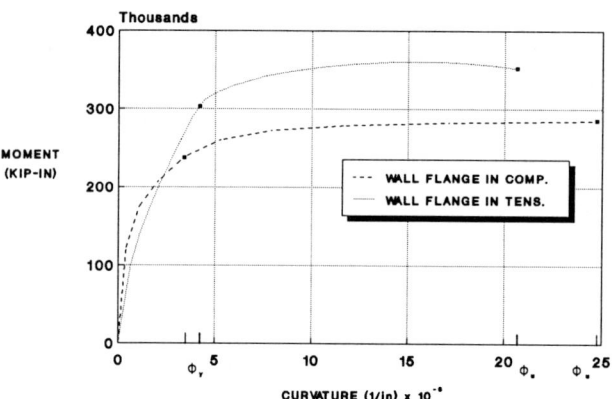

Fig. 4 Moment Curvature Plot for L-Shaped Shear Wall

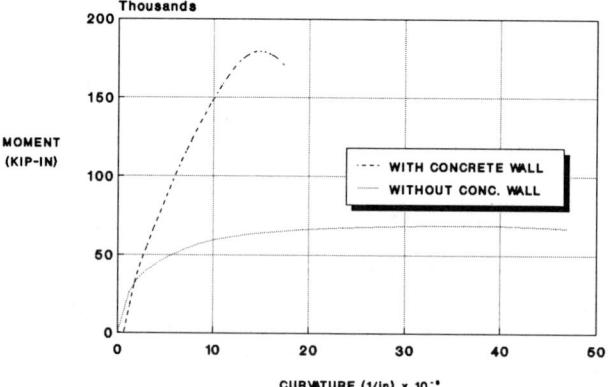

Fig. 5 Moment Curvature for Rectangular Wall (West Load)

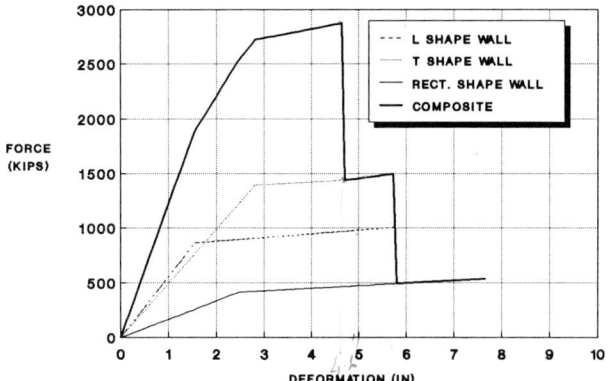

Fig. 6 System Load Deflection Performance for East Load
(T Wall Flanges in Tension, Rect Wall/w Shear Wall Flange)

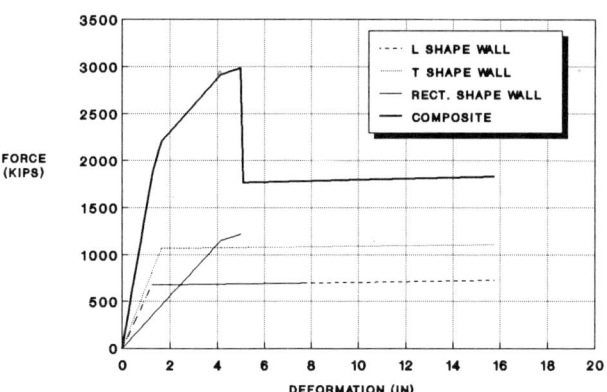

Fig. 7 System Load Deflection Performance for West Load
(T Wall Flanges in Comp., Rect Wall/w Shear Wall Flange)

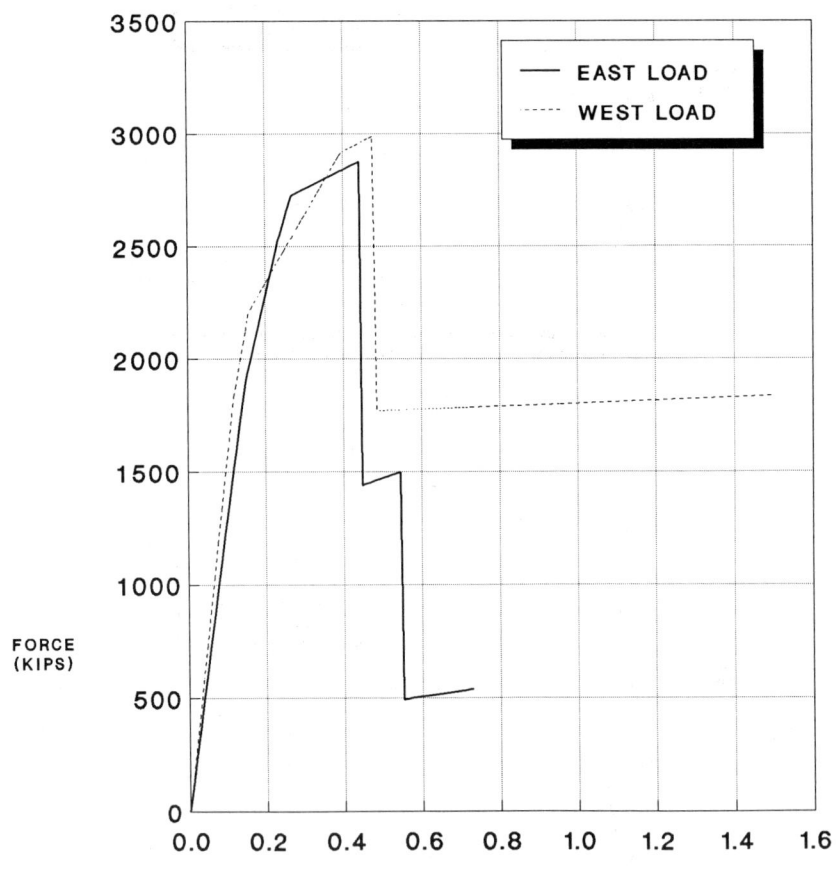

DRIFT RATIO (%)

Fig. 8 System Load Deflection Performance

154

Proceedings of the Second Conference on Tall Buildings in Seismic Regions
55th Regional Conference
May 16 and 17, 1991, Los Angeles, California

The Developpment of Full Scale Active Mass Damper System

Masahiko Higashino, Satoru Aizawa and Yutaka Hayamizu

Takenaka Corporation; Tokyo, JAPAN

1. Introduction

The structral design of architectures has been done mainly in viewpoint of the safty of the structure. However, people recently began to require the "habitability" of the buildings, the confort by means of vibrational errvironment, supported by the growth in economy and the progress in technology. For highrise buildings, several approaches have been done to improve the vibrational condition subjected to wind and small earthquake excitation.

To improve the habitability of a highrise building, the authors have been developping Active Mass Damper (AMD) system. Using a hightech of the controling by computer, the AMD has the advantage that it will control higher modes in addition to primary mode theoreticaly, beside the passive Tuned Mass Damper (TMD) could only control the tuned mode.

The theoretical study of active control technology have been pursued aggressively in United States during the period of 1970's through 1980's. We must notice valuable approaces done by T. T. Soong, S. F. Masri, etc.. The remarkable advancement in computer technology and control technology recently have made it possible to realize the Active Control of practical strucrure.

The authors began the experimental study of prototype model using 12 kg moving mass in 1987 to determine the basic performance of the device in comparison with the theory. After the comfirmation of the efficiency of AMD from primary test, we scaled up the model to middle scale (200 kg mass) and we came up to full scale model (6 ton mass) now.

This report describes about the behavior of this 6 ton mass AMD through the conduction of shaking table test and the earthquake observation of full scale structure.

2. Outline of Control Theory

We design the AMD to reduce the displacement at the top floor of a buildings by 1/2 compared to that of uncontrolle state. The control is

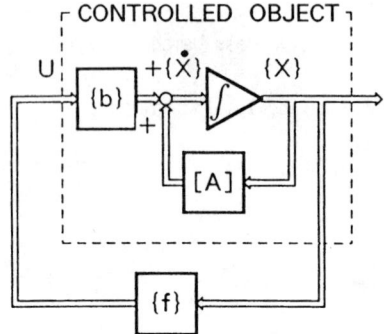

Fig. 1 Block Diagram of
Control Algorithm

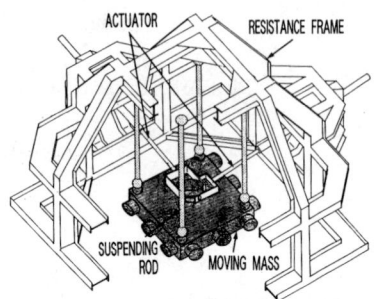

Fig. 2 Construction of AMD

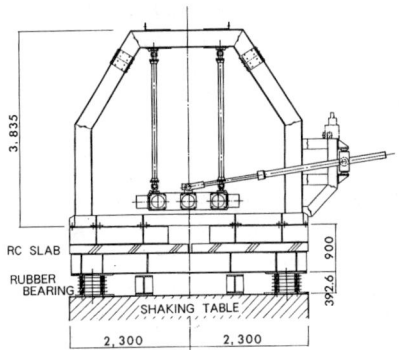

Fig. 3 Model for
Shaking Table Test

Table 1 Dimensions of AMD

Control Direction	2 (X,Y)
Weight of Moving Mass	6 tonf
Length of Suspending Rod	245 cm
Stroke	95 cm
Maximum Velocity	300 cm/sec
Maximum Control Force	10 tonf

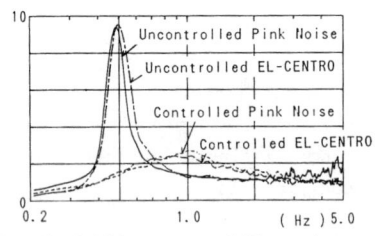

Fig. 4 Difference of Transfer
Functions between
Controlled and Uncontrolled

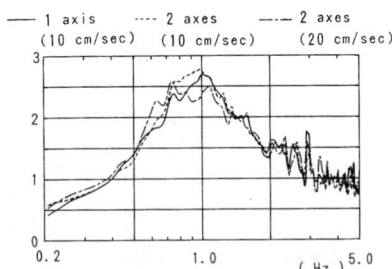

Fig. 5 Comparison of Transfer
Functions due to The
Difference of Input
Level and Axes Number

expressed in the following state equation.

$$\{\dot{x}\} = [A] \{x\} + \{B\} u + \{z\}$$ (1)

where
　{x} : State vector of the structure and AMD.
　[A], [B] : Matrix which is determined depending on property of building and AMD.
　{z} : External force vector
The control force of u is provided by the following equation.

$$u = \{f\}^T \{x\}$$ (2)

The feedback vector of {f} is decided so that the following evaluation function can become as small as possibie.

$$J = \int_0^\infty [\{x\}^T [Q] \{x\} + ru^2] dt$$ (3)

where
　[Q], r : Weight
Based on optimal contorol theory, the control is performed by reducing the relative displacement of the structure. Parameters signaled are the relative displacement and velocity between the foundation and uppermost levels, and movement of the mass. Fig. 1 is a block diagram represent ation of the control algorithm. The dimensions of this full scale AMD is summerized in Table 1.

3. Outlines of AMD

The outline of this large scale AMD is given in Fig. 2. The 6-ton moving mass is suspended (T=3.1 sec) so that it will respond instantaneously and the responses of the structure will be controlled in two directions by electrohydraulic servo actuators which are orthogonally installed. A weight of 6 tons translates to one percent of the actual weight (600 tons) of the observation model to be described later.

4. Results of Shaking Table Tast

We carried out shaking table test. Prior to an earthquake obsevation using a full scale observation model, to verify the vibration control performance of AMD. As shown by Fig. 3, laminated rubber bearings were adapted to constitute one lumped mass system. Simulating a high rise building, we designed the natural period of this single mass system as 2 seconds. The weight is 24 ton without moving mass.

Table 2 Maximum Response Values in Variation of Input Earthquake and Level

Input wave		Pink Noise			EL CENTRO 1940 NS				HACHINOHE 1968 EW			
Level (cm/sec.)		6.5			10			20	10			20
input axes		1 axis		2 axes	1 axis		2 axes		1 axis		2 axes	
Control		OFF	ON		OFF	ON			OFF	ON		
X	Input acc. (Gal)	132.0	131.7	133.1	107.0	103.6	105.6	216.3	65.8	72.6	70.6	146.3
	Frame absolute acc. (Gal)	23.1	67.1 (2.9)*	75.7	65.8	131.4 (2.1)	123.9 (1.9)	279.9	53.2	86.6 (1.5)	77.1 (1.4)	14.0
	Frame relative vel. (cm/sec.)	10.8	9.5 (0.88)	9.4	23.9	17.8 (0.77)	18.6 (0.79)	36.5	19.5	15.1 (0.70)	12.6 (0.44)	26.2
	Frame relative disp. (cm)	2.4	1.8 (0.75)	1.7	7.6	3.5 (0.47)	3.9 (0.52)	9.1	6.1	3.2 (0.48)	2.9 (0.44)	5.9
	Mass stroke (cm)	—	20.0	12.8	—	29.8	33.6	72.5	—	23.4	22.8	49.0
	Cyl. thrust (ton)	—	0.60	0.68	—	1.12	1.21	2.7	—	0.64	0.63	1.3
Y	Input acc. (Gal)	—	—	125.5	—	—	62.8	117.3	—	—	53.8	99.0
	Frame absolute acc. (Gal)	—	—	88.2	—	—	81.0	166.2	—	—	86.9	165.0
	Frame relative vel. (cm/sec.)	—	—	10.7	—	—	14.1	26.7	—	—	17.3	31.0
	Frame relative disp. (cm)	—	—	1.4	—	—	3.3	6.4	—	—	3.6	6.7
	Mass stroke (cm)	—	—	9.3	—	—	23.2	42.3	—	—	26.0	47.9
	Cyl. thrust (ton)	—	—	0.61	—	—	0.66	1.24	—	—	0.60	1.2

* : ( ) = ON/OFF   The ratio calculated by the response level of no control.

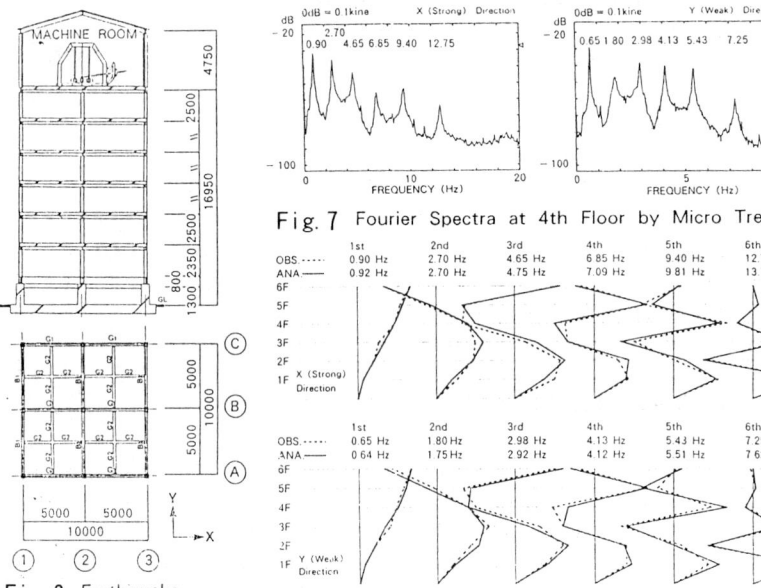

Fig. 6 Earthquake Observation Model

Fig. 7 Fourier Spectra at 4th Floor by Micro Tremor

Fig. 8 Mode Shape of Model Structure

Table 2 shows a part of the maximum response values which were obtained from the results of various types of tests. The relative displacement in X axial direction at 10 cm/sec level is reduced to 0.44-0.75 compared with the uncontrolled state. On the other hand, the acceleration increased to 1.5-2.9. The result of the test in which the two axes were applied for the input was almost the same as the result which was obtained from the test in which the wave was input to one axis. The validity of dual axis control was confirmed. The result of the response at the input level of 20 cm/sec was approximately two times that of the response at the input level 10 cm/sec, and the linear property was confirmed. Wherer, the thrust of the cylinder in this test was 2.7 tons (EL-CENTRO, 20 cm/sec) at a maximum.

Fig. 4 and 5 show the displacement transfer functions (Frame relative displacement/Table displacement) calculated from the results of several kinds of excitations. In comparison with the uncontrolled state, the main peak changes from 0.5 Hz to approximately 1.0 Hz, and the height of the peak (the response magnification ratio) became 1/5. The damping ratio (h), which was calculated from the peak value was approximately 0.2. Furthermore, without regard to the difference of the input level and that of the input direction, a significant difference could not be recognized in the transfer function.

5. Observation Model

Fig. 6 is the schematic drawing of a full scale model. This obervation design was modelled after the vibration character of comparatively tall buildings. The AMD is installed on the top floor. This structure is provided with a floor area of 10m x 10m, 6 levels of steel frame structure, and on each level the weight is 100 tons, including RC slabs.

Fig. 7 gives the fourier spectra on the 4th floor in micro tremors. Both charts show peaks clearly over the entire spectra from the 1st to 6th modes; the 1st dominant frequencies of the model are 0.9 Hz in the X (strong) direction and 0.65 Hz in the Y (weak) direction. Fig. 8 shows the comparison between the modal shapes found by micro tremor measurement and by six lumped mass analyses. In both X and Y directions, there are good similarties between the frquencies and the modal shapes found by analyses and observation; it therefore supports the validity of our analytic model proven by micro tremor level.

Figs. 9 and 10 give a comparison between analyzed and observed responses of the model without control during Tokyo Bay earthquake on 10th August, 1989. Let the damping ratio be 1% in the X direction and 0.3% in the Y direction (the values of the analyzed maximum relative displacement figures closest to the observation) for every modal order. In the analyzed waveforms under the conditions described above, some phase shifts are observed in the latter

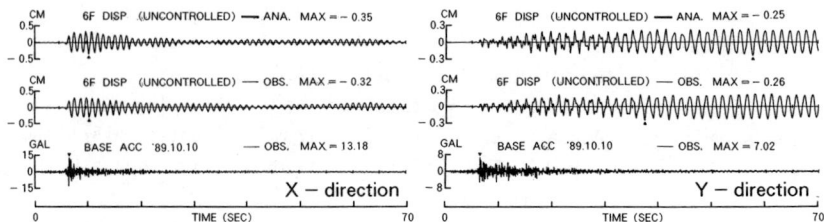

Fig. 9 Observed and Analyzed Responses of Uncontrolled Structure

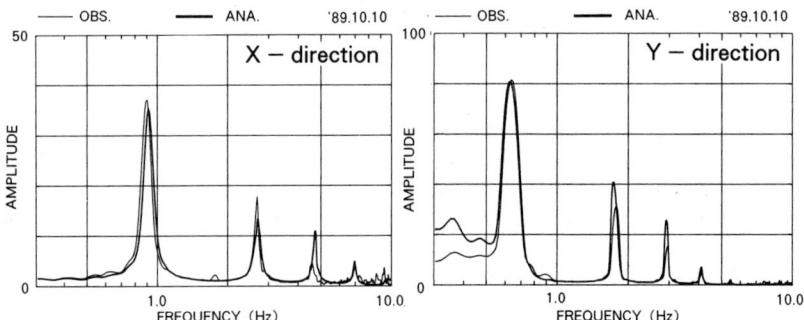

Fig. 10 Observed and Analyzed Transfer Functions of Uncontrolled
Structure

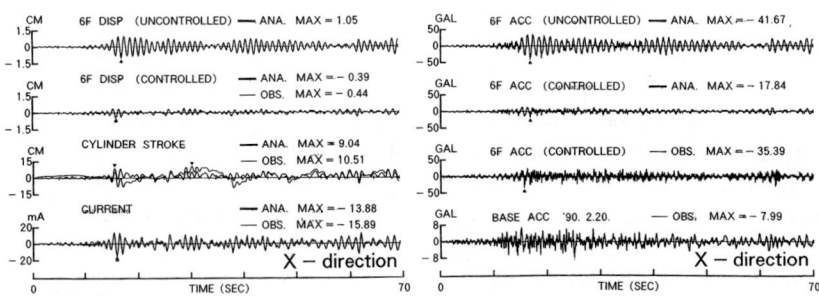

Fig. 11 Comparison of Responses between Controlled and
Uncontrolled State

part. However, there is much similarity between analyses and observation in the waveforms and transfer function shapes.

6. Earthquake Observation

Using a full scale observation model, the control effect of AMD was examined during the Izu-Oshima earthquake on 14th October, 1989 and on 20th February, 1990. Fig. 11 illustrates the data of the former earthquake in both X and Y directions. Additionally, analyzed responses, obtained when AMD is not activated, are given in this figure. In comparison with responses in an uncontrolled state, it was observed that the relative displacement was reduced by 1/2 to 1/3 in those of a controlled state. Although the controlled acceleration responses are combined with the components of high-frequency, the maximum values are almost the same as those of the uncontrolled state. This proves that the displacement response of the structure can effectively be reduced without causing acceleration to increase with AMD. Fig. 12 shows X direction data of the latter earthquake; it reinforces the effect obtained above.

Fig. 13 gives transfer functions of both earthquakes. It proves that, when the frequency is lower than about 2 Hz, the damping effect is greater in comparison with the uncontrolled transfer functions of those observed on 10th October, 1989. However, in ranges over 3 Hz, an uncontrolled peak cannot be completely reduced. The X direction in particular, shows higher values than those in an uncontrolled state when the range is higher than 5 Hz. In this range, the higher mode is more frequent in comparison with those of the Y direction; there is a possibility that the means of set point control parameter, phase delay in control and or friction cause some excitation. These are the problems yet to be solved.

7. Simulation Analyses of a Frame Structure with AMD

Figs. 11 and 12 illustrate the results of simulation, as well as observation data. Both observed displacement responses and controlled currents are almost the same as the analyzed results. This proves that AMD has principally achieved its design control. The comparison of acceleration responses with the experimental model data is given by Fig. 13, based on data recorded on 20th February, 1990. Although the waveforms combine more high-frequency components in comparison with the simulated data, the movement in low-frequency components are almost identical. It may be a result of the same reasons as described in the transfer functions. With regard to the stroke of the cylinder, it is almost indentical. However, the waveforms observed do have some waviness; it will be necessary to improve the controllability of AMD in very low frequency region.

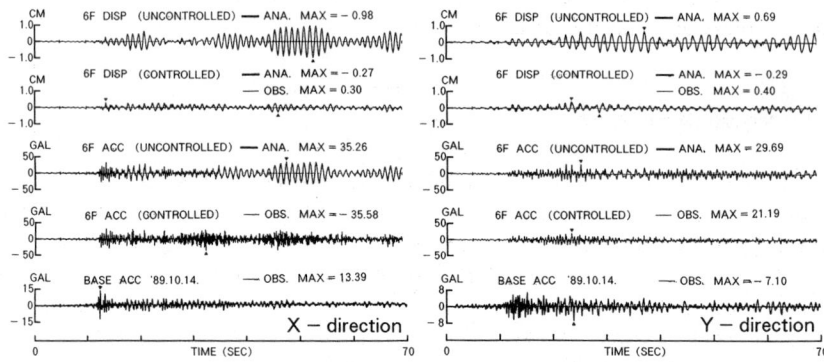

Fig. 12 Responses of AMD and Sturacture

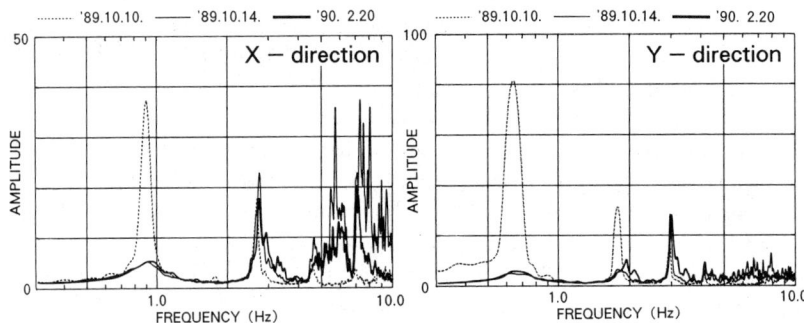

Fig. 13 Comparison of Transfer Functions in Each Earthquake

8. Conclusion

This report is devoted to the findings obtained through the shaking table test of a bidirectionally controlled AMD and the properties of the swaying of full scale observation model caused by a small earthquake. In a frequency range lower than 3 Hz, desired performance can be practically obtained. However, in higher ranges, there are some problems to be solved. AMD is a device which is technically on a proper course of development, based on its stability and reliability. Therefore, our research will continue in the improvement of its performance.

References

1) AIZAWA, S. et al, An Experiemental Study On Active Mas Damper Proceeding of Ninth WCEE, 1988, Tokyo-Kyoto, JAPAN
2) Reinhorn, A. M. et al, Experiments on Active Structural Control Under Seismic Loads, ASCE Structure Congress, 1989, San Francisco
3) Fukao, Y. et al, Experiments of Active Mass Damper Control Under Seismic Loads, ASCE Structure Congress, 1990, Baltimore
4) Masri, S. F., G. A. Caughy, On-Line Control of Nonlinear Flexible Structures, ASME Jour. Applied Mechanics, 1881, pp. 877-884.
5) Soong, T. T., Cang, J. G. A, Structural Control Using Active Tuned Mass Dampers, ASCE Jour. EM6, 1982 pp. 1091-1098.

Proceedings of the Second Conference on Tall Buildings in Seismic Regions
55th Regional Conference
May 16 and 17, 1991, Los Angeles, California

DESIGN AND DEVELOPMENT OF DAMPING AMPLIFIER SYSTEM
FOR PASSIVE RESPONSE-CONTROLLED STRUCTURE

[1]Osamu Hosozawa and [1]Osamu Tsujita (M. JSCA)

[1]Building Design Department, Taisei Corporation,
Shinjuku-ku, Tokyo, Japan

ABSTRACT

To reduce the response of tall buildings subject to
earthquake motions, there is increasing interest in the
development of response control systems.

This paper describes the design of a passive system
which, using a simple lever-based yoke mechanism,
amplifies the damping effect on the structure.

To investigate the effectiveness of this system, two
fundamental tests were carried out. One was a dynamic
loading test of the damper elements, from which it was
observed that the damping force was almost proportional
to the damper's velocity. The other test used a
three-story steel-frame model on a shaking table to
determine the response characteristics of a structure
fitted with this system. This test demonstrated that
the response of the structure was significantly reduced.

This system is most suited for use in a tall buildings.
Initially, an analysis was carried out on a ten-story
steel structure to compare its behavior with and without
the damping amplifier system under dynamic seismic
forces. Results show that if the damping factor of the
structure is tuned to 0.2 using this system, the maximum
response displacements and accelerations are reduced by
up to 40%. Similarly, the maximum base shear force on
the structure is also reduced by about 30%, thus
allowing structural members to be reduced in size.

In conclusion, it is evident that this damping amplifier
system offers a practical and cost-effective solution to
the response control of tall buildings subject to
seismic and wind forces.

INTRODUCTION

Buildings suffer from vibration as a result of external
dynamic disturbances such as earthquakes, wind, etc.
Various response-controlled systems have been proposed
and investigated and some have been put to practical
effect.

Response-controlled systems installed in buildings to combat vibration can be roughly classified as passive types, in which the damping effect is obtained without supplying energy to the system, and active types, in which energy is supplied in the form of an external force to control the vibration. While the active type has the advantage that it can cope in a more flexible way to unforeseen external disturbances such as earthquake motions, many problems need to be solved before putting such a system into practical use. These include the time lag arising in the feedback system, the extra energy that has to be input from outside and the higher initial cost. With regard to the passive type, although the response of the buildings depends on the external disturbance, one advantage is that a significant effect can be expected using a relatively simple system.

This report summarizes the results of experiments and analyses on the effectiveness of a passive response-controlled system in which a large damping effect is achieved by increasing the damping resistance using the lever principle.

OUTLINE OF RESPONSE-CONTROLLED SYSTEM

Fig. 1 is a conceptual outline of the damping amplifier system. The system is designed to increase the small damping resistance of the building by using the principle of the lever. The system comprises an inverted T-shaped lever with supporting framework and a pair of dampers.

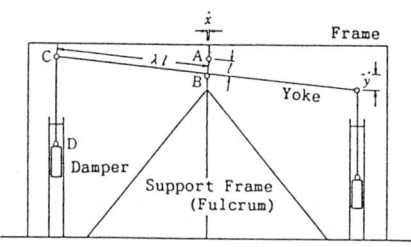

Fig. 1 Schematic Illustration of System

Defining the length ratio of the lever arms as λ, a horizontal storey deflection x is converted to the dampers as an up-and-down movement y, which is λ times larger due to the action of the lever ($y = \lambda \cdot x$, $\dot{y} = \lambda \cdot \dot{x}$)

If the damping coefficient for each damper is Cy, its resultant force is $Cy \cdot \dot{y} = Cy \cdot \lambda \cdot \dot{x}$
For equilibrium, the equivalent damping coefficient at point A, Cx is given by $Cx \cdot \dot{x} = 2\lambda \cdot Cy \cdot \dot{y}$
    i.e. $Cx = 2\lambda^2 \cdot Cy$ (1)

In other words, Cx is amplified by $2\lambda^2$ times Cy.

166

PERFORMANCE VERIFICATION TESTS

Summary of tests

Two sets of tests were carried out to verify the performance of the damping amplifier system.

1)   Individual test of damper --- to check the basic performance of a single damper as used in the damping amplifier system.

2)   Shaking table test of 3-storey model building --- to investigate the vibration characteristics of a structural steel frame incorporating the damping amplifier system.

Element test of damper

A dynamic force application test was carried out on the dampers used to check their individual characteristics (dependence of damping effect on velocity). Photo 1 shows the appearance of the sample dampers. They consist of a piston and cylinder and give a damping effect due to pressure loss arising when the piston moves up and down.

The tests consisted of changing the following variables and making comparisons:

(1)   The gap between the piston and cylinder (0.5 mm, 1.0 mm, 2.0 mm, 3.0 mm), (2)   the cylinder bore (89.1 mm, 200.0 mm), (3)   the length of the piston (100 mm, 150 mm, 200 mm, 300 mm), and (4)   the working fluid (water, water glycol).

Photo 1   Loading Set-up and Test Specimen
(Element Tests)

The tests were performed on a loading device with a 5-ton actuator and vibrations were applied by exciting the piston portion using a sine excitation. The input frequency and amplitude were pre-set, and a fixed number of cycles was allowed in order to stabilize the results.

Fig. 2 shows an example of the relationship between
velocity of vibration and the resultant damping force.
The figure also shows calculated values obtained by
evaluating pressure loss P using the following
equation:

$$\Delta P = \Delta P_1 + \Delta P_2 \qquad (2)$$
$$\Delta P_1 = \Lambda \, l r v^2 / 2dg$$
$$\Delta P_2 = \zeta r v^2 / 2g$$

(Fluid:Water-Glycerol)
(Diameter of cylinder
 in the clear: 89.1mm)

where,
ΔP_1: pressure loss due to
        fluid friction [kg/cm²]
ΔP_2: pressure loss due to
        change in bore cross-
        section [kg/cm²]
Λ: coefficient of friction
        of rough pipe
        $= 0.016 + 1.13/Re^{0.41}$
        : Landau's equation
Re: Reynolds number = vd/ν
ν: coefficient of kinetic
        viscosity [cm²/sec]
l: length of piston [cm]
r: specific gravity of
        fluid [kg/cm³]
v: flow velocity [cm/sec]
d: equivalent bore of
        close-fitting section [cm]
g: acceleration due to
        gravity [cm/sec²]
ζ: Coefficient of loss
        when cross-section of
        pipeline changes rapidly
        $= \zeta_1 + \zeta_2$
ζ_1: small diameter to large
        diameter
ζ_2: large diameter to small
        diameter

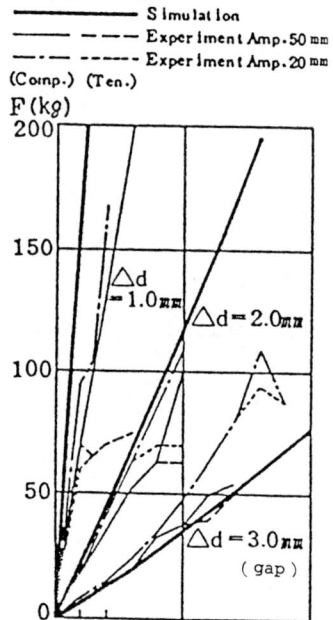

Fig. 2 Plot of Damping Force
versus Velocity
(Element Test)

Fig. 2 shows that for a fluid of high viscosity,
the gap between the cylinder and piston has a sensitive
effect and it can be inferred that a large damping force
will result. The tests verify this.

Shaking table test of model building

To check the vibration characteristics of a structure
fitted with this damping amplifier system a shaking
table test was performed using a structural model.

Fig. 3 shows the tested damping resistance amplifier system, Fig. 4 a rough sketch of the structural model and Photo 2 a complete view of the model. The lever length ratio, λ was made adjustable by changing the pivot position, i.e. the length of A-B in Fig. 1. The dampers used in this system

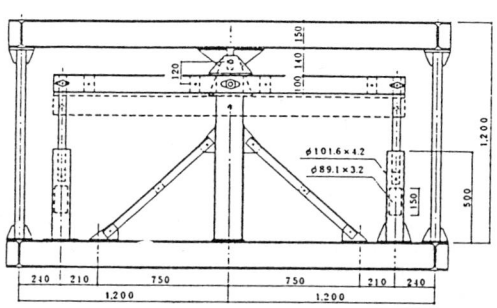

Fig. 3 Test Specimen of Damping
Resistance Amplifier System

were of a fluid-filled type as used in the element test, with one attached to each end of the lever. The steel structural frame consisted of three-storey in a three-dimensional layout with a single span and four columns. Each storey was fitted with the damping resistance amplifier system as shown in Fig. 3.

Fig. 5 shows the frequency response curve resulting from a uni-directional sine wave step sweep excitation in the horizontal direction.

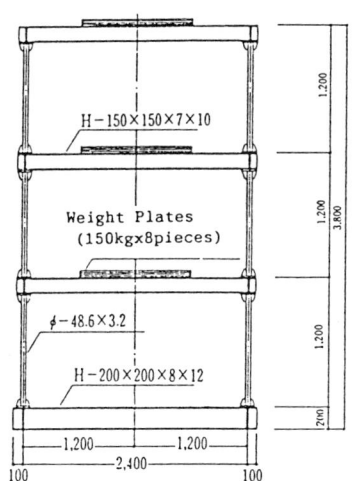

Fig. 4 Structural Model
(Shaking Table Test)

Photo 2 Three Storey Model

This figure shows that the primary response ratio of the top of the frame when excited by the sine wave was reduced to about 1/10 by installing the damping system, and the secondary and tertiary peaks almost disappeared. Thus there is a tendency for not only the primary, but also higher orders to be controlled.

Fig. 6 shows the response at the top of the structural model during a seismic wave excitation and Table 1 is a comparison of maximum response values. The results confirm that the maximum response acceleration was reduced to about 1/2 of that in a model without the equipment.

Fig. 7 is a comparison of acceleration response as calculated in a simulation. A two-dimensional column and beam model including damping resistance was also analyzed.

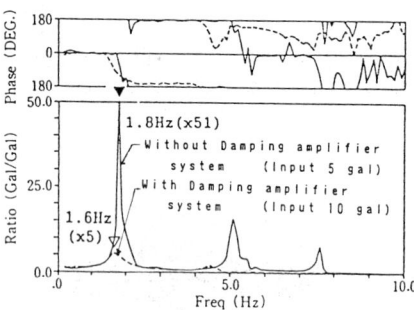

Fig. 5 Frequency Response at the Top
of the Model

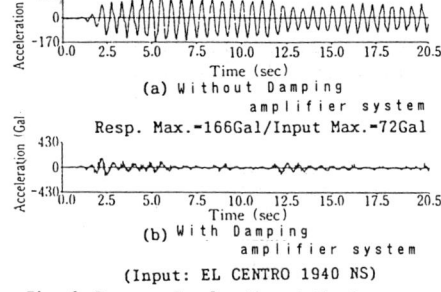

(Input: EL CENTRO 1940 NS)

Fig. 6 Response Acceleration at the Top
of the Model

Table 1 Maximum Response Values

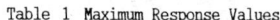

Type	Maximum Response	EL CENTRO 1940 NS input			HACHINOHE 1968 NS input	
		10Gal	30Gal	80Gal	30Gal	80Gal
Without	Top Acc. (Gal)	69	165		128	
	(Top/Table)	(4.9)	(5.7)		(4.6)	
	3rd Fl. Acc. (Gal)	56	137		107	
	2nd Fl. Acc. (Gal)	43	97		67	
	Table Acc. (Gal)	14	29		28	
With	Top Acc. (Gal)	35	96	165[179]	70	152[150]
	(Top/Table)	(2.7)	(3.5)	(2.2)	(2.3)	(2.0)
	3rd Fl. Acc. (Gal)	34	112	278[128]	76	242[141]
	2nd Fl. Acc. (Gal)	37	96	155[104]	95	145[112]
	Table Acc. (Gal)	13	27	74	31	77
	Damper Load (kg)	3	11	25	6	19
	Damper Disp. (mm)	1.3	15.4	41.8	8.9	33.0

[ ] : Results through Low-Pass Filter (10Hz)

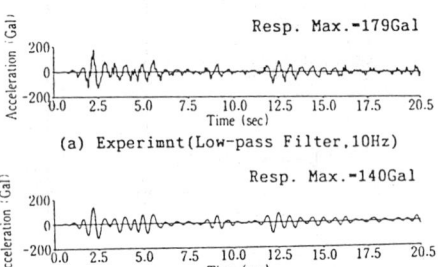

(Input: EL CENTRO 1940 NS, 66Gal)

Fig. 7 Comparisons of Experimental Results
and Simulated Results for Response
Acceleration of Structural Model

The analysis model of the amplifier
system is shown Fig. 8, and the
damping force was modeled by using
equation (2) since its dependence
on velocity was demonstrated in
the element test of the damper.
Measured and simulated results
correlate well.

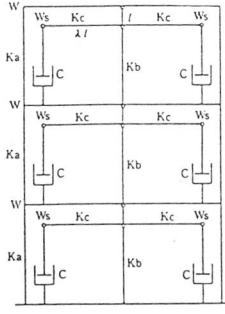

Fig. 8 Analysis model
for 3-storey building

SEISMIC RESPONSE OF STRUCTURE
WITH DAMPING AMPLIFIER SYSTEM

Analysis model

The seismic response was analyzed on model frames of
representing a 10-storey steel building with the damping
resistance amplifier fitted and on a typical rigid frame
structure without the amplifier. The effects of the
system were examined by comparison. Fig. 9 shows a
typical floor plan. The damping characteristics of the
damper were designed to be proportional to velocity so
that the damping coefficient would be proportional to
the rigidity of the frame. In other words, assuming
that $c_i = \alpha \cdot k_i$ and the constant of proportionality
$\alpha = 2h/\omega_0$, damping would be equivalent to a general
proportionally rigid type when each storey's h is
constant. (h is the damping factor)

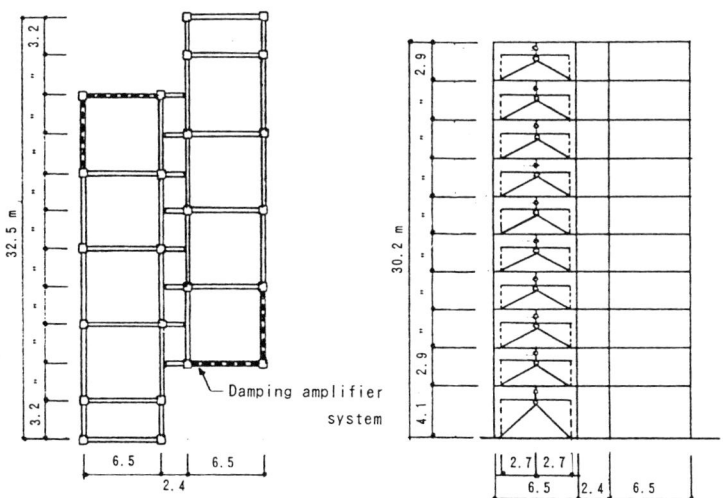

Fig. 9 Typical Floor Plan

Fig. 10 Analysis Model for
10 - Storey Building

171

The response was compared in the two models. For the
seismic wave, EL CENTRO 1940 (NS direction) was used as
the example and the maximum acceleration was set up at
256 gal (25 cm/sec.: velocity). Two dampers were
assumed on each storey, so the damping was additive, and
cases of 5%, 10%, 20%, 30%, 40% and 50% damping were
analyzed.

Fig. 10 shows the analysis model in which $\lambda = 6$.

Analysis results

Fig. 11(a), 12(a) and 13(a) show the response of the
storey due to the difference in additional damping, and
Fig. 11(b), 12(b) and 13(b) compare additive damping
with the response at 2nd, 6th and Roof level.

Response displacement at the top fell to about 60% when
additive damping was 20% and the response fell to a
minimum point when additive damping was around 20-30%
(Fig. 11). The acceleration at the top also decreased to
about 70% for additive damping of 20%, a similar change
to that seen in displacement (Fig. 12). However, as
additive damping was increased to 50%, the responses no
longer decreased.

As can be seen from Fig. 13, the shear force
distribution in the storey followed the same tendency as
the acceleration distribution.

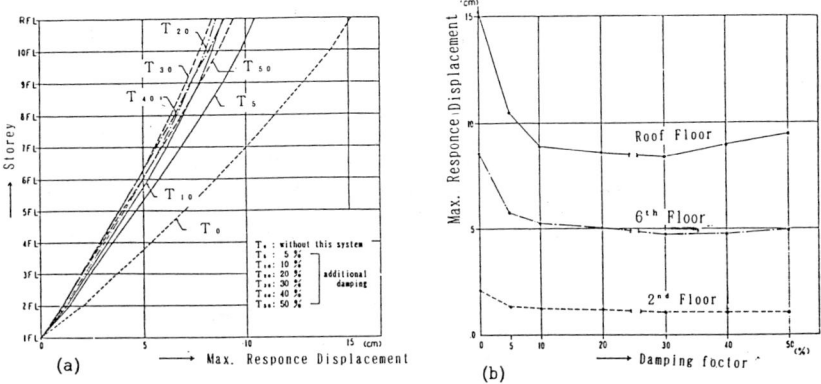

Fig. 11 Maximum Response Displacement

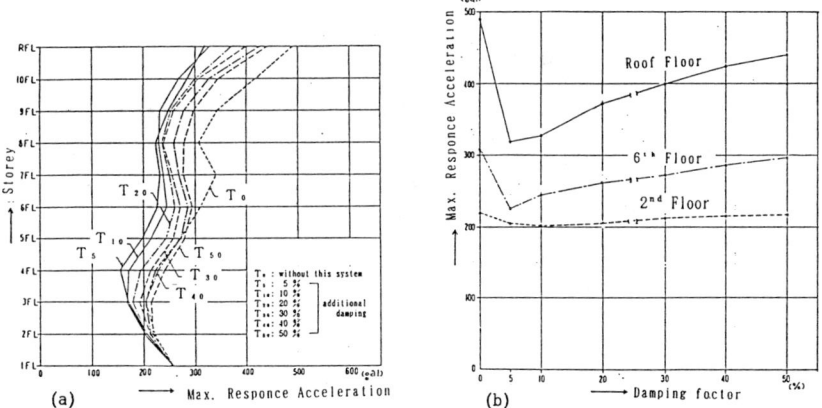

Fig. 12 Maximum Response Acceleration

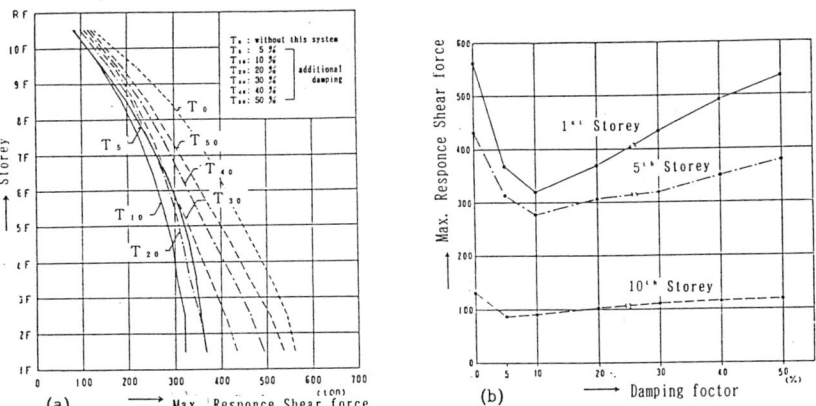

Fig. 13 Maximum Shear Force

173

CONCLUSIONS

The conclusion are as follows.

(1) In an element test of the damper, the damping
     force was shown to depend on velocity and
     results matched those of a calculation based
     on a theoretical equation taking into
     consideration the pressure loss of the working
     fluid.

(2) A shaking table test of the frame model
     demonstrated that by adding suitable damping
     with the damping resistance amplifier system,
     the response of the structure could be reduced
     by a large amount. By adopting the system on
     more than one floor, not only the primary mode
     response, but also higher order responses can
     be reduced.
     In a simulation of the shaking table test,
     both these effects were seen indicating the
     adequacy of our response analysis technique.

(3) The seismic response analysis of a building
     fitted with the damping resistance amplifier
     system verified that by adding damping of
     about $h = 0.2$ to each storey, the
     displacement,and acceleration at the top and
     the base shear could be reduced to about
     60% - 70% of their values in a building not
     fitted with the system.

Reference

Mizuo Yamada, et al., 1990
A STUDY ON RESPONSE CONTROL BY DAMPING AMPLIFIER SYSTEM
AIJ. Abstract

Proceedings of the Second Conference on Tall Buildings in Seismic Regions
55th Regional Conference
May 16 and 17, 1991, Los Angeles, California

THE RESPONSE OF TALL BUILDINGS TO WIND AND ITS MITIGATION

S. Islam[1], G. Hart[2], R. Englekirk[3], J. Raggett[4], S. Huang[5]

Abstract: This paper summarizes the results of a study of the sensitivity of wind-induced acceleration to various structural parameters. Serious shortcommings of using drift limits to limit undesirable building motion perceived by building occupants is also discussed. Finally, a brief discussion of various methods available for controlling wind-induced dynamic motions is presented.

INTRODUCTION

Wind-induced dynamic response which is measured in terms of both lateral imposed loads and acceleration induced at the upper floors, is a very important factor in the design of tall buildings. Even in regions of high seismic activity, the serviceability and thus the cost of a tall building is strongly influenced by wind. Use of high-strength and light-weight materials and light exterior facades that contribute little to the structural strength and stiffness have made modern buildings much more flexible and lightly damped than those of the past. As a result, modern buildings are much more sensitive to wind-induced dynamic excitation than their predecessors. This has created an ever increasing need for more precise descriptions of wind forces on tall buildings, their response to such forces and finally, to find effective means of reducing building dynamic response in order to increase building safety and reduce occupant discomfort and non- structural damage.

The dynamic response of tall buildings to wind is significantly influenced by factors such as the site, over which we have little or no control, and the following building-related parameters: shape and height, vibration periods, mode shapes, mass and mass distribution, lateral stiffness and stiffness distribution, and structural damping. The following section describes the results of a comprehensive study of the sensitivity of wind-induced response of buildings to changes in some of these structural properties. Floor acceleration is addressed primarily because for high-rise buildings in active wind climate, it is the acceleration experienced at the upper floors which is of

---

1. Engineer, Englekirk, Hart & Sabol, Inc., Los Angeles, California
2. Professor, Department of Civil Engineering, University of California, Los Angeles, and Principal, Englekirk, Hart & Sabol, Inc., Los Angeles, California
3. Adjunct Professor, Department of Civil Engineering, University of California, Los Angeles, and Chief Executive Officer, Englekirk, Hart & Sabol, Inc., Los Angeles, California
4. President, West Wind Laboratory, Carmel, California
5. Vice President, Englekirk, Hart & Sabol, Inc., Los Angeles, California

major concern. Serious shortcommings of using drift limit criterion for limiting such undesirable building motion is also discussed. Finally where preliminary design assessment indicates potential motion problem, it is useful to know the various options available for reducing motions. The final section, therefore, discusses the different methods available for controlling the dynamic motions of tall buildings.

ANALYSIS OF STRUCTURAL RESPONSE

A general formulation based on random vibration theory was developed and used for analyzing the three-dimensional dynamic response of tall buildings subjected to wind forces (Islam et al, 1989a). The building was modeled with three-degrees-of-freedom: two perpendicular translational motions and one rotational motion. A linear mode shape normalized to a value of unity at the top of the building was used for all three degrees of freedom. Denoting alongwind, acrosswind and torsional motion of the top floor by x, y and θ (Figure 1), the peak resultant acceleration at a point N (u,v) on the top floor may be approximated as

$$a_R(u,v) = \beta\,(a_x^2 + a_y^2 + (u^2 + v^2)a_\theta^2)^{1/2} \tag{1}$$

in which a_x, a_y and a_θ are the mean peak alongwind, acrosswind and torsional acceleration and β is the "joint action factor" which accounts for the fact that, in general, individual peaks of the three acceleration components do not occur simultaneously. It may be pointed out that the use of joint action factor is simply a convenient mathematical tool. In reality, the case may be that the maximum resultant acceleration at any point occurs when one of the components is at the maximum and the others contribute to it. The values of joint action factor used varies anywhere from 0.7 to 1.0 (Islam et al 1989b). The individual peak accelerations a_x, a_y, and a_θ are obtained by multiplying the corresponding root-mean-square acceleration by a peak factor whose value varies between 3.5 to 4.0.

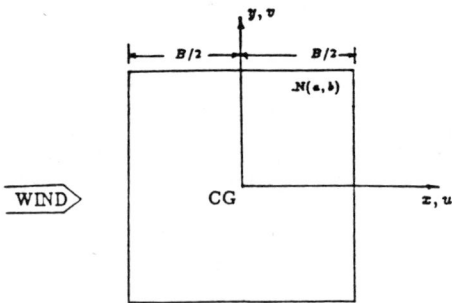

Fig 1: Plan of Building Analyzed

Wind forces for this study were obtained from wind tunnel experiments performed on a rigid model of an isolated square building in an urban boundary layer. A detailed description of the development of wind force spectra is given elsewhere (Islam et al. 1990a). The generalized alongwind

(SFd), acrosswind (SFl) and torsional (SFm) force spectra associated with respective linear fundamental mode are ploted in non-dimensional form in Figure 2. The normalized forces are plotted as a function of reduced frequency, nB/U_H, where n = frequency (Hz); B = the building plan dimension; and U_H = mean hourly design wind velocity at height H (i.e., at the top of the building). Once the design wind speed at an elevation corresponding to the top of the building is selected, the associated spectra of modal forces can be obtained from Figure 2.

Figure 3 shows how the alongwind, acrosswind and torsion-induced acceleration at the corner of the building vary with damping. For this and all subsequent analyses, a square building 591 ft high and 98 ft wide having modal damping 1% of critical and torsional period 1.2 times the translational period was used along with a ten year return period wind speed of 58 mph at the top of the building. The result shows that acceleration is approximately inversely proportional to the square root of the damping for all three response components.

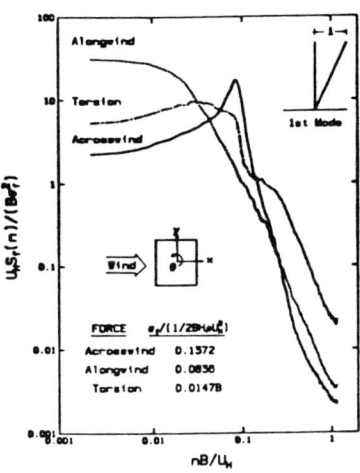

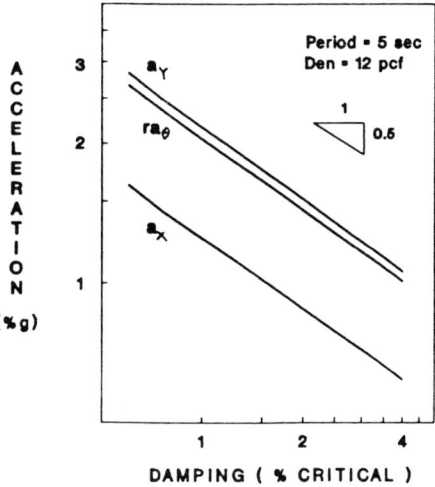

Fig 2. - Modal Force Spectra   Fig 3. - Response .vs. Damping

Figure 4 shows the effect of increasing modal mass with a proportional increase in stiffness (i.e., the modal period remains unchanged). Modal mass which is defined as

$$Modal\,Mass = \sum_{i=1}^{n} m_i \phi_i^2 \qquad (2)$$

in which m_i = mass of the ith floor and ϕ_i = ordinate of the fundamental mode at the ith floor, can be increased either by increasing the floor mass directly or by changing the mode shape of the building (or both). Here the increase in modal mass is shown in terms of increase in building density. Note that typical modern high-rise steel buildings have building densities in the range of 8 pcf to 10 pcf. The results show that acceleration is inversely proportional to modal mass. This may explain why buildings like Empire State Building which is approximately twice as heavy as modern buildings of equal height never had motion problems.

However, if mass is added throughout the building without adding to the stiffness, the increased mass has a twofold effect on the response. Although the added mass results in a direct decrease in response (as before), increasing the mass also results in a decrease in fundamental frequency, shifting it into the range where there is more energy in the force spectra and thus resulting in an increase in response. The net result is shown in Figure 5. The stiffness was kept constant by limiting the roof top deflection to H/500 (Islam et al 1989a). The accelerations in this case are found to be not as sensitive to variation in mass as before.

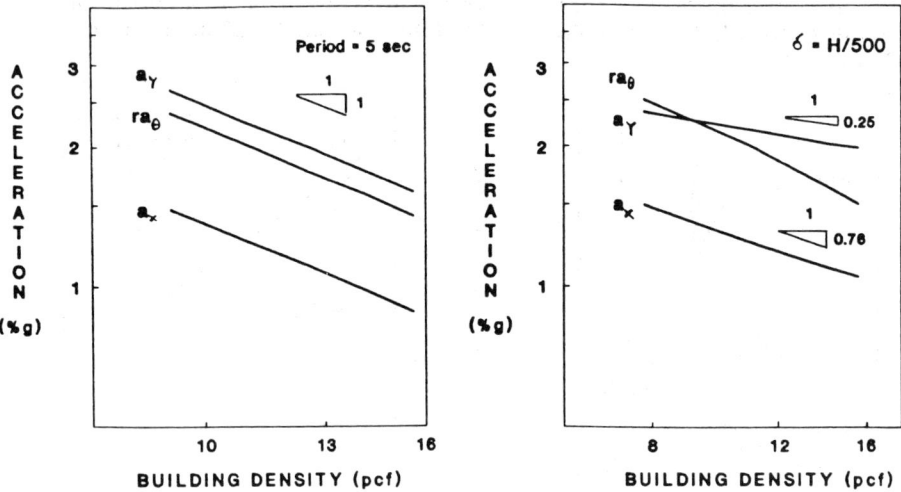

Fig 4. - Response .vs. Mass
Constant Period

Fig. 5 - Response vs Mass
Constant Stiffness

In Figure 6, the acrosswind, alongwind and torsion-induced acceleration are plotted as a function of lateral drift index H/δ (height/tip deflection), which is a direct measure of stiffness. The results suggests that although the torsion-induced acceleration is only weakly dependent on stiffness, the alongwind and particularly the acrosswind acceleration is quite sensitive to variation in stiffness. Figure 7 shows the same results in somewaht different form. Here the accelerations are plotted as a function of fundamental translational period. Since the building mass was held constant while the period was varied, the parameter actually being varied was the modal stiffness. The results once again shows how sensitive acrosswind acceleration is to changes in stiffness. Also note the clear dominance of acrosswind acceleration at long periods. This is very important because for very tall buildings, the acrosswind response is not only the major contributor to the resultant acceleration but also results in lateral imposed loads which are significantly greater than those obtained for the alongwind direction.

To place the accelerations in perspective, the National Building Code of Canada (NBCC) criterion for office buildings is also shown in Figure 7. NBCC recommends a 10-yr return period peak acceleration of 0.03g as a guideline for comfort of occupants for office buildings (the criterion for apartment buildings is much more restrictive). It is important to point out that besides contributing to the total acceleration, the torsional acceleration may also cause building occupants to perceive a rotating horizon and thus enhance their awareness of motion.

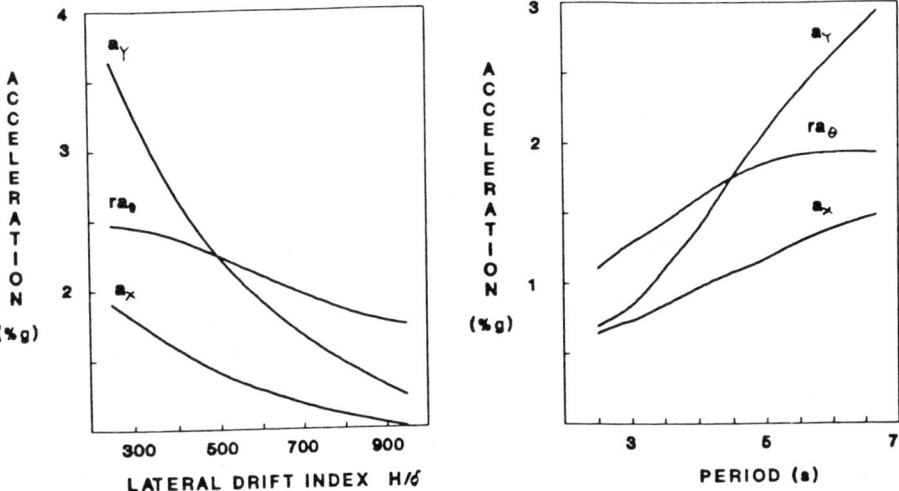

Fig 6. - Response .vs. Drift Index   Fig. 7 - Response .vs. Period

Table 1 summarizes the dependence of alongwind and acrosswind acceleration on damping, mass, stiffness, building period and wind speed.

TABLE 1
DEPENDENCE OF ACCELERATION ON VARIOUS
STRUCTURAL PARAMETERS AND ON WIND SPEED

PARAMETER	ALONGWIND RESPONSE	ACROSSWIND REPSONSE
Damping (ζ)	$\zeta^{-1/2}$	$\zeta^{-1/2}$
Modal Period (T)	$T^{0.74}$	$T^{1.54}$
Stiffness (K)	$K^{-0.37}$	$K^{-0.77}$
Modal Mass (M_E):		
- Mass added with proportional increase in stiffness (T constant)	M_E^{-1}	M_E^{-1}
- Mass added without adding to stiffness	$M_E^{-0.25}$	$M_E^{-0.57}$
Wind Speed (U_h)	$U_H^{2.74}$	$U_H^{3.4}$

179

DRIFT LIMITS AS A SERVICEABILITY CRITERION

Traditionally, the serviceability of tall buildings have been ensured by limiting the lateral deflection at the top of the building to a fraction of the height under the statically equivalent design wind load. Even today, many practicing engineers use drift limits not only as a mean to control nonstructural damage to cladding and partitions but also to limit undesirable motion perception by the building occupants. Although, most specifications and codes in the United States do not give specific drift limits, values of 0.0017H (H/600) to 0.0033H (H/300), in which H is the overall height of the building, are commonly used in practice (ASCE Task Committee on Drift 1988). Interstory drifts are typically limited to the range 0.0015h to 0.0025h, in which h is the story height. Such limits on total or interstory drift control stiffness or curvature of the structural frame, thereby preventing damage to nonstructural panels and partitions, cracking of windows and separation of cladding (Ad Hoc Committee on Serviceability 1986). These limits, however, have been found to be misleading as an indicator of expected performance of buildings with regard to human response to wind-induced motions. For example, the occupants of a tall building designed in 1970 for a conservative drift limit of less than 0.0014 (H/700) were known to have been uncomfortable (Tall Building Monographs Vol. SB 1980). These suggests that even conservative limits like the one mentioned above may not be sufficient to spare occupant from discomfort. This should not come as a surprise because drift limit is only equivalent to a stiffness constraint (the stiffness being directly proportional to the drift ratio). Whereas, occupants comfort is closely related to building acceleration which is a dynamic problem and therefore, besides stiffness, is also affected by other parameters like damping, mass and building period. Simple limits on drift which addresses the static problem therefore, cannot by itself be a good parameter to use for controlling human comfort. This is clearly evident from the results presented in Figure 8.

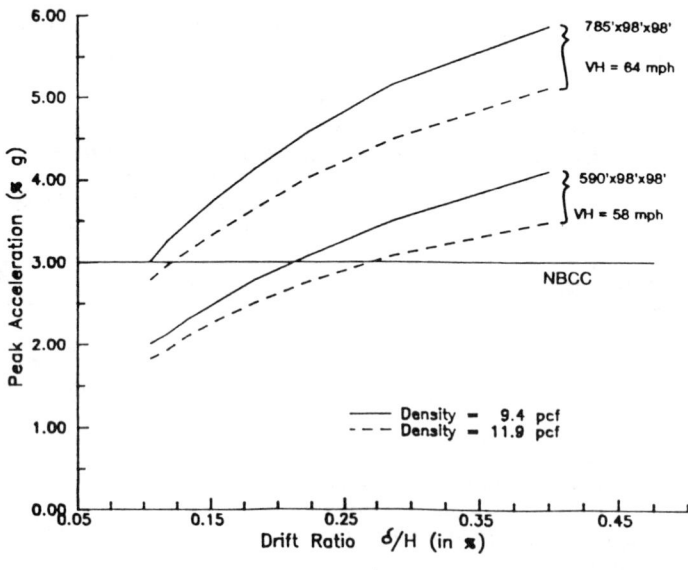

Fig. 8 - Response as a Function of Drift Ratio

Figure 8 shows the resultant peak acceleration at the corner of two generic tall buildings as a function of lateral drift ratio (tip deflection/height), δ/H (Islam et al. 1989a). The acceleration response was computed using the method described in the previous section. It is evident from the results that for the same drift limit, the wind-induced response of the 785 ft high building is significantly greater than that of the 590 ft high building. This is attributed primarily to the fact that even though the two buildings are designed for the same drift ratio, the 785 ft high building because of its longer period is in the range where there is more energy in the force spectrum and thus subjected to higher dynamic response.

An additional reservation regarding the use of simple "drift" criteria as a sole indicator of building serviceability stems from the fact that the design wind loads used by the engineers to compute the drift vary considerably. In some cases, an arbitrary uniform lateral load anywhere between 20-25 psf is applied over the height of the building, while in other cases, the wind load is the unfactored ANSI-A58 or Uniform Building Code or the wind tunnel load. For example, in a recent poll carried out by the ASCE Task Committee on Drift Control of Steel Bulding Structures (1986), a hypothetical 380-ft high 30-story office building with lightweight cladding was given to structural engineers of several firms who were asked to state the design drift limit and the associated average wind load in psf. The responses to this question are shown in Figures 9 and 10 (reproduced from ASCE Task Committee on Drift Control 1986). Figure 9 shows wind pressure versus drift ratio. Fifty seven percent of the respondents recommended a drift limit of 0.0025H (H/400) for wind pressures from 20 to 35 psf. Figure 10 shows wind pressure versus lateral stiffness. This is a measure of the absolute stiffness of each design, obtained by dividing the pressure by the drift index. Note that even though the same drift crieria is used by most engineers, use of different wind loads for computing the drift has resulted in different effective stiffness of the building. Recall that the lateral response to wind has two components: static drift and dynamic fluctuations about the static position (quite often the major component for tall buildings), and it is the effective stiffness that is used for computing the fluctuating component. Therefore, it follows that the wind-induced response of the building designed for the same drift criteria will be different when designed by different structural firms. This plus the fact that buildings in different geographical locations are expected to have different wind loads acting on them, further suggests that a single drift limit is likely to be a misleading indicator of the expected performance of the structural system.

METHODS OF REDUCING BUILDING MOTION

Quite often wind analysis of tall flexible buildings indicate potential problem due to very high accelerations experienced at the top several floors of the building. The obvious question that arises then is how to reduce the building motion so that the response is within acceptable limits ? The results presented earlier and summarized in Table 1 suggests that this can be achieved in several ways. For most buildings it is possible to reduce bulding motion by careful engineering at the planning stages of structural design. This may take the form of adjusting mass and stiffness of the building. As shown earlier, an increase in effective mass is benificial from the point of view of reducing wind-induced motion. Although changes in mass is rarely possible,

181

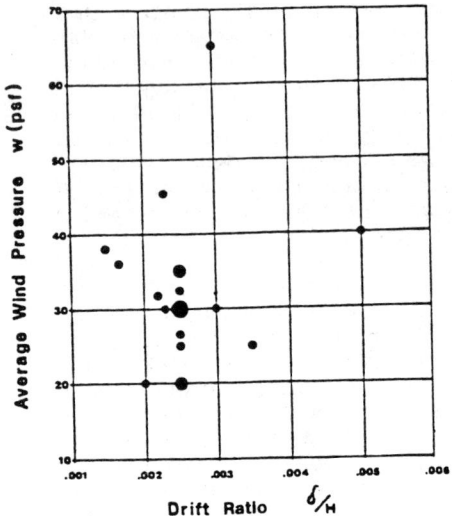

Fig. 9 - Wind Pressure vs. δ/H

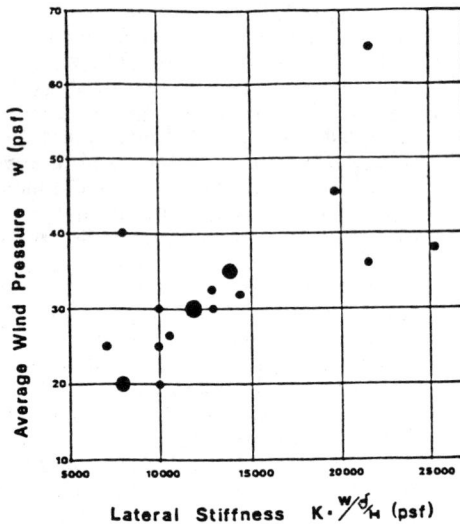

Fig. 10 - Wind Pressure vs. Stiffness

the effective (modal) mass can be increased with minor readjustment of structural stiffness and thereby modifying the mode shape of the building. Architectural design features such as tall free spanning lobbies and distinctive building tops can and have been judiciosly used by structural engineers to increase the effective mass. For example, this concept was used in reducing the top floor acceleration of the Allied Bank Tower in Dallas and Bell Atlantic Tower in Philadelphia (Banavalkar 1989). However, it may be noted that if the building happens to be in active seismic area, a delicate balance needs to be achieved between conflicting requirements for wind and seismic loads. The conflict occurs because an increase in effective mass results in an increased seismic lateral load on the building.

It is also possible to reduce the building motion by increasing the building stiffness. For example, in the case of the 72-story Interfirst Plaza Building in Dallas, design studies (including aeroelastic model tests) showed that it was more cost-effective to control wind induced motions by adding structural stiffness than installing a Tuned Mass Damper (Irwin et al 1988). Stiffness was also increased in the design of the Atlantic Palace Building Condominium in New Jersey in order to reduce wind-induced acceleration.

For most very tall buildings an increase in damping can be especially effective in reducing building motion because the inherent damping in such structures is so small. The results presented earlier suggests that by merely doubling the structural damping, one can reduce the dynamic response by 29 percent. Methods for reducing building dynamic response through an increase in damping includes the use of devices such as viscoelastic dampers and tuned mass dampers. Viscoelastic dampers are designed so that part of the mechanical energy of the building motion is dissipated by converting it into heat which results in a reduction of the amplitude of vibratory motion. For effective transfer of energy, the viscoelastic material is located such that it undergoes pure shear deformation. Examples of buildings using viscoelastic dampers are the Twin Towers of the World Trade Center in New York where the dampers are believed to have increased the damping from 1% to 2.5% of critical, Columbia SeaFirst Building in Seattle and Two Union Square Building also in Seattle.

The Tuned Mass Damper (TMD) essentially consists of a secondary mass of the order of 1% of the buildings generalized mass and is attached to the main structure near the top using pneumatic springs and viscous dampers. The natural period of the secondary mass-spring-damper system is tuned to that of the building it is supposed to provide damping. If properly tuned, the effect of the TMD is essentially equivalent to increasing the viscous damping of the main structure several fold. Tuned Mass Dampers have so far been used in Citicorp Building in New York, John Hancock Building in Boston, Sydney Tower in Sydney and CN Tower in Toronto. However, it should be pointed out that since the floor space required for a TMD is usually a large part of one upper-floor, the loss of rentability space is always a concern for the owner. Besides, the TMD itself might cost $4M to $5M illustrating that the cost is not insignificant.

Other concepts for TMD's, although not yet implemented on buildings, have also been examined. These include masses hung by rigid arm and using pendulum action to provide damping (Grossman 1990). Another type of TMD concept that has been proposed uses the damping produced by liquid sloshing. This concept which is similar to that used in ships to reduce rolling, relies on the viscous action of the fluid and wave breaking action to dissipate energy. There is also this whole area of "active structural control" which although primarily limited to research looks very promising. In brief, this consists of controlling or modifying the motion of the structure by means of the action of a control system which uses external energy supply. An excellent description of different types of damping systems and their advantages and disadvantages is given by Wiesner (Wiesner, 1986).

Finally, wind-induced building motion can also be controlled by altering the wind flow pattern around the building i.e., by aerodynamic treatment of the buildings which has the effect of reducing the wind forces. Deliberately misaligning the directions of preferred motion of the structure with the directions of the maximum aerodynamic excitations have also been suggested as a potential method (Isyumov et al, 1989).

CONCLUDING REMARKS

It is demonstrated that providing externally applied dampers is the not the only way to reduce building motion. Structural engineers can achieve this objective by careful engineering at the planning stages of the structural design. For most highrise buildings this may be adequate. However, for very tall buildings, it usually is more beneficial to increase damping then to adjust the mass and stiffness of the building. Many of our tall buildings have been provided with external damping systems. The primary purpose of these systems is to increase comfort of building occupants i.e., reduce building accelerations. In fact, we think of using damping systems or other means of reducing wind induced building motion only and only if the acceleration at the upper floors exceed the comfort limit. But going beyond the current practice why not think of reducing wind induced response even when acceleration is not a problem. This will certainly result in reduced lateral imposed wind loads which in turn will reduce the building cost. None of the available damping systems can be presently considered reliable enough to be counted upon for strength design of the building, but they can certainly be counted upon for serviceability related design issues which governs the design. Besides, they will also provide an added factor of safety to the strength of the building.

In conlusion, there is surely great savings to be realized by designing tall buildings intelligently for wind. A conscious effort to reduce wind induced response, therefore, should become more commonplace as we design our tall buildings.

ACKNOWLEDGEMENT

The writers are grateful to Dr. Mukund Srinivasan for his help in preparing the manuscript.

REFERENCES

Ad Hoc Committee on Serviceability Research (1986); "Structural Serviceability: A Critical Appraisal and Research Needs," J. Struct. Engr., ASCE, Vol. 112, Dec. 1986, pp.2646-2664.

ASCE Task Committee on Drift Control of Steel Building Structures (1988), "Wind Drift Design of Steel Framed Buildings: State of the Art Report," J. Struct. Engr., ASCE, Vol. 114, Sept. 1988, pp. 2085-2108.

Banavalkar, P.V. (1989), "Structural Systems to Improve Wind Induced Dynamic Performance of High Rise Buildings," Proc. 6th National Conference on Wind Engr., Houston, Texas, pp. C1.21-C1.32.

Grossman, J.S. (1990), "Slender Concrete Structures - The New Edge," ACI Structural Journal, Vol. 87, No. 1, Jan.-Feb. 1990, pp. 39-52.

Irwin, P.A. et al (1988), "Wind Induced Motion of Tall Buildings," Proc. Symposium/Workshop on Serviceability of Buildings, Vol. 1, May 16-18, 1988, Univ. of Ottawa, Ottawa, Canada, pp. 200-213.

Islam, M.S., Ellingwood, B. and Corotis, R.B. (1990a), "Transfer Function Modeling of Dynamic Wind Loads on Buildings," J. Engr. Mech., ASCE, Vol. 116, No. 7, July 1990, pp. 1473-1478.

Islam, M.S., Ellingwood, B. and Corotis, R.B. (1989a), "Dynamic Response of Tall Buildings to Stochastic Wind Load," J. Struc. Engr., ASCE, Vol. 116, No. 11, Nov. 1990, pp. 2982-3002.

Islam, M.S., Hart, G.C. and Raggett, J. (1989b), "Serviceability of Tall Buildings: Occupant Comfort," Technical Report 1989-12, Englekirk, Hart & Sabol Consulting Engineers, Los Angeles, Calif., Dec. 1989.

Isyumov, N. et al (1989), "Effects of the Orientation of the Principal Axis of Stiffness on the Dynamic Response of Slender Square Building," Proc. 6th U.S. National Conference on Wind Engineering, Houston, 1989, pp. B5.39-B5.48.

Tall Building Criteria and Loading, Vol. SB (1980), Monograph on Planning and Design of Tall Buildings, ASCE, New York, 1980.

Wiesner, K. (1986), "The Role of Damping Systems," Proc. 3rd International Conference on Tall Buildings:"The Second Century of Skyscraper", Chicago, Illinois

Proceedings of the Second Conference on Tall Buildings in Seismic Regions
55th Regional Conference
May 16 and 17, 1991, Los Angeles, California

WIND RESISTANT DESIGN OF A TALL BUILDING
WITH AN ELLIPSOIDAL CROSS SECTION

By
Masaru Itoh*

Recently there is a tendency towards constructing odd shaped-tall buildings in Japan. Although Japan is located in a high seismic zone, there are many cases where the wind load is larger than the earthquake load for buildings with heights over 180m, as is the case for the ACT Tower under study. With the results from experiments in the wind tunnel it has been possible to check the building structural resistance to winds. And as for the possibility of people's uncomfortableness due to the vibration of the building, when there are strong winds many times a year, Antivibration Systems have been installed to solve the problem.

1. WIND DESIGN

1.1 Design Procedure

The flow chart in figure 1 presents the wind design procedure and the wind design load.

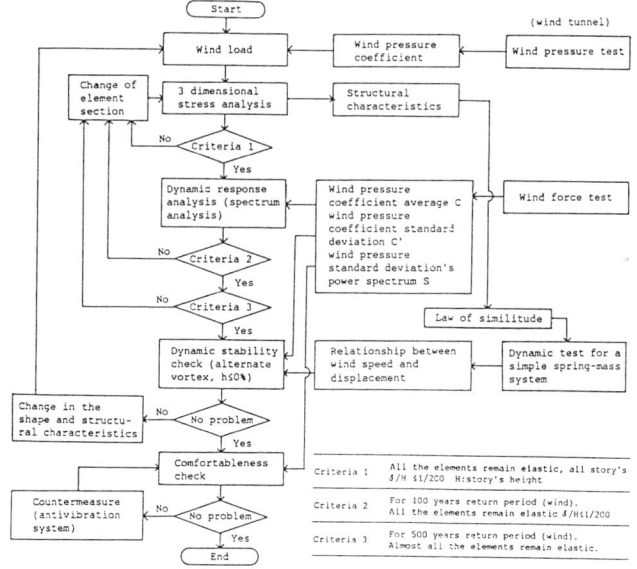

Figure 1 Wind Design Flow Chart

*Nihon Sekkei, Inc.
Tokyo, JAPAN

There are two methods to determine the wind design load in Japan:
The Japanese Building Standard Code (JBSC):

$$P = C.q.A \qquad q = 60\sqrt{h} \ (h \leq 16m), \ 120\sqrt[4]{h} \ (h \geq 16m)$$

Where,

P: wind pressure (kg)   q: velocity pressure (kg/m^2)
C: wind pressure coefficient   h: height (m)
A: face area (m^2)

The Architectural Institute of Japan (AIJ) method, which is based on the buffeting theory:

$$P = C.q.Gf.A \qquad q = (1/2) \ \rho \ V^2$$

Where,

ρ: air density $(=0.125kg.s^2/m^4)$
V: wind velocity (= R.E.V10 m/s)
R: return period factor (for 50years:1.0, 100 years:1.07, 500 years:1.23)
E: wind velocity distribution $(=0.8(h/10)^{0.2}$, for exposure type III)
V10: basic wind velocity at 10m height
Gf: gust factor

The return period is set and its corresponding wind velocity is determined. Having the building site exposure type is possible to get the wind velocity distribution and then to calculate the velocity pressure. The gust factor is determined from the building natural frequency, the wind velocity power spectrum and, the turbulence scale which depends on the face area.

For the ACT Tower design the wind pressure obtained by the Building Standard Code was larger and therefore the one we used.

1.2 Building Outline

ACT Tower is in Hamamatsu city, Shizuoka prefecture. It has an ellipsoidal cross section and consists of 46 floors, two basement floors and, there are a media tower and a heliport on the penthouse roof. The total height is 211.9m and the total area is 152871m². The cross section changes at 116.3m height, having 76.8m length and 45.3m width at the bottom part (offices, typical floor), and 72.0m lengths and 30.8m width at the top one (hotel). The media tower has a cylindric shape with a 25m radius (see figure 2).

The basement part is used as parking area and machine rooms, and the upperstructure is used for banquet rooms, stores, offices, restaurants and hotel.

The basement is a composite structure (Steel-Reinforced Concrete SRC) and the upperstructure is a Steel one.

The Construction of this building will start on May 1991 and be finished by March 1994.

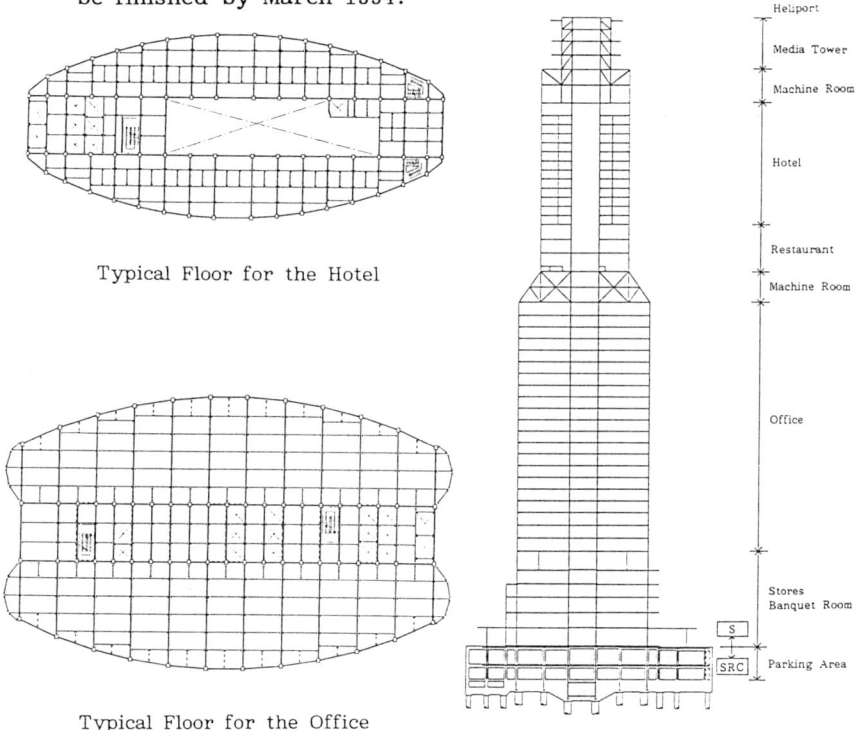

Typical Floor for the Hotel

Typical Floor for the Office

X Frame Elevation

Figure 2 Plan and Elevation of Act Building

Table 1 Building Characteristics

	X Direction	Y Direction	Torsional Direction
Natural Period(s)	4.52	4.73	3.80
Effective Mass	3.285×10^{4} kgf s^2/cm	3.004×10^{4} kgf s^2/cm	1.682×10^{3} kgf cm s^2

189

1.3 Wind Tunnel Tests

The wind tunnel is a boundary layer closed-jet tunnel and has 3.1m width, 2m height. The wind flow velocity distribution is a function of the (height)$^{1/5}$.
1/500 is the scale used for the test model which also contains an 800m radius portion of the building neighborhood.
Three wind tunnel tests were performed: Wind pressure test, wind force test, and dynamic test.

i)   Wind Pressure Test

This test is done to evaluate the wind pressure coefficient. The model is divided in ten levels, each one of them has ten holes where pressure tubes are set in order to do the test measurements.
The wind pressure has been measured at all these holes, then by integrating these data the wind pressure coefficient can be obtained.
On figure 3   corresponds to the wind pressure coefficient calculated using the wind speed at the top and,   is the wind pressure coefficient calculated using the wind speed at each one of the ten level heights.
The wind pressure coefficient used in the present design is shown in this figure.
Figure 4 shows the wind load and seismic load corresponding to the ACT Tower design.

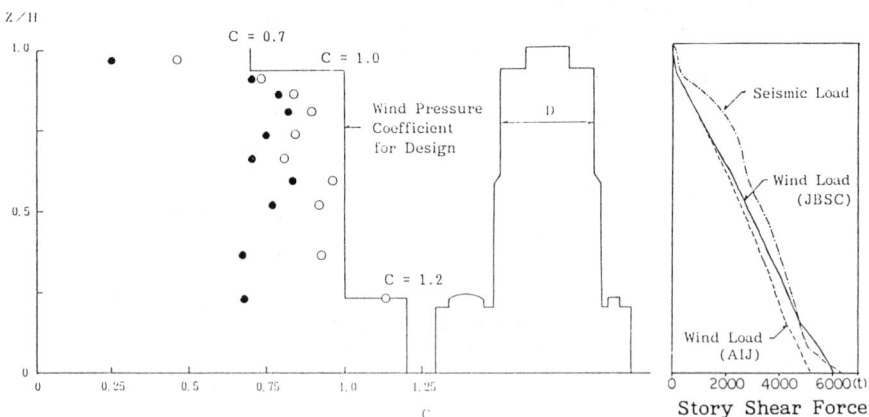

Figure 3 Wind Pressure Coefficient   Figure 4 Wind Load and
                                                        Seismic Load
                                                        for Design

ii) Wind Force Test

The horizontal force, overturning moment and torsional moment are measured directly through this test. The mean wind pressure coefficient and its standard deviation are shown in figure 5.

The mean wind pressure coefficient is calculated using the wind speed at the top and a representative width of the building. The normalized power spectrum for the wind force standard deviation is shown on figure 6.

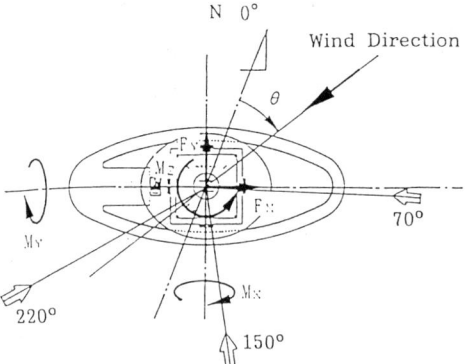

Orthogonal Coordinates and Wind Direction

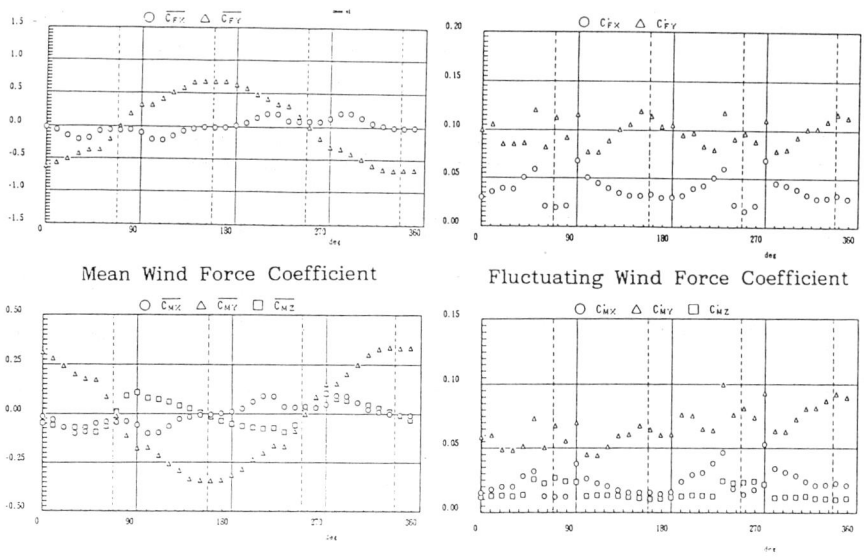

Mean Wind Force Coefficient   Fluctuating Wind Force Coefficient

Mean Moment Coefficient   Fluctuating Moment Coefficient

Figure 5   The Mean wind pressure coefficient
and its standard Deviation

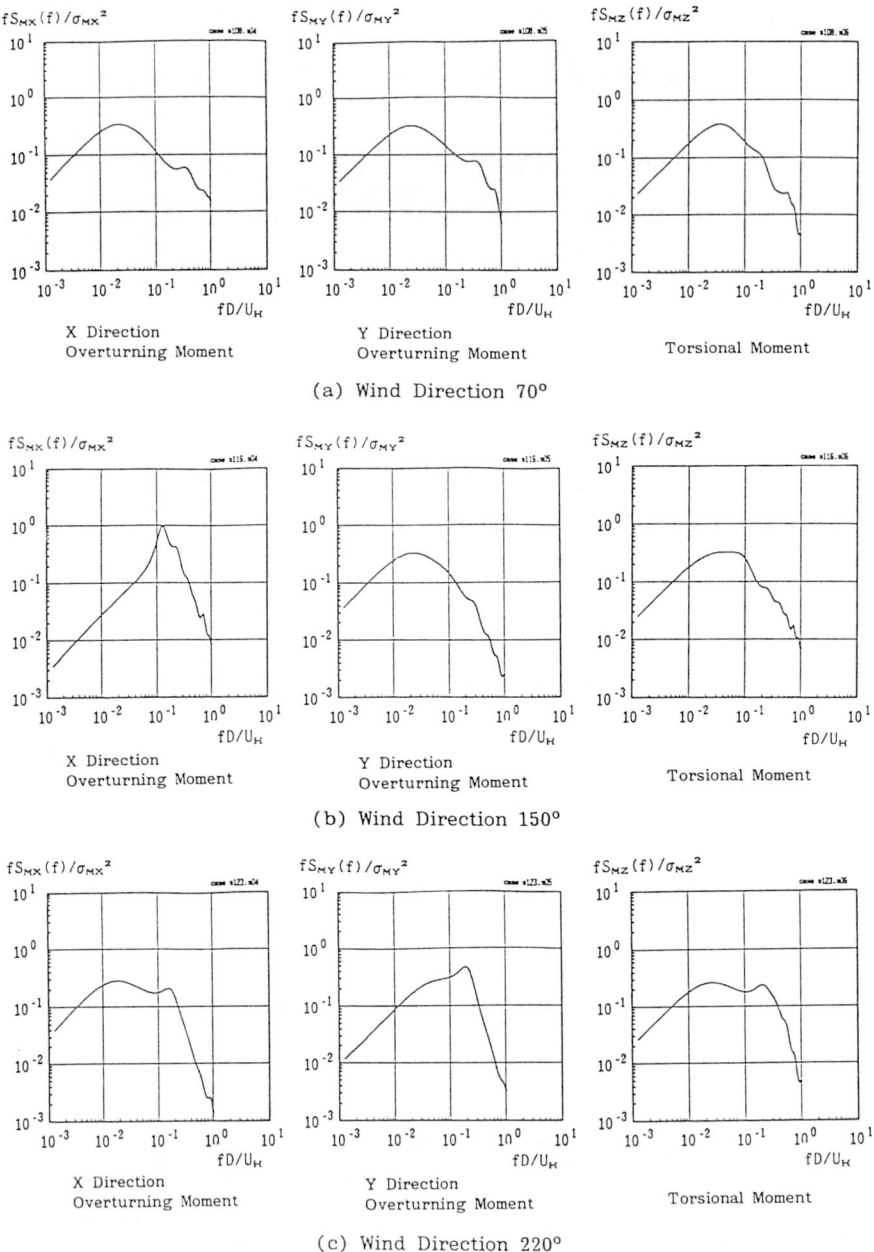

(a) Wind Direction 70°

(b) Wind Direction 150°

(c) Wind Direction 220°

Figure 6   Normalized Power Spectrum for the Fluctuating Wind Force

iii) Dynamic Test

The test model should have mass parameters, damping factor and normalized wind velocities that are similar to the real building.

Figure 7 shows the displacement and wind speed relationship on the wind direction where the maximum displacement occurs, it contains the mean, standard deviation (rms) and maximum displacement.

The line drawn in the figure is the result of the response analysis that will be discussed later.

As for the mean values, the test results are larger than the analysis results. And for the cases of rms and maximum values, the test results are slightly smaller than the analysis results.

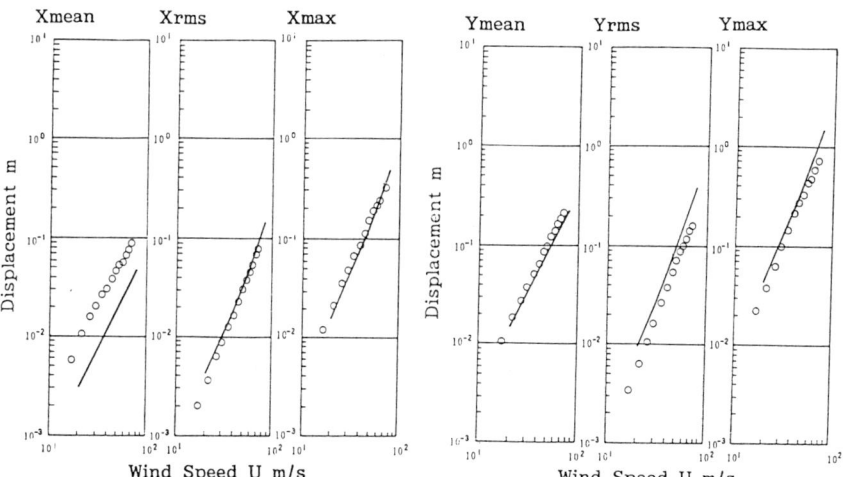

Figure 7 Displacement and Wind Speed Relationship

1.4 Dynamic Analysis

The dynamic analysis was performed using the mean and standard deviation, and power spectrum of the overturning moment and torsional moment coefficients obtained in the wind force test.

The building response spectra is obtained by combining the wind spectra (for X, Y and Θ directions) and the magnification factors vs. frequency curve. (See figure 8).

As for the response results the following combination has been performed in order to get the maximum values in X, Y and Θ directions used in the design:

(the maximum value in the direction of analysis)

+

(the mean and rms values in the other directions)

By using these maximum values (displacements and rotations) it has been possible to check that the building remains in the elastic stage.

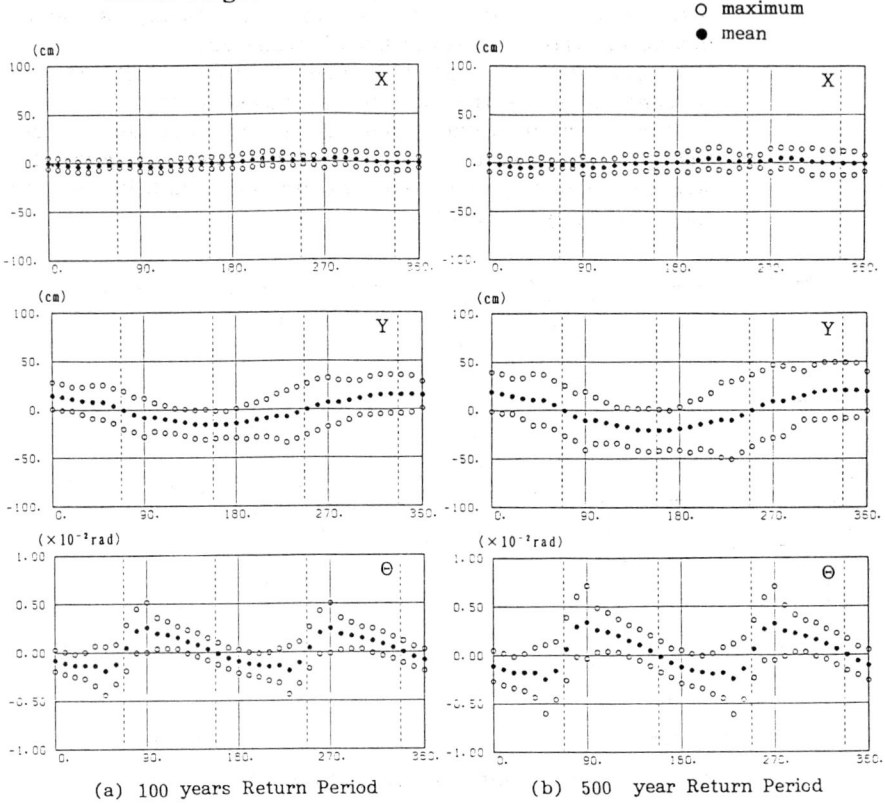

(a) 100 years Return Period   (b) 500 year Return Period

Figure 8   Response Results

1.5 Dynamic Stability Check

The resonance wind speed due to alternate vortex is calculated using the strouhal numbers $Stx=0.13$, $Sty=0.18$ obtained from the wind tunnel tests and are equal to 115,87m/s respectively. As this velocity is larger than the 500 years return period wind speed at the top of the building equal to 54.5m/s, it is possible to state that there is no unstability problem.

When performing the dynamic test it has been also checked that there is not a possibility of galloping.

1.6 Comfortableness Check

The comfortableness estimation has been done using the ISO 6897 and a draft recommendation for comfortableness estimation published by the Architectural Institute of Japan.

On figure 9 is shown the rms of the acceleration response for 5 years return period. And on figure 10 is presented the maximum response acceleration for 1 year return period, where H-3 is a standard curve for buildings and H-4 corresponds to the maximum allowable values for buildings.

From the response results it is possible to say that there is not problem for X direction, however Y direction has to be treated in a special way as it will be explained later.

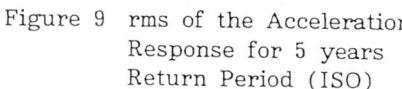

O X direction acceleration at the 45F center of gravity.
● X direction acceleration at the 45F corner frame.
△ Y direction acceleration at the 45F center gravity.
▲ Y direction acceleration at the 45F corner frame.

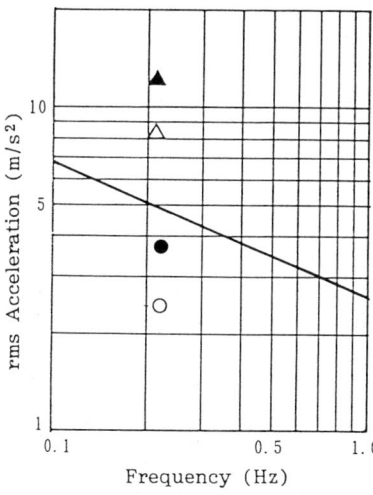
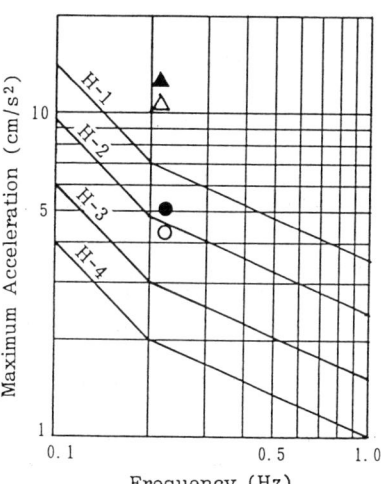

Figure 9   rms of the Acceleration
Response for 5 years
Return Period (ISO)

Figure 10 Maximum Response
Acceleration for 1 year
Return Period (AIJ)

2. Anti Vibration System

As explained in section 1.6 strong winds can occur several times in a year causing uncomfortable building vibration in terms of human sensitivity. In order to avoid this problem an antivibration system has been installed.

This antivibration system is such that when the building acceleration is larger than $3cm/s^2$ the damping effect will start to work, and when the displacement of the building corresponds to a 5 years return period the damper will become passive, and when the displacement of the building corresponds to a 100 years return period the system will be stopped.

By using this antivibration system the rms of the acceleration in Y direction, for 5 years return period, is 40% of the value when this system is not used.

Table 2. Antivibration System Characteristics

	Description	Remark
Type	Pendulum Tuned Mass Damper	The computer will acti-vate the system only for Y direction
Number of Machines	2	In order to control the possibility of torsion
Mass	90t/machine	60% of the Effective mass

3. Conclusions

• Using the data from the wind tunnel tests.
It has been possible to evaluate the building response, and then to check the building structural resistance and safety to winds.

• Seismic and Wind design have been carried out for the ACT Tower design.
For the case of seismic load, the building elements are design to reach the plastic range. However for the case of wind loads, as the wind occurrence time can be quite long (1 hour or more), the building elements are designed to remain in the elastic range.

• Antivibration systems have been installed to avoid people's uncomfortableness due to the vibration of the building when strong winds occur for return periods that are less or equal than 5 years.

Proceedings of the Second Conference on Tall Buildings in Seismic Regions
55th Regional Conference
May 16 and 17, 1991, Los Angeles, California

Three-Dimensional Inelastic Analysis of High Rise
Buildings: Micro-Computer Application

J.C. Jeing

T.Y. Lin International
San Francisco, CA. USA

SUMMARY

Current engineering practice uses 2 popular methods for evaluating the ultimate strength of
buildings subjected to strong earthquake loading: 1) Step-by-step incremental static lateral
force analysis and 2) Step-by-step direct integration time history analysis. The shortcoming of
the static analysis method is its inaccurate assumption that earthquake loads are static because
earthquake loads are in fact, dynamic. The time history analysis method, on the other hand,
has been restricted by the large quantities of computations required and the difficulty of
interpreting the results. This paper presents a simplified step-by-step incremental Peak Ground
Acceleration (PGA) spectrum analysis method that uses a micro-computer to approximate the
inelastic capacity of a building subjected to strong earthquakes. This spectrum analysis method
incorporates the earthquake loads into structural vibration mode shapes and requires a smaller
computer size to perform an inelastic analysis of high rise building.

INTRODUCTION

In current engineering practice, a time history analysis is used to simulate the behavior of
buildings in an earthquake using a time history of ground motion and mathematically
calculating the deformation and stress state of the mathematical model at repeated instants of
time. Under low level loadings the structures tend to behave elastically. Under the large stress
and deformation levels experienced in a strong earthquake, however, structures behave
inelastically.

The mathematical techniques and computer software required for performing time history
analysis are available. However, most of the software packages tend to be impractical for
complex structures due to the large size of computer memory and great length of time required
to perform the calculations, as well as the difficulties of interpreting the results. In addition,
the results obtained from such analyses are highly dependent on the particular form of ground
motion time history used. It is possible to perform several analyses for different ground motion
histories, all with the same peak amplitudes of motion, and obtain substantially different
findings as to structural adequacy. Since earthquake time histories vary significantly and can
only be approximated, the validity of this approach is somewhat limited.

Although ground motion time histories vary considerably from earthquake to earthquake, research has shown that, for any given site, response spectrum curves can be developed to encompass the array of time histories likely to be experienced at the site. The response spectrum analysis can be used to adequately reflect the real response of a building. The size of the computer required to perform the response spectrum analysis and the time required for the analysis are significantly smaller and shorter than required using a time history analysis method.

Spectrum Capacity Method

The approximate inelastic analysis procedures using the spectrum method are as follows:

(1)   Perform a static analysis of vertical dead loads and save the structural displacements and member forces. This is the initial stress level of the structure before the earthquake.

(2)   Perform a modal analysis, and determine the level of excitation causing first major yielding of the structure: Review the stress ratio of all structural elements by adding the modal analysis forces to the initial dead load analysis forces. Find the maximum stress ratio of the overstressed element, or select a group of stress ratios that are within 10 percent of the highest overstressed elements. Calculate the reduction factor for the highest overstressed member so as to reduce the stress ratio to the yield point of this member. The factor is a percentage of modal analysis results which add to initial analysis results to make this element reach the yield point. Multiply this factor to reduce all the structural elements in the modal analysis results and add to the initial results. This is the first yielding level of the structure caused by the earthquake.

(3)   Revise the stiffnesses of those yielded elements to represent the plastic hinges.

(4)   Calculate an additional reduction factor for the next new modal analysis result using the same technique in item (2) above. This is the second yielding level of the structure caused by the earthquake.

(5)   Repeat the same procedure above until the combined results reach an ultimate limit (e.g., a mechanism, instability, or excessive distortion)
Step (n-1) + Step (n) * Factor (n) = Yield (n).

(6)   Sum up the reduction factors. The total reduction factor is the percentage of the demand earthquake response spectrum (e.g., Peak Ground Acceleration) which the building can resist inelastically. If the total reduction factor is less than 1.0, the building can not resist the demand earthquake without collapse. If the total reduction factor is greater than 1.0, the inelastic capacity of the building is strong enough to resist the earthquake.

(7)   Accumulate the lateral deformations and compare the story drift limits.

Micro-Computer Application

A computer program SCM3D was developed for PC/386 micro-computer to perform the inelastic analysis of high rise buildings. The structural idealization for the analysis of a framed building is essentially identical to the ETABS (reference 1) idealization. It assumes that each story of a building has a horizontal rigid diaphragm to link with one or more three-dimensional space frame(s), consisting of beam, column, bracing, and shear wall members. The program formulates structural stiffness and calculates the stress of structural members story by story. This one-story-at-a-time concept requires substantially less memory, and hence, is practical for micro-computer application in performing both static and dynamic analysis of high rise buildings.

Example of Analysis

The example shown in Fig. 1 is a 54 - story apartment complex building. the structure of this building is a steel space frame with bracings at the elevator core and stairway areas. The structural members consist of Box Columns 800 x 800 x 50 x 50 to 500 x 500 x 30 x 30, I - Beams 900 x 400 x 40 x 25 to 700 x 300 x 20 x 15, and Tube Braces 400 x 400 x 20 x 20 to 200 x 200 x 10 x 10.

The structure was modeled in one 3-D space frame, and a demand earthquake spectrum of 0.2g PGA with 5 percent damping was applied in the diagonal direction of 45 degrees. The elastic responses of the building are shown in Table 1. The states of structural yielding are shown in Fig. 2. The structural members start to yield in frame I-9 at 70 percent of the demand PGA, and the first yielding to occur in the next frame F-6 is the 7th step at 110 percent of demand PGA. The yielding occured in the third frame J-A and the 4th frame 1-10 in the 9th step at 140 percent of demand PGA.

The roof displacement and base shear of the structure are shown in Fig. 3. Both are gradually decreased after the structure yielded when compared to the condition of same structure as a fully elastic mode. Fig.4 shows the demand earthquake spectrum curve and the building capacity curve based on the periods and spectral accelerations for the fundamental mode of vibration. The analysis results demonstrate that the ultimate inelastic capacity of this building can resist twice the demand of this earthquake. In other words, the building will be damaged but will not collapse in a 0.4g PGA level of earthquake. The story drift limit of this building is governed by wind loads.

Conclusion

The ability of structures to resist the excessive accelerations and deformations of severe earthquakes is not directly proportional to the equivalent design seismic forces or to the amplitudes of the peak ground accelerations of earthquakes. Fig.5 shows the inelastic reaction of a structure subjected to the demand strong earthquake. The responses of a high rise building tend to decrease while the responses of a low rise building, on the other hand, tend to increase after the initial yielding of the structure.

References

1) E.L. Wilson, J.P. Hollings and H.H. Dovey: "Three Dimensional Analysis of Building Systems (Extended Version)" Report No. EERC 75-13 University of California, Berkeley, California.

2) "Seismic Design Guidelines For Essential Buildings" Departments of The Army, The Navy and The Air Force USA.

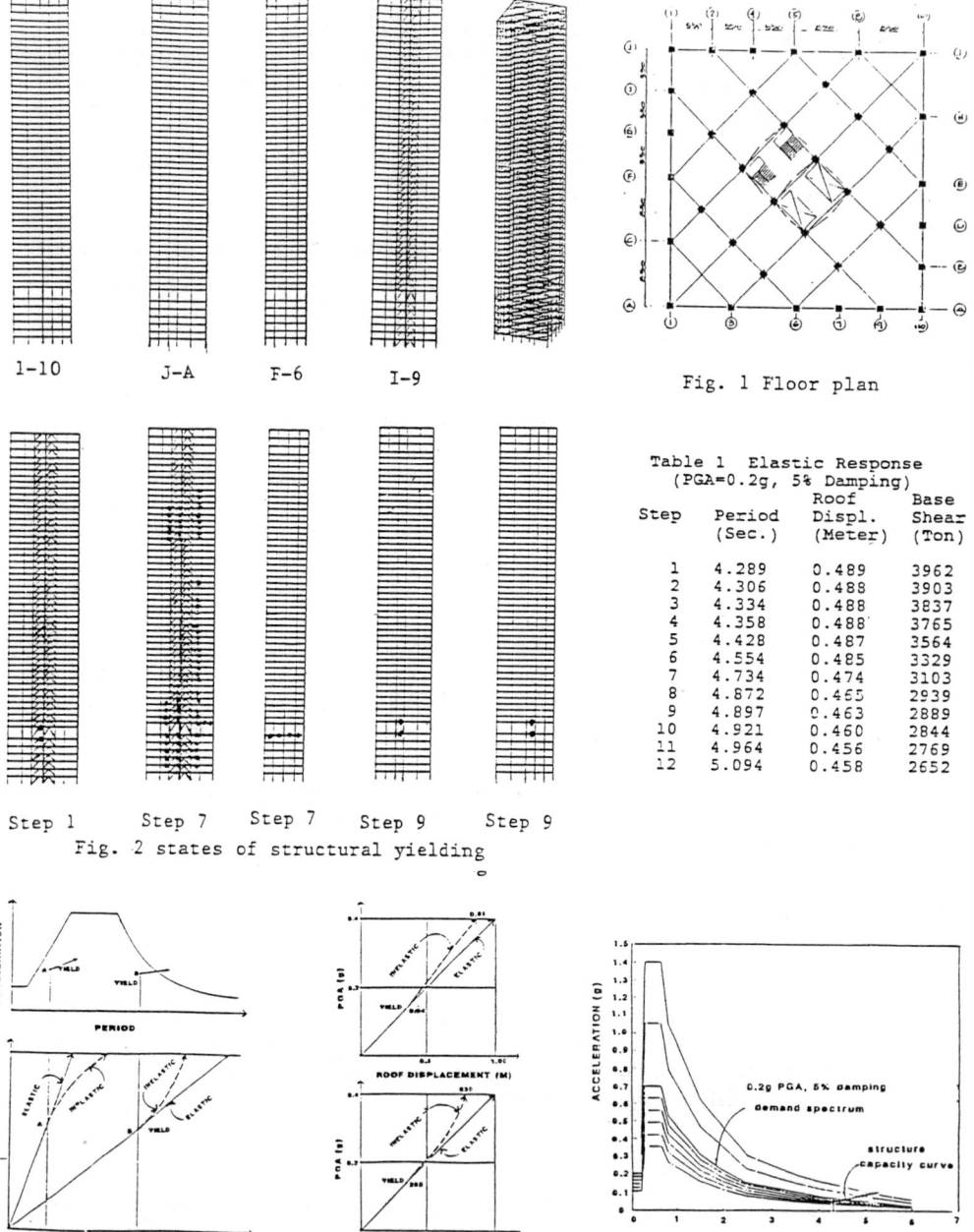

1-10   J-A   F-6   I-9

Fig. 1 Floor plan

Step 1   Step 7   Step 7   Step 9   Step 9

Fig. 2 states of structural yielding

Table 1   Elastic Response
(PGA=0.2g, 5% Damping)

Step	Period (Sec.)	Roof Displ. (Meter)	Base Shear (Ton)
1	4.289	0.489	3962
2	4.306	0.488	3903
3	4.334	0.488	3837
4	4.358	0.488	3765
5	4.428	0.487	3564
6	4.554	0.485	3329
7	4.734	0.474	3103
8	4.872	0.455	2939
9	4.897	0.463	2889
10	4.921	0.460	2844
11	4.964	0.456	2769
12	5.094	0.458	2652

Fig. 5 Base shear   Fig. 3 shear, displacement   Fig. 4 Structure capacity

Proceedings of the Second Conference on Tall Buildings in Seismic Regions
55th Regional Conference
May 16 and 17, 1991, Los Angeles, California

TALL BUILDINGS IN THE PUBLIC REALM

R. Scott Johnson
Johnson Fain and Pereira Associates
6100 Wilshire Boulevard
Los Angeles, California 90048

This panel discussion is held by five important participants in the process of development, entitlement and realization of major buildings in the Southern California area and beyond. While environmental concerns, political approval processes, and the race to rebuild the central city continue to put the design and construction of tall buildings in the public spotlight, California leads the nation in this evolving climate of realization. As the 80's have seen fierce competition for quality and urban amenities in the design of major commercial structures, the economic slowdown marking the opening of the 90's promises to bring more discussion and more competition.

R. Scott Johnson, Architect and Designer of major buildings in the Los Angeles area, will lead the one hour discussion. **Steve Meixner**, Vice President and Regional Director of JMB/Urban Development Company, has developed and continues to plan major projects in the area. **O'Malley Miller**, Partner at Allen, Matkins, Leck, Gamble & Mallory, specializes in land use and zoning issues in the entitlement process. **John Kaliski** is Principal Architect for the Community Redevelopment Agency for the City of Los Angeles and oversees all design and construction for projects in its downtown area. **William Fain, Jr.**, Managing Partner and Chief Urban Designer for Johnson Fain and Pereira Associates, is currently masterplanning major sites throughout the region.

Proceedings of the Second Conference on Tall Buildings in Seismic Regions
55th Regional Conference
May 16 and 17, 1991, Los Angeles, California

EARTHQUAKE AND WIND RESISTANT DESIGN
OF
THE SHINJUKU PARK TOWER

Takuji Kobori [*], S.Ban , T.Yamada
S.Muramatu , Y.Takenaka . T.Arita , T.Tujimoto

Kobori Research Complex Inc.
6-5-30, Akasaka, Minato-ku, Tokyo 107, Japan

INTRODUCTION

The Shinjuku Park Tower, located in the Shinjuku business area of the cosmopolitan city of Tokyo, will be one of the 21st century's monumental high-rise beautiful intelligent buildings (Photo-1). It will function as a center for information media and for creative cultural activity. Architectural design is by Kenzo Tange Associates Urbanists-Architects, and structural design is by Kobori Research Complex Inc.

STRUCTURAL PLANNING

The Shinjuku Park Tower comprises an underground structure of 5 basement floors and a superstructure rising to 52 stories (235 meters). The underground portion is a composite structure of steel and reinforced concrete, and the upper portion is a rigid steel frame. Show rooms and halls are planned for the 3rd through the 7th floors, offices for the 8th to 37th floors, and hotels and banquet rooms for the 39th to 52nd floors.

This tall building has a unique configuration composed of three tower blocks. The south tower is 52 stories, the middle one is 47 stories and the north one is 41 stories. Each tower basically comprises a square (32x32 meters) unit section with additional window space (19.2x6.4 meters) protruding from it. The three towers stand together with some overlap between them, and this constitutes the final design configuration of this high rise building (Fig.1).

The primal structural components are *mast columns*, *a frame tube*, and *a belt truss*. These structural units constitute the whole frame system. The complex planning and proportioning of this tall building achieve 1) wide open working space for the office areas 2) compatible vertical composition between the office areas and the hotel areas in one building.

In addition to these structural devices, vibration controllers or mass drivers are installed to attenuate unpleasant vibration response caused by frequently occurring moderate earthquakes and strong wind gusts. Furthermore, vertical vibration of slabs supported by long span members is prevented by using lattice beams with high stiffness.

---

[*] President of Kobori Research Complex Inc., Prof. Emeritus of Kyoto University

Photo-1 Exterior View of the Shinjuku Park Tower (Craft Photograph)

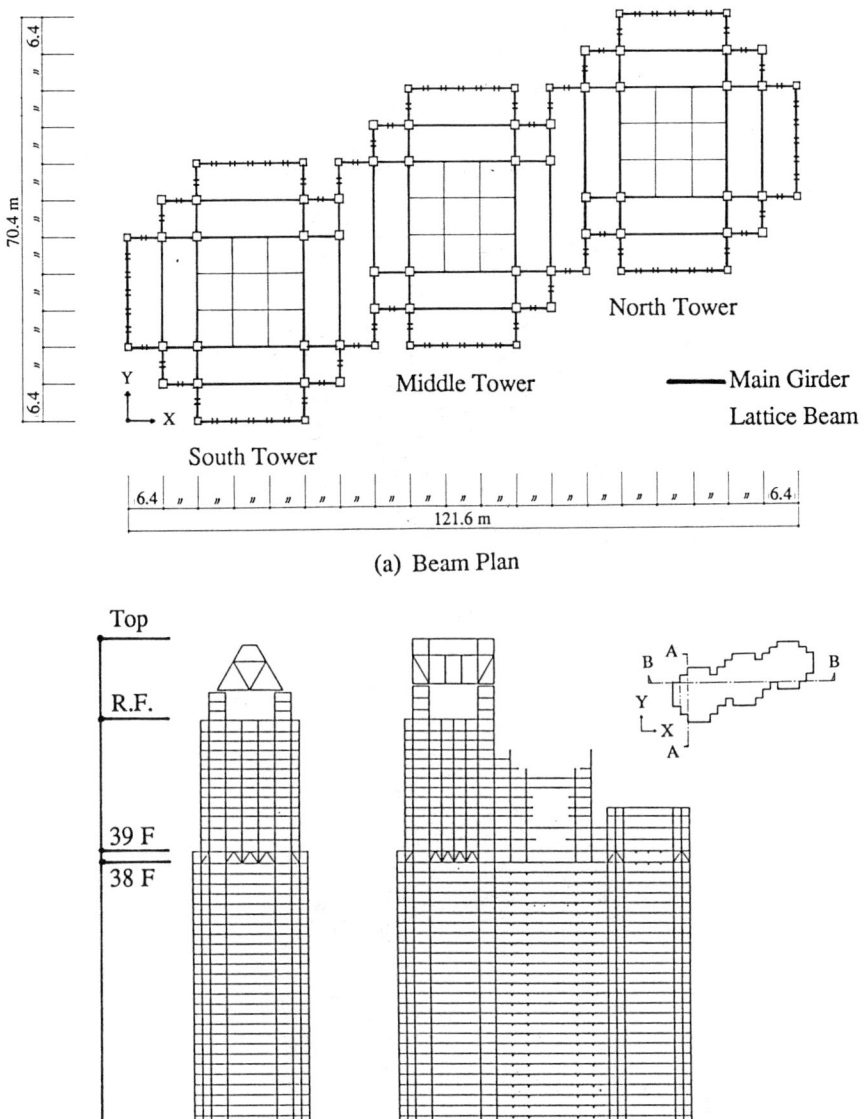

(a) Beam Plan

70.4 m

6.4

6.4

North Tower

Middle Tower

Main Girder

Lattice Beam

South Tower

Y

X

6.4   121.6 m   6.4

Top

R.F.

39 F

38 F

G.L.

(b) A - A Framing Elevation   (c) B - B Framing Elevation

Fig.1 Beam Plan and Framing Elevation

205

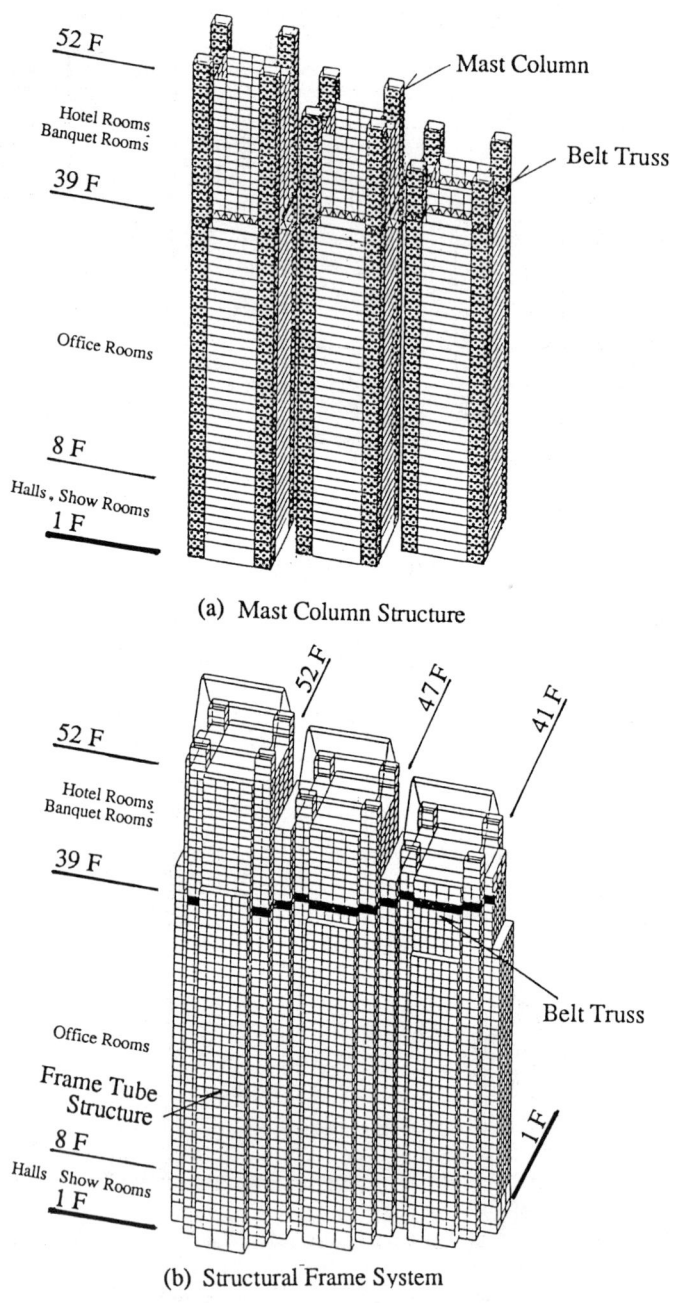

52 F

Hotel Rooms
Banquet Rooms

39 F

Office Rooms

8 F

Halls, Show Rooms
1 F

Mast Column

Belt Truss

(a) Mast Column Structure

52 F

Hotel Rooms
Banquet Rooms

39 F

Office Rooms

Frame Tube
Structure

8 F

Halls Show Rooms
1 F

52 F

47 F

41 F

Belt Truss

1 F

(b) Structural Frame System

Fig.2   Structural Frame

206

MAST COLUMN STRUCTURE

The mast column structure , which virtually constitutes this high rise building, realizes a large open space (44.8x19.2 m) with no internal columns.   Each tower has a *mast column* at each corner of its 32-meter-square plan.   One mast column is composed of four large columns bound together by stocky beams.   Twelve mast columns, in all, extend from the underground level to the top of each tower and support the whole structure (Fig.2).

FRAME TUBE STRUCTURE

Medium sized columns are additionally placed at 3.2-meter intervals circumscribing the three towers.   This enhances the torsional stiffness of the whole structure to resist the expected extreme response in case of large earthquakes and strong wind forces. These columns together create a tube-like structure, or *frame tube structure*. Above the 39th floor level, the column spacing is increased to 4.8 meters in the x-direction and 6.4 meters in the Y-direction (Fig.2).

BELT TRUSS STRUCTURE

The 38th story is a buffer zone between the office areas and the hotel areas. The whole 38th story space is occupied by a *belt truss structure* comprising a large girder truss combination which connects the twelve mast columns at this level and creates a rigid base for the structure above (Fig.2).

The *belt truss structure* is expected to absorb the inconsistent stresses and deformations smoothly between the differently spaced column structures: 3.2-meter intervals in the lower portion and 4.8- and 6.4-meter intervals in the upper portion. At the same time it enhances the lateral stiffness of the upper structure and minimizes the large lateral deformations that would be caused by large earthquakes and strong winds.

Table-1 Member Size

Column Member   Material : SM50B(t >40mm), SM50A(t ≤40mm)

	Mast Column Members	Frame Tube Members	
		Corner Column	Middle Column
Size	□ - 850x850x28 ≀ □ - 850x850x80	□ - 500x500x28 ≀ □ - 500x500x80	H-414x405x18x28 ≀ BH-518x467x80x80

Girder Member   Material : SM50A

	2nd~41st Floor	42dn~Roof Floor
Size	BH-900x300x19x22 ≀ BH-900x450x19x36	BH-700x300x14x22 ≀ BH-700x400x19x28

Belt Truss Member   Material : SM50A

	Compression and Tension Members	Lattice Members
Size	BH-900x300x16x19 ≀ BH-900x600x22x40	H-300x300x10x15 ≀ H-428x107x20x35

STRUCTURAL DESIGN

Both static and dynamic analyses, including eigen value calculation, of this complex structure for wind disturbances and ground excitations were carried out using an accurate 3-dimensional model created by " 3 dimensional FAPP " : an original computer program for high rise structures. The ground excitation level was 30 cm/sec for this linear analysis. The results of eigen value analysis are given in Table 2 and the modal shapes are shown in Fig.3.

Then elasto-plastic non-linear dynamic analysis with 50 cm/sec level ground excitation was carried out according to the simplified 3-dimensional model shown in Fig. 4. This model is based on the following assumptions.

1. The mast columns are assumed to be large cantilevers which resist both bending moments and shear forces, the enhanced stiffness due to the surrounding beams and girders is taken into account. The medium sized columns constituting the *frame tube structure* are hardly influenced by the perpendicular direction beams, so they are just assumed to be isolated members independently carrying bending moments and shear forces. These members are connected to each other at every floor level, and they are treated as a rigid body.

2. The elasto-plastic characteristic curve of the simplified 3-dimensional model was determined according to the yielding strength of each original member. A 'Normal Tri-Linear' model was used to determine the shear-deformation curve for each mast column and a 'Bi-linear' characteristic model was used for the girders that connect a pair of two mast columns.

Table 2 Modal Period

Order of Mode	1st	2nd	3rd	4th	5th	6th
Modal Period (sec)	5.24	4.50	3.98	1.70	1.56	1.38
Principal Direction	Y	X	φ	Y	X	φ

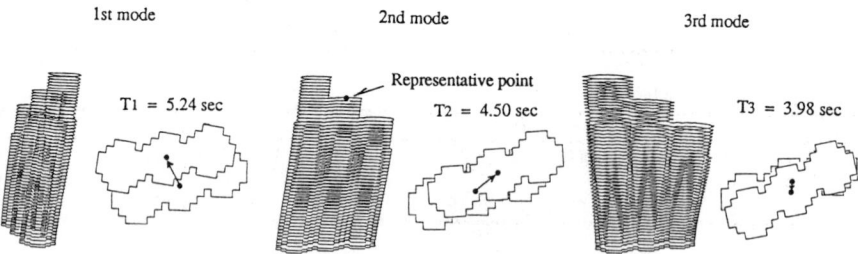

Fig.3 Modal Shape and Modal Period

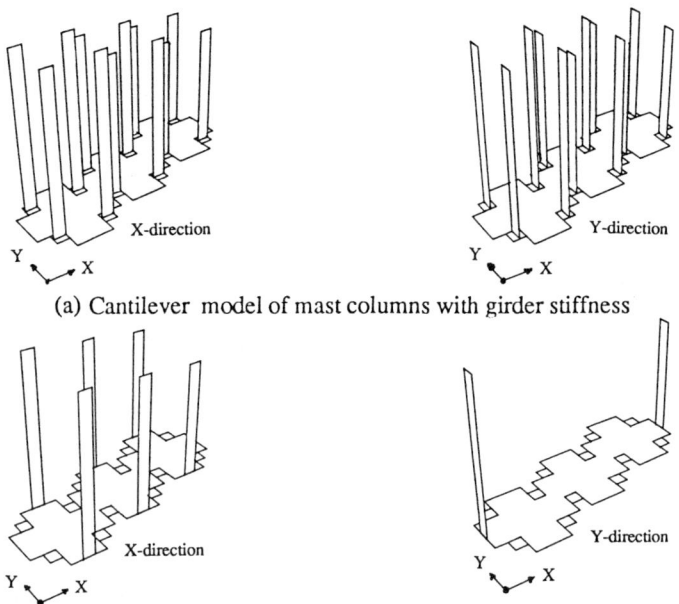

(a) Cantilever model of mast columns with girder stiffness

(b) Cantilever model of frame tube columns

Fig.4   3-Dimensional Simplified Elasto-plastic Analysis Model

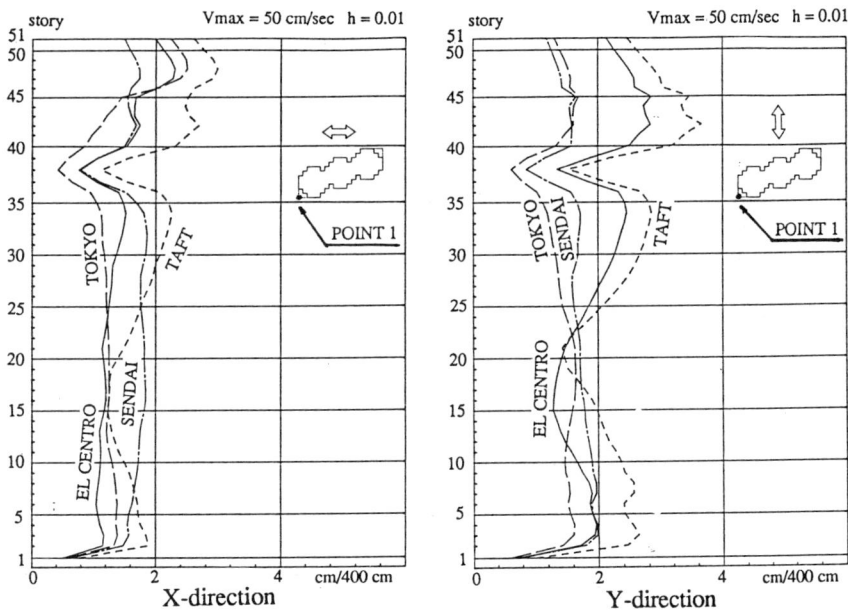

Fig.5   Maximum Response Story Drift (ground excitation 50 cm/sec)

EARTHQUAKE RESISTANT DESIGN

Earthquake resistant design is basically a feed-back process. The sizes of the structural members are selected after considering the scale, usage, and structural system of the building. Then dynamic numerical analyses for various ground excitations are executed to check the stress and deformation of each member so that its size can be modified to achieve the required rigidity, ductility and strength.

The design base shear coefficient was determined to be 0.055, the vertical distribution of shear forces was determined according to the Ai distribution pattern designated by the building standard law enforcement order of Japan ($C_B = 0.045$) and the additional lateral point load ($C_B = 0.010$) at the top of the model. The dynamic responses were checked for two cases: level 1 and level 2 according to the guideline of the Building Center of Japan. Level 1 corresponds to a probable big earthquake occurring during the expected life of the building, and level 2 corresponds to the largest earthquake that has ever occurred.

Level 1 : The maximum ground excitation level is 30 cm/sec at the ground floor. Every response story drift is less than 1/170 and the expected response stress of each member is less than the allowable material stress according to the building standard law enforcement order of Japan.

Level 2 : The maximum ground excitation level is 50 cm/sec at the ground floor. Some members may go into the plastic region. However, they should not deform beyond the excessive deformation criteria and also the expected story drift should not exceed 1/100.

The maximum ductility factor of the frame is 1.04 ~ 1.48 in x-direction, and less than 1.40 in Y-direction. The maximum response story drifts for the level 2 analysis are shown in Fig.5.

WIND RESISTANT DESIGN

The directions of wind gusts and reaction vectors on the structure are naturally expected to be related but not always identical. Indeed, the experimental wind tunnel study, which used a miniature structural model and measured the bending moments and shear forces at the base point, clarified the relation between the wind force and the reaction force vector directions. The design wind load was evaluated considering three components of wind forces : with wind, across wind and torsional (Fig.7).

The wind resistant design of this structure followed the "Recommendation of design load AIJ 1981" to determine the load criteria by considering the experimental results. A static analysis was carried out to check the stresses and deformations of the members to ensure structural safety under large wind forces. The wind speed that determines the design wind load is selected according to the expected return period. The design load is based on the following expected return period levels and criteria.

Level 1 : Expected return period is 100 years. The stresses in the structural members under this design load should be less than the allowable material stress. The maximum story drift should be less than 1/170, after considering the deformations induced by the torsional effect.

Level 2 : Expected return period is 500 years. The bending moments of the structural members determined by the linear static analysis according to this design load should be

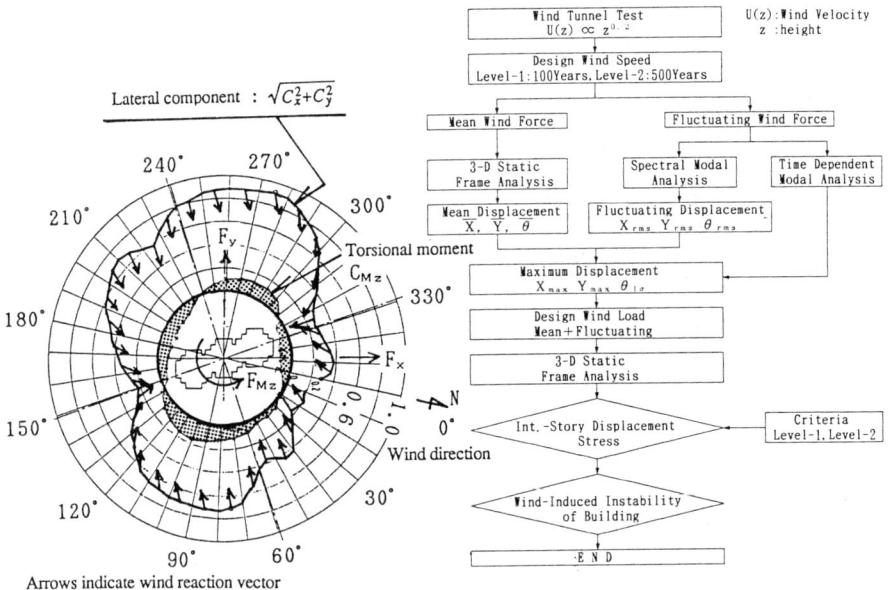

Lateral component : $\sqrt{C_x^2 + C_y^2}$

Arrows indicate wind reaction vector

Fig.6 Mean Wind Force Coefficient
and Mean Wind Direction

Wind Tunnel Test
$U(z) \propto z^{0.2}$

$U(z)$: Wind Velocity
z : height

Design Wind Speed
Level-1:100Years, Level-2:500Years

Mean Wind Force

Fluctuating Wind Force

3-D Static
Frame Analysis

Spectral Modal
Analysis

Time Dependent
Modal Analysis

Mean Displacement
$\overline{X}, \overline{Y}, \overline{\theta}$

Fluctuating Displacement
$X_{rms} Y_{rms} \theta_{rms}$

Maximum Displacement
$X_{max} Y_{max} \theta_{1\sigma}$

Design Wind Load
Mean + Fluctuating

3-D Static
Frame Analysis

Int.-Story Displacement
Stress

Criteria
Level-1, Level-2

Wind-Induced Instability
of Building

E N D

Fig.7 Flow Chart for Wind Resistant Design

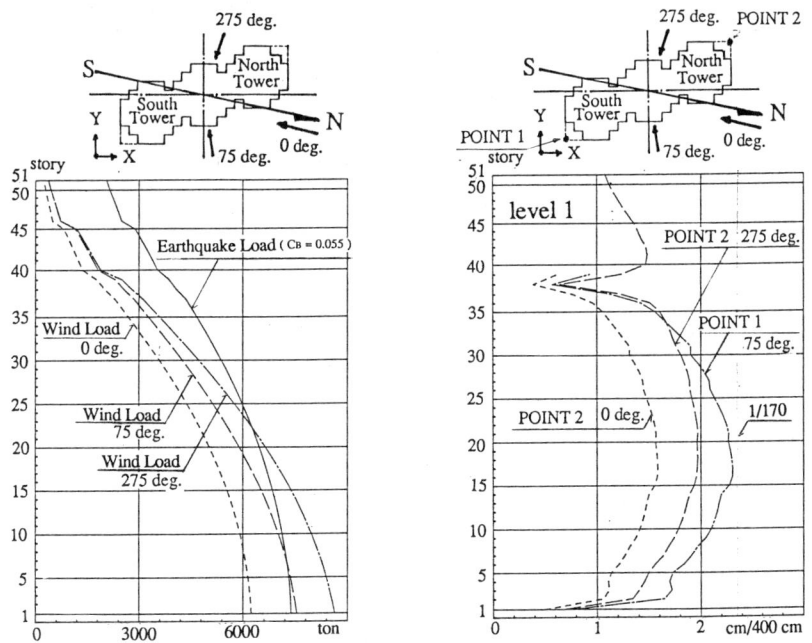

Fig.8 Design Loads

Fig.9 Maximum Story Drift (Wind Load)

less than the plastic moments at the members. The maximum story drift should be less than 1/100 after considering the deformations induced by the torsional effect. The resonant wind speed which causes the aerodynamic instability should be higher than the design wind speed.

Specifically, level 1 design load in the long span direction or X-direction was determined by the north wind or 0 degree wind, while the design load in the short span direction or Y-direction was determined by the 75 degree wind and 275 degree wind. Because of the different torsional effects according to the wind directions, two directions are considered for the short span direction design forces. Level 2 design load was determined by 0 degree, 65 degree, and 280 degree wind forces.

The base shear coefficient for level 1 and level 2 are 0.064 (275 degree wind) and 0.088 (280 degree), respectively. The design load distributions for the earthquake and level 1 wind forces are shown in Fig.8. The maximum story drift of the level 1 design wind forces are plotted in Fig.9 (0 degree wind, 75 degree wind and 275 degree wind).

VIBRATION CONTROL SYSTEM

It is understood from the previous data that the response vibration to wind disturbances in the short span direction of this building is more significant than that to the earthquake excitations. Although structural damage is hardly expected, even under such very severe wind forces the trembling response of the structure will not always be insignificant. Furthermore, because this tall building has hotel rooms and banquet rooms in the upper portion, the response motion in this area is a more sensitive problem than in other parts of the building. Therefore careful attention must be paid to this problem. To minimize the unpleasant tremors, it was determined to install vibration controllers on the 38th floor just above the belt truss structure. They are intentionally placed at an eccentric location away from the gravity center of the floor so that they are expected to reduce both lateral motion and torsional one simultaneously.

The design wind velocity for this system is 13.3 m/sec, which is the average value over a 10-minute wind blow with a return interval of 5 years at 10 meters above ground level. The aim is that nobody will not feel any unpleasant trembling. It is anticipated that the response vibration at the 52nd floor level will be attenuated by half. This vibration control system is also expected to function in earthquakes to reduce the fear of people inside.

SUMMARY

The structural configuration of the Shinjuku Park Tower is achieved by several unique features : *mast columns, frame tube structure,* and *belt truss structure.* These features restrict torsional deformation, upper portion lateral deformation, and excessive stress concentration on local frames caused by the building complex configuration. Additionally installed is a mass driver system that attenuates unpleasant vibration, thus enhancing the overall amenity of this giant structure.

REFERENCES

1.Katagiri,J.,Nakamura,O., (1990) "Method of spectral-modal analysis for wind-induced lateral-torsional motion of a tall building" Proceedings of 11th National Symposium on Wind Engineering, Tokyo, Japan.
2.Kobori,T.,Tsujimoto,T. (1990) "Evaluation of design wind load for a high rised building with stepped height" Proceedings of 11th National Symposium on Wind Engineering, Tokyo, Japan.

Proceedings of the Second Conference on Tall Buildings in Seismic Regions
55th Regional Conference
May 16 and 17, 1991, Los Angeles, California

Shaking Table Experiment and Practical Application of Active Variable Stiffness(AVS)System

Takuji Kobori[*]
Motoichi Takahashi, Tadashi Nasu,
Naoki Niwa, Narito Kurata,

Kobori Research Complex, Kajima Corporation
KI BLDG. 6-5-30 Akasaka, Minato-ku, Tokyo, 107, Japan

Junichi Hirai, Katsura Ogasawara

Kajima Institute of Construction Technology
2-19-1 Tobitakyu, Chofu-shi, Tokyo, 182, Japan

ABSTRACT

The authors have already proposed the Active Variable Stiffness (AVS) System. This is a seismic-response-controlled system. It is designed to reduce the response of buildings to earthquakes, by actively controlling their stiffness to prevent resonance with unpredictable earthquake motions.

In this paper, the results of the experiment using a shaking table, and their application to an actual building, are described. The experiment was carried out with AVS devices, and actual measuring and controlling setup, which were developed with the purpose of applying them to actual buildings. This system can efficiently respond to large earthquakes with minimal energy supply to control devices. A feed-forward control algorithm was used for a three-story steel model structure in this experiment.

The characteristics of the respective components and the effectiveness of this system were confirmed. By comparing the maximum response value between the controlled and uncontrolled cases, we showed that the controlling effect reduced the acceleration response of the model structure by 25-50%, without increasing the displacement response.

Furthermore, an application of this system to an actual three-story steel building, which is required to obtain the fundamental data, is shown. The building was completed in October 1990, and the complete AVS System was in full operation with observation of the earthquake records from early 1991.

1.Introduction

The authors have already proposed the Active Variable Stiffness (AVS) System. This is a seismic response-controlled system which is designed to reduce the response of buildings by actively controlling their stiffness to prevent resonance with unpredictable earthquake motions[1,2].

A simple variable stiffness seismic control system has been practically applied for the first time, and by subjecting a one-story steel model specimen to shaking table

----

*) Prof. Emeritus of Kyoto Univ., Executive Vice-President of Kajima Corp.

experiments[3], its effectiveness has been verified. However, a variable stiffness device aimed at conserving the electrical energy required to drive it has been newly developed, together with a measurement and control device that uses practical hardware. The shaking table experiment for this latest system has been conducted by using a 3-story, 2-frame steel structure and its applicability to actual buildings has been studied.

Also, a device has been developed, which can store data to be used in the practical application of the AVS system. The system was applied to a steel-frame 3-story building. This paper outlines its concept.

2. Method of Experiment

2.1 Test specimen frame with variable stiffness brace attached

The test specimen is a 3-story, 1-span, 2-frame steel structure as shown in Fig.1. On each story, an inverted V shaped brace was installed in the excitation direction. Variable stiffness devices were installed in series between the brace head and the beam, so that the variable mechanism of the brace could alter the specimen stiffness. The newly developed variable stiffness device comprises a two-rod-type hydraulic cylinder (cylinder lock device). A valve is installed in the connecting tube that joins the cylinder chambers at both sides of the rod. This value is opened or closed by an external signal (electric voltage 12V) which either allows oil to flow from one end of the tube to the other (unlocked condition) or stops oil from flowing (locked condition). Thus, it is possible to engage or disengage the joint in the brace. The time required for the change-over is only about 30msec.

2.2 Equipment for measuring and controlling

Fig.2 shows equipment for measuring and controlling. This equipment comprises sensors for sensing motions, analyzers for analyzing input motion and computers for judging and instructing. The computers comprise a plural system for simultaneously instructing many elements. These hardware parts must be designed to fit actual buildings. Therefore, equipment with confirmed characteristics can be utilized in real buildings.

2.3 Control method

There are a total of eight combinations of connecting conditions in the variable stiffness braces installed at various floors. For the experiment herein, three typical combinations were selected : Type 1 (all floors unlocked), Type 2 (only first floor locked) and Type 3 (all floors locked). These were selected because their frequency change due to stiffness alteration can be presented rather clearly. The initial stiffness condition of control was set at Type 3.

The method of control was the earthquake motion analyzer feed-forward control[3] as shown in Fig.3. Non-resonance was realized as follows. 1.Input acceleration to the specimen was analyzed in real time by the earthquake motion analyzer. 2. The stiffness type with small resonance was selected. 3. The connecting condition of the variable stiffness brace was changed over to produce less resonance, 4.The stiffness condition of the specimen was altered to the desired state.

The earthquake motion analyzer was composed of plural band-pass filters complying with the passing nature of the response transmission characteristics according to the various stiffness types. From the various filter outputs of incident seismic waves, the responses of the uncontrolled stiffness types were evaluated, and by instant comparison of the frequencies the stiffness type with least resonance was selected.

2.4 Method of excitation

The experiment was carried out with an in-plane horizontal shaking of the variable stiffness brace. The input motions applied were the swept-sine wave(frequency 8Hz ~ 0.8Hz), the Mexico Wave (1985 CDAO-NS) with the time axis scaled down to 32% so that the effect of the variable stiffness control could be clearly confirmed, and the Taft wave (1952 EW). The time histories of the input motions are shown in Fig. 4.

3. Experimental Results

3.1 Vibration characteristics of the specimen

To establish the filter passing characteristics of the earthquake motion analyzer and to identify the vibration characteristics of the specimen, a swept-sine wave excitation experiment was conducted with various stiffness types and the vibration characteristics at the uncontrolled stage were examined. The response acceleration transfer function at the top of the specimen is shown in Fig.5.

3.2 Control experiment results

(1) Time history of response at specimen top

Figs.6 (a) ~ (c) show the time histories of the uncontrolled response accelerations and displacements at the top of the specimen compared with the controlled values, based on earthquake motion analysis in real time. The figure also shows the stiffness condition in the controlled state.

Swept-sine wave input All of the stiffness types in the uncontrolled state enter resonance in response to the predominant frequency of the input motions, and parts with increasing responses can be seen. In the controlled state, the stiffness is changed twice to avoid resonance. Thus, the response is reduced significantly.

Mexico wave input The various stiffness types in the uncontrolled state have generally large acceleration and displacement responses. However, in the controlled state, these large responses are skillfully avoided by altering the stiffness type a total of 9 times and the responses are significantly reduced.

Taft wave input In the uncontrolled state, the acceleration response of stiffness Type 3 is generally large. However, the difference in displacement response among various stiffness types are relatively small. Thus, the effect of control by stiffness change cannot be seen as clearly as with the above two waves.

(2) Comparison of maximum response

Figs.7 (a) and (b) show the maximum acceleration and displacement responses of the specimen at each floor in various uncontrolled states and the controlled state. With the maximum response acceleration in the uncontrolled state stiffness Type 3 shows a large value. However, in the controlled state this is reduced by approximately 1/4 to 1/2. Moreover, with the maximum response displacement in the uncontrolled state, stiffness Type 1 shows the largest value. However, in the controlled state, for the swept-sine wave and the Mexico wave these are reduced by approximately 1/4 to

2/3, but for the Taft wave no noticeable decrease was seen. These figures indicate that the lower the stiffness the smaller the acceleration response, and in such cases the displacement response tends to be large. However, attention is drawn to the fact that with the control, the acceleration response is reduced without increasing the displacement response.

(3) Input energy of earthquake motion

$$\int_0^T \{\dot{y}\}^T [M]\{\ddot{y}\} dt + \int_0^T \{\dot{y}\}^T [C]\{\dot{y}\} dt + \int_0^T \{\dot{y}\}^T [K]\{y\} dt = -\ddot{y}_0 \int_0^T \{\dot{y}\}^T [M]\{1\} dt$$

Table 1 shows the value of the total seismic input energy as defined on the right side of the above equation. Also, Fig.8 shows the time history of input energy to the specimen for the Mexico wave excitation. When the passage of time of the predominant frequency change is clear, such as with the Mexico Earthquake, in the controlled state, the input energy at various points of time is at a minimum. Thus, it can be clearly seen that both acceleration and displacement responses are greatly reduced in the controlled state.

(4) Acceleration response spectrum

Fig.9 shows the acceleration response spectra at the top of the specimen with the Mexico wave input. This figure indicates that the peaks of the spectrum in the uncontrolled state are decreased to a value close to the lower limit of the uncontrolled state, by executing the controlled state. It can also be seen that non-resonance is possible in the frequency domain.

4. Summary of Experiment

The newly developed variable stiffness device was installed in a 3-story specimen. Thus, for the very first time, a variable stiffness seismic control system for a multi-layer structure was demonstrated using a hardware consisting of measurement and control devices . The real time feed-forward algorithm with earthquake motion analyzer that reflects the characteristics of the various system components was developed and the experiments on the variable stiffness control were conducted using a shaking table. The obtained results are described below.

1) It was verified that variable stiffness system could be realized by applying the newly-developed cylinder-lock-type variable-stiffness device. This device has an excellent response and power conserving function and can conduct high-speed lock and unlock change-over with only 12V - 20W of electrical power.

2) To heighten the reliability of the entire system, the measurement and control device for the processes were diversified and experiments were conducted by coordinating in real time (for the first time) a plural number of control computers. Consequently, the applicability of this measurement and control device to an actual building was determined.

3) When the control effect is evaluated from the maximum response value, it is possible to reduce the response acceleration when placed in the controlled state, by about 1/4 to 1/2 without increase in displacement response even compared with Type 3, which has the highest stiffness (uncontrolled).

4) For earthquake waves, the more clearly defined is the transition of the predominant frequency (i.e. Mexico Earthquake), the easier it is to realize non-resonance with this control system, and the greater the reduction in response.

5. Application to Actual Buildings.

To amass the various data which is necessary to develop the AVS system for practical use, an trial application of this system to an actual building was performed. An example of the application is introduced below.

5.1 Outline of building

The AVS system was installed in the control building for a large-scale shaking table facility, newly constructed within the compound of the Kajima Institute of Construction Technology (KICT) of the Kajima Corporation, Tokyo. This building was completed in November 1990. It has a structural steel framing and contains a control room and several research rooms. The building has a total floor area of 440m². It is 7m in wide (1 span), 21m (3 span) long, and 12m high, and comprises 3 stories. Fig.10 shows the floor plan and the framing elevation where the variable stiffness devices are installed. The entire shaking table building is shown in Photo 1. The shaking table building, built adjacent to the control building on the same foundation, is structurally independent.

5.2 Outline of system

To control the building in the width direction, a variable stiffness device was installed on each floor of both end frames at the joint of the brace head and beam (Photo 2).

The control system (Fig.11) is composed of [1] A measurement sensor for detecting earthquake ground motions, [2] A earthquake motion analyzer for determining the earthquake characteristics, [3] A control computer for selecting the building's stiffness, [4] A variable stiffness device for changing the building's stiffness, and [5] A non-suspension electrical power source device that can continue the control function even when the public power service is interrupted.

5.3 Special features

The AVS system possesses special features for practical application to buildings as follows:

[1] The variable stiffness device can operate at high speed. It takes 5/100sec to change from 0 to 70-ton locking force and only requires 20W of electrical power per unit. This enables a great energy conservation compared with conventional active seismic control systems.

[2] An emergency non-suspension electrical power source device that can supply power to the building for 30 minutes is provided. Thus, if the public power supply is interrupted, it is possible to continue the stable operation of the control computer and the variable stiffness system.

[3] The variable stiffness device is small enough to be installed inside a ceiling. Therefore, no special area need be provided for its installation.

[4] An operation monitoring function is provided to safe-guard the system.

5.4 Simulation

Ground motions of the Izu-Oshima Kinkai Earthquake of October 14, 1989, were recorded within the KICT site. These seismic waves were magnified to 250Gal and a simulation analysis was conducted based on the assumption of a large earthquake. The results, shown in Fig.12, indicate that by changing the stiffness of the building according to the seismic motion characteristics, the response of the building can be reduced significantly even if the earthquake is very severe.

Henceforth, earthquake observations will be continued to confirm the actual seismic control effect. In this way, various measured data, such as vibration of the building during an earthquake and movement of the system, can be collected and analyzed. Further improvements to the AVS system described herein are planned.

REFERENCES
1) T.Kobori, and et al.: New Development to Seismic Resistant Structure, Approach to Dynamic Intelligent Building. AIJ Conference Lecture Summary, August 1986.
2) T.Kobori, and et al.: Trial to Dynamic Intelligent Building - DIB Possessing Variable Stiffness Function, AIJ Conference Lecture Summary, August 1986.
3) T.Kobori, and et al.: Experimental Study on Active Variable Stiffness System - Active Seismic Response Controlled Structure -, FOURTH WORLD CONGRESS OF COUNCIL ON TALL BUILDINGS AND URBAN HABITAT, November 1990.

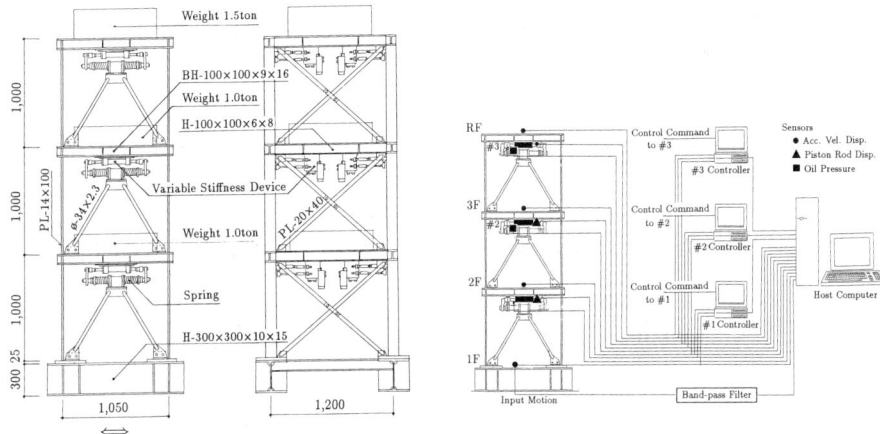

Fig.1 Outline of Test Specimen

Fig.2 Measurement and Control Setup

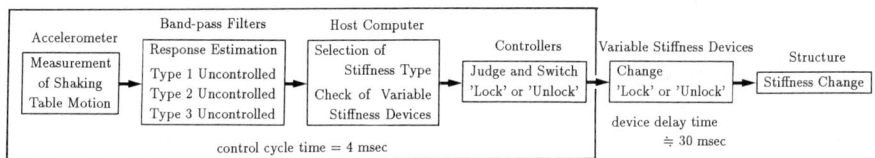

Fig.3 Feedforward Control Scheme

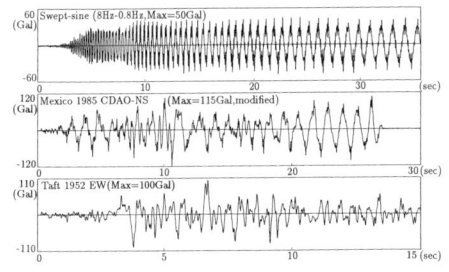

Fig.4 Input Motion

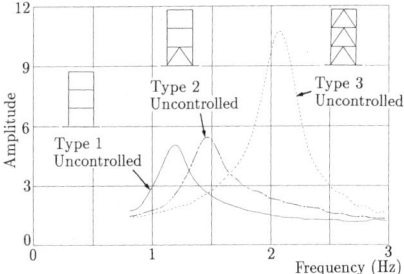

Fig.5 Acceleration Transfer Function

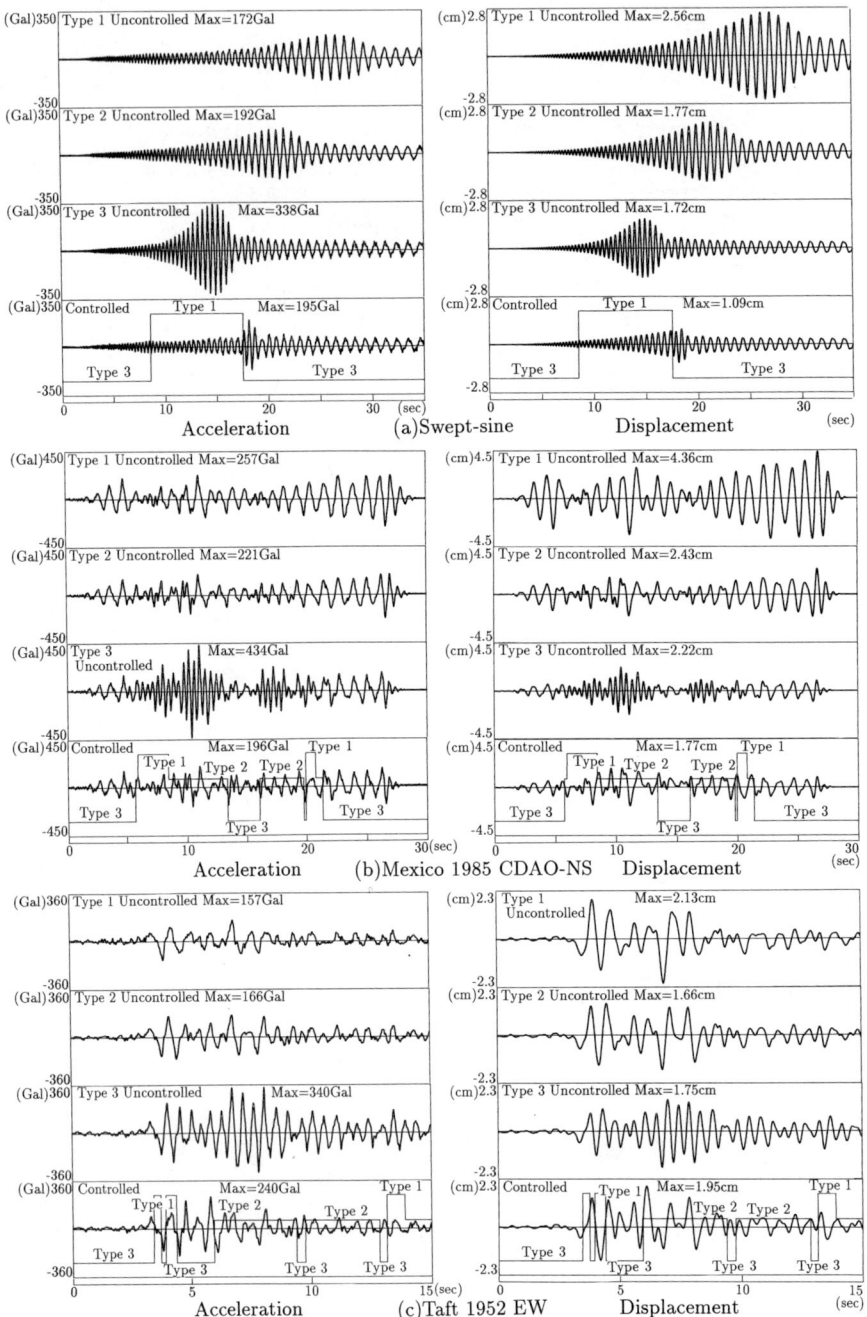

(Gal)350 Type 1 Uncontrolled Max=172Gal
-350
(Gal)350 Type 2 Uncontrolled Max=192Gal
-350
(Gal)350 Type 3 Uncontrolled   Max=338Gal
-350
(Gal)350 Controlled   Type 1   Max=195Gal
Type 3   Type 3
-350
0   10   20   30   (sec)
Acceleration   (a)Swept-sine

(cm)2.8 Type 1 Uncontrolled Max=2.56cm
-2.8
(cm)2.8 Type 2 Uncontrolled Max=1.77cm
-2.8
(cm)2.8 Type 3 Uncontrolled Max=1.72cm
-2.8
(cm)2.8 Controlled   Type 1   Max=1.09cm
Type 3   Type 3
-2.8
0   10   20   30   (sec)
Displacement

(Gal)450 Type 1 Uncontrolled Max=257Gal
-450
(Gal)450 Type 2 Uncontrolled Max=221Gal
-450
(Gal)450 Type 3 Uncontrolled   Max=434Gal
-450
(Gal)450 Controlled   Max=196Gal   Type 1
Type 1   Type 2   Type 2
Type 3   Type 3   Type 3
-450
0   10   20   30(sec)
Acceleration   (b)Mexico 1985 CDAO-NS

(cm)4.5 Type 1 Uncontrolled Max=4.36cm
-4.5
(cm)4.5 Type 2 Uncontrolled Max=2.43cm
-4.5
(cm)4.5 Type 3 Uncontrolled Max=2.22cm
-4.5
(cm)4.5 Controlled   Max=1.77cm   Type 1
Type 1   Type 2   Type 2
Type 3   Type 3   Type 3
-4.5
0   10   20   30   (sec)
Displacement

(Gal)360 Type 1 Uncontrolled Max=157Gal
-360
(Gal)360 Type 2 Uncontrolled Max=166Gal
-360
(Gal)360 Type 3 Uncontrolled   Max=340Gal
-360
(Gal)360 Controlled   Max=240Gal   Type 1
Type 2   Type 2   Type 2
Type 3   Type 3   Type 3   Type 3
-360
0   5   10   15(sec)
Acceleration   (c)Taft 1952 EW

(cm)2.3 Type 1 Uncontrolled Max=2.13cm
-2.3
(cm)2.3 Type 2 Uncontrolled Max=1.66cm
-2.3
(cm)2.3 Type 3 Uncontrolled Max=1.75cm
-2.3
(cm)2.3 Controlled   Type 1   Max=1.95cm   Type 1
Type 2   Type 2
Type 3   Type 3   Type 3
-2.3
0   5   10   15   (sec)
Displacement

Fig.6 Response Time History at the Top of Test Specimen

220

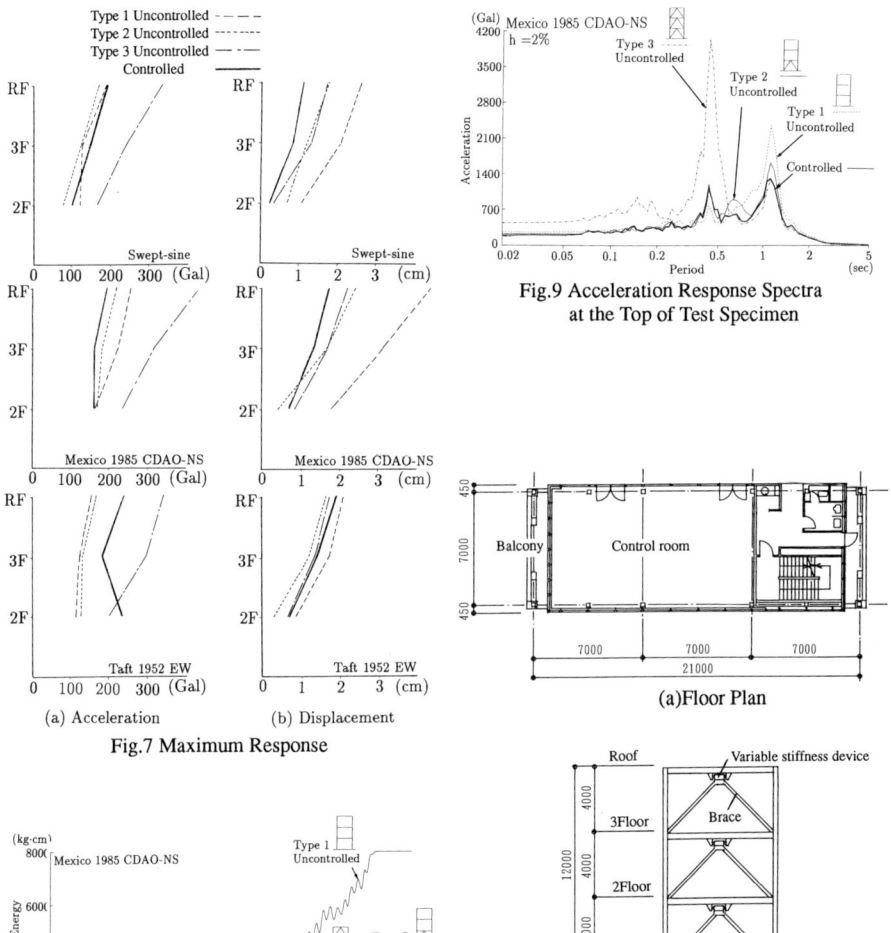

Type 1 Uncontrolled — —
Type 2 Uncontrolled -------
Type 3 Uncontrolled — · —
Controlled ———

RF 3F 2F
Swept-sine
0 100 200 300 (Gal)

RF 3F 2F
Swept-sine
0 1 2 3 (cm)

RF 3F 2F
Mexico 1985 CDAO-NS
0 100 200 300 (Gal)

RF 3F 2F
Mexico 1985 CDAO-NS
0 1 2 3 (cm)

RF 3F 2F
Taft 1952 EW
0 100 200 300 (Gal)

RF 3F 2F
Taft 1952 EW
0 1 2 3 (cm)

(a) Acceleration   (b) Displacement

Fig.7 Maximum Response

(Gal)
4200 Mexico 1985 CDAO-NS
      h =2%
3500
2800   Type 3
              Uncontrolled
2100
              Type 2
              Uncontrolled
1400   Type 1
                  Uncontrolled
700   Controlled ———
0
0.02 0.05 0.1 0.2 0.5 1 2 5
              Period   (sec)

Acceleration

Fig.9 Acceleration Response Spectra
at the Top of Test Specimen

(kg·cm)
8000 Mexico 1985 CDAO-NS
                              Type 1
                              Uncontrolled
6000
                          Type 2
                          Uncontrolled
4000   Type 3
          Uncontrolled
2000
                          Controlled
0   10   20   30
                      (sec)

Seismic Input Energy

Fig.8 Time History of Seismic Input Energy

450 7000 450
Balcony   Control room
7000 7000 7000
21000

(a)Floor Plan

Roof   Variable stiffness device
3Floor   Brace
2Floor
12000
1Floor
4000 4000 4000
7000

(b)Framing Elevation

Fig.10 Outline of the Building

Table.1 Total Seismic Input Energy

		Swept-sine	Mexico	Taft
Un-controlled	Type 1	6108	7955	1541
	Type 2	3652	3117	1474
	Type 3	2418	2932	1812
Controlled		1383	2177	1514

221

Photo.1 Entire view of the Shaking Table Building

Photo.2 Installation
of the Variable Stiffness Device

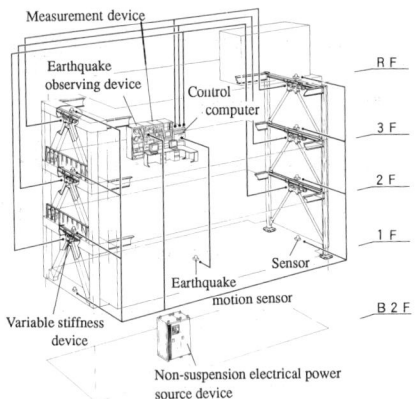

Fig.11 Composition of the Control System

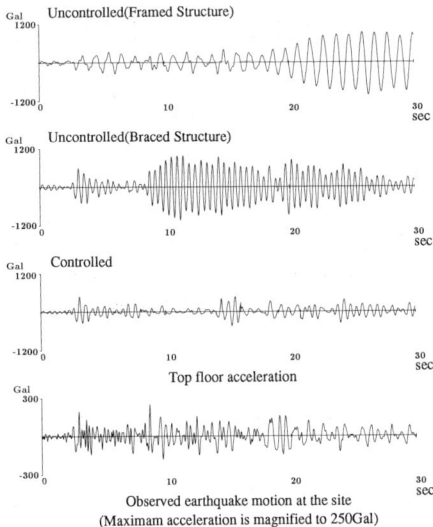

Fig.12 Simulated Control Effect under Izu-Oshima
Kinkai Earthquake on Oct. 14, 1989

Proceedings of the Second Conference on Tall Buildings in Seismic Regions
55th Regional Conference
May 16 and 17, 1991, Los Angeles, California

Structural Design and Response Control
on Dynamic Intelligent Building 200

Takuji KOBORI[1],   Toshihiko KUBOTA[2],
Norihide KOSHIKA[3],   Kazuhiko YAMADA[4]

Kobori Research Complex, Kajima Corp.
KI Bldg., 6-5-30 Akasaka, Minato-ku, Tokyo 107, Japan

ABSTRACT

Super-high-rise buildings will play an increasingly important role in enhancing the social environment, especially in large cities such as Tokyo where soaring real estate prices demand effective land utilization. Society in the near future will become increasingly dependent on information. Therefore, these buildings will be required to possess much better facilities for maintaining information processing functions in disasters. They should also provide a more comfortable working environment with improved amenities.

The Dynamic Intelligent Building (DIB) promoted here utilizes a structural system based on a new concept called the 'response-controlled structure for earthquakes and typhoons'. Buildings utilizing this concept spontaneously react to external excitations as if animate, thus canceling them out. A small-scale building based on this concept was realized in August, 1989 and its effectiveness in reducing building sway during several earthquakes and typhoons has been confirmed.

The DIB-200 plan proposed here is a 200-story super-high-rise building utilizing the DIB concept. This plan integrates twelve 50-story cylindrical units. However, other combinations of units can be utilized, depending on the situation. In design, response control systems for earthquakes and typhoons are indispensable in ensuring high-quality performance.

This paper introduces the concept of structural design and the role of the vibration response control system.

1.INTRODUCTION

Many years have passed since the appearance of the first 100-story high-rise building. In Japan, a 70-story high-rise building will soon be constructed in Yokohama bay. This follows Sunshine-60 and the Tokyo City Hall Tower. It has already been confirmed, 13 years ago (in 1978), that a 100-story high-rise building would be feasible not only technologically but economically, even in Japan where earthquakes and typhoons often occur.

In the past ten years, people, facilities and information have become increasingly concentrated in Tokyo. This has led to severe urban problems such as a shortage of land causing soaring land prices. One of many proposals to solve these problems is to construct super-high-rise buildings, since they utilize land very efficiently. This is the present challenge for the engineers. These buildings will express both symbolic and monumental aspects of modern technology. To adequately reflect these factors, it will be necessary to conduct basic investigations to realize super-high-rise buildings.

----

1) Prof.Emeritus of Kyoto University, Executive Vice-President of Kajima Corp.
2) Assistant Manager
3) Senior Research Engineer
4) Research Engineer

In our highly information-based society, the rapid development of electronic equipment has enabled information of high value to be accumulated. If this accumulated information were to be destroyed by a disaster, there is a serious danger that the loss will never be recovered. Therefore, the function of buildings in protecting such information should be of higher quality than at present. Buildings are now designed simply to avoid complete destruction and protect human life. To guarantee the stable of communication and accumulation of information, super-high-rise buildings will play a key role since they can accommodate a whole society in a self-contained space. Furthermore, a high level of amenities and safety will be demanded for office activity and living comfort.

Based on these requirements, we aim here to construct a super-high-rise building in a modern city. The design and engineering investigated here is based on the assumption of a 200-story, 800m-high building, but it can be altered to apply to a 50, 100 or 150-story building as required.

The most fundamental point in this plan is to determine whether we can work and live comfortably in such a building, regardless of the safety in a disaster. Although there is much experience with buildings in the 100-story class, there are no answers yet to several problems posed by taller buildings. However, we handle this problem positively by promoting our concept of a 'Dynamic Intelligent Building' which actively copes with earthquakes and typhoons to restrain vibrations and sway.

Our design, structure and response-control method are introduced below.

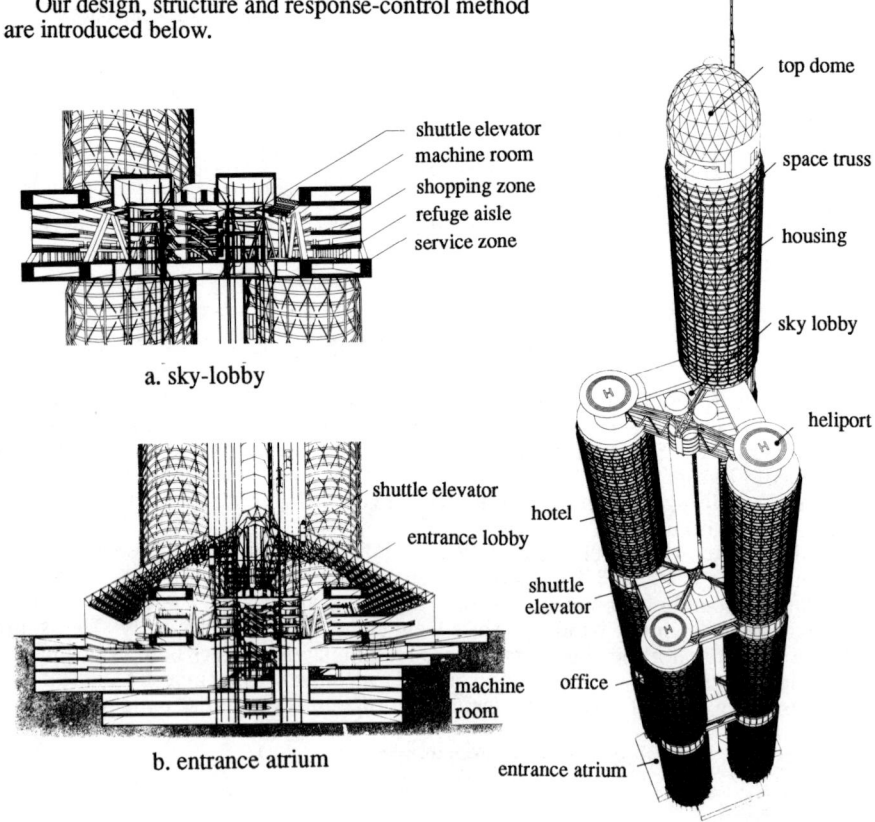

a. sky-lobby

b. entrance atrium

c. perspective view

Fig.1 Design outline

2.DESIGN

As representative of architecture in the coming century, super-high-rise buildings will be required to upgrade living comfort, health and safety. The following basic points must be settled.

High density : For high land utilization, it is assumed that the ratio of volume to ground area is 2000%. Thus, for a total volume of 1,350,000m^2, 260m-square of ground area is necessary. The ratio of construction area to ground area is about 32%. Thus, open green area occupies about 70% of the total ground area.

Diversity : Since several units can be set in three-dimensional space, various functions can be accommodated such as office, hotel and living.

Combination : This is the most unique characteristic of our design (Fig.1). Twelve cylindrical units 50m in diameter, 2000m^2 in area per floor and 50 stories in height, are combined into a super-high-rise building, comprising four long cylinders. The units are connected at sky-lobby floors. One cylinder comprises 200 stories, two comprise 150 stories and the other comprises 100 stories. The sky-lobbies are where shuttle elevators stop and are also where the cylinders are interconnected, not only for access but also for stiffening the structure as a whole. In addition, a 40m-deep foundation structure is necessary to support the upper structure. This combination can be altered depending on the required scale.

Plan shape and view from outside : A cylindrical system was chosen to adequately express the design concept and to easily accommodate several functions such as office, hotel and living. The structural body is directly seen from outside, and its massiveness is moderated by a diversified appearance. Furthermore, the shape of the building's outer profile reduces residents fear through the provision of indirect outer space.

Equipment : To minimize the influence of the infra-equipment present, self-demand-supply systems are utilized where possible. The systems include co-generation systems and a rain and used water system. In addition, a self-back-up and a co-back-up system are introduced to ensure reliability at a high elevation.

The designed super-high-rise building has following merits:
a) People can move very freely. Everyone can easily understand where he is in a huge space.
b) The outer facing area is large, thus providing good lighting and spectacular views.
c) Various functions are included in the sky garden and the space ground.
d) Refuge gardens are located on each roof and refuge aisles between units are located in each sky-lobby, offering safety and reliability, by separating the units.
e) The influence of earthquake and wind loads is reduced by the combination of the different heights of the cylinders and the circular shape of the units.
f) The image of a huge mass is softened because the view from outside alters depending on the viewer's position. It will be readily accepted in its environment.

3.STRUCTURE

Design criteria : With the recent rapid development of an information-based society, higher quality buildings will be demanded in the near future. These buildings will require increasingly complex information functions. Housing will also need to provide greater comfort because the individual living will become more luxurious. Therefore, stricter criteria will be assigned in the design of super-high-rise buildings than presently exists for high-rise buildings as shown in Table 1.

Framing : Super-framing is introduced for structural design (Fig.2). Four cylinders are connected at each 50-story level by large truss girders which stiffen the structure to restrain deformations. Cubic truss elements surrounding the outer columns of each cylinder also maintain high stiffness reducing residents' fear. Wind streams into windows on the high floors are also reduced because of the building's profile.

Material : High-strength materials are adopted. Columns consisting of cylindrical units are made of confined-steel pipe reinforced concrete. The steel frame is comprised of large steel girders.

Construction : It is expected to take about seven years to construct such a building. If this is achieved, the building will meet social requirements. Moreover, it is not necessary to wait seven years before the building is occupied because each 50-story unit can be utilized on its completion.

Table.1 Load and criteria for design

level		I	II	III
period		frequent	possible	rare
design criteria	structure	linear	linear	linear
	living comfort	comfortable (≤ 10cm/s^2)	endurable	no panic
	equipment	in general	in general	maintain function
	response controlling	normal-system working	event-system working	final-system working
load	wind	V_S=27m/s	V_S=70m/s	V_S=80m/s
	earthquake	small	large	severe
*ordinal design criteria	structure	linear	linear	nonlinear, not destroyed

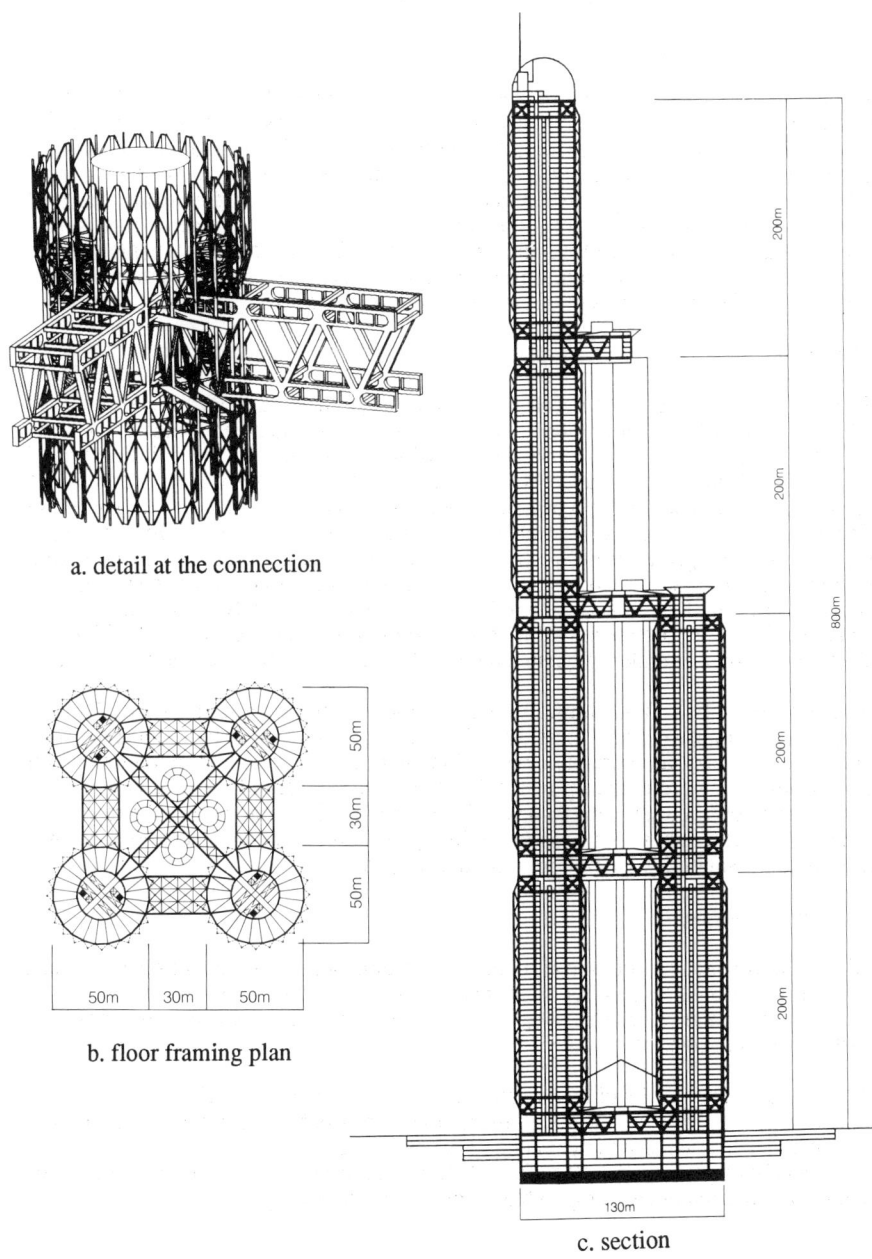

a. detail at the connection

b. floor framing plan

c. section

Fig.2 Structure of the building

227

4.RESPONSE CONTROL

(1) Installation of systems

Response control systems are installed to restrain sway and vibrations of buildings subjected to earthquakes and such strong winds as typhoons. Passive systems, for example damping elements, are first attached to structural elements to add a damping factor to the whole structure. Next, active systems are installed to cope with responses which cannot be reduced by passive systems. These active systems are mainly designed for frequent earthquakes and winds, although they are partially effective in rare, strong earthquakes and winds. Therefore, not only is external power saved by using passive systems but highly effective reduction can also be expected by introducing active systems. Fig.3 shows the concept of various response control systems.

(2) Disturbance

Wind load : The building comprises such slender cylinders that dynamic responses to wind loads cannot be ignored. To measure dynamically changing wind force, wind tunnel tests were conducted. The attained wind forces around the building model were converted into wind loads (Fig.4), and dynamic analysis was executed. The experimental results show that vortexes especially affect the top 200m of the 800m cylinder, because it extends beyond the other parts. In the lower 600m, the wind streams around the group of cylinders. Thus, vortexes do not occur and fluctuating wind forces are reduced. However, since the average wind force is strong, the connection of each element and each unit should be rigid. For dynamic analyses, the average wind load velocities at 600m above ground level are assumed to be 27m/s to confirm living comfort under frequent events and 70m/s for a 100-year returned period.

Seismic load : The first natural period of the building is about 10 sec. Design earthquakes for high-rise buildings, based on earthquake records by the usual strong motion accelerometer, are unreliable over a long period. Therefore, design motions for large earthquakes were made up of a sub-experimental method. With this method a hypothetical record of a large earthquake, Sakata is created by integrating the records of several small earthquakes (Fig.5b). In addition, measurement by reliable equipment is also adapted for small design motions, Koutou (Fig.5a).

(3) Response analysis

Analytical model : An analytical model was established for wind and seismic loads, as shown in Fig.6. The passive response control systems are assumed to add 4% damping factor to the whole structure. The Active Mass Driver system were chosen as the active response control system.

Eigen value : The results of eigen value analyses are shown in Fig.7.
Wind response: Fig.8 and Fig.9 show response results for wind loads which input into each node. The wind was assumed to flow in Y-direction.
Earthquake response : Fig.10 to Fig.12 show response results for earthquake loads which input into each node along both directions simultaneously.

(4) Control effect

The resulting figures without the response control systems infer the possibility of damage to sensitive equipment and offenses to living comfort. However, it is confirmed that the response control systems restrain accelerations and maintain reliability and safety, thus satisfying the design criteria.

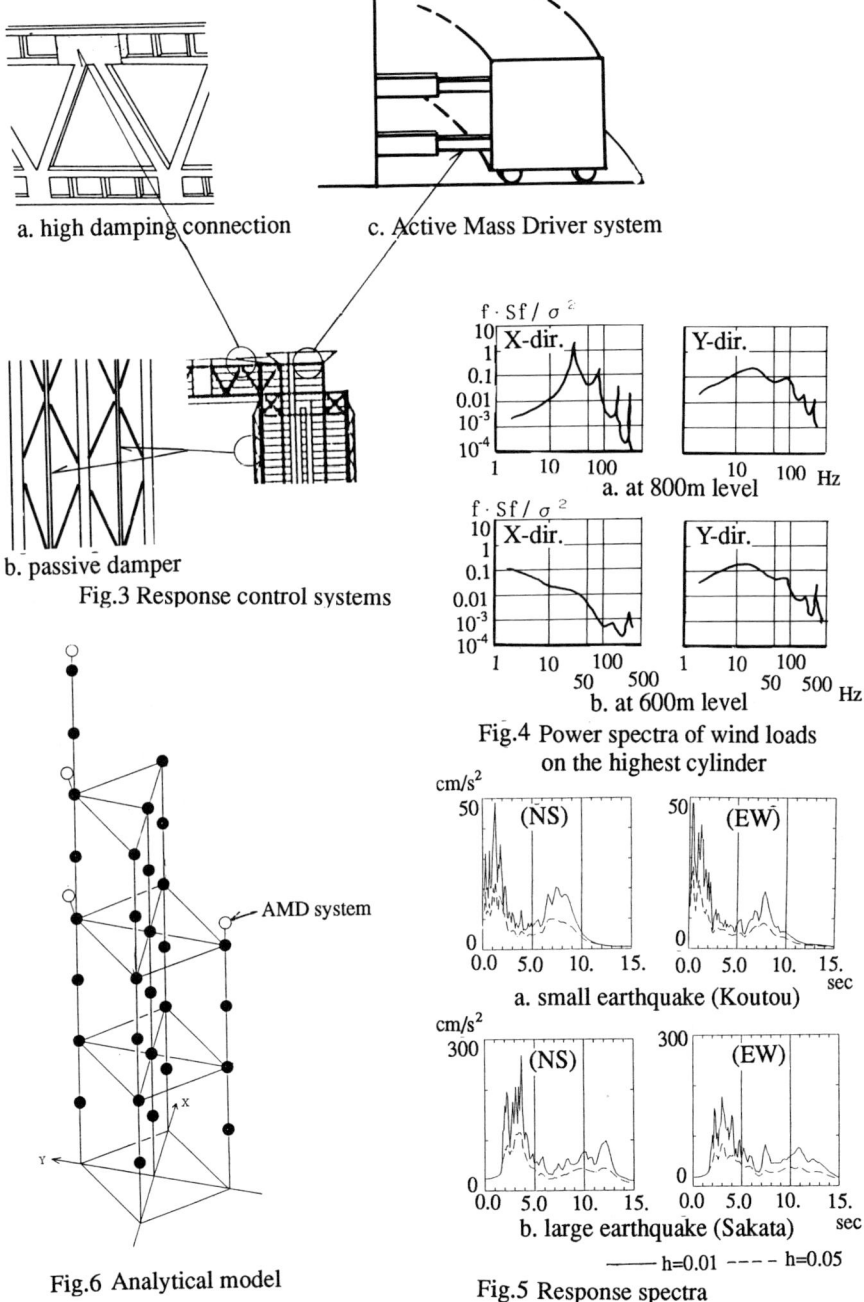

a. high damping connection

c. Active Mass Driver system

b. passive damper

Fig.3 Response control systems

$$f \cdot Sf / \sigma^2$$

X-dir.   Y-dir.

a. at 800m level

$$f \cdot Sf / \sigma^2$$

X-dir.   Y-dir.

b. at 600m level

Fig.4 Power spectra of wind loads
on the highest cylinder

cm/s^2

(NS)   (EW)

a. small earthquake (Koutou)

cm/s^2

(NS)   (EW)

b. large earthquake (Sakata)

——— h=0.01 ---- h=0.05

Fig.5 Response spectra
of design earthquakes

AMD system

Fig.6 Analytical model

229

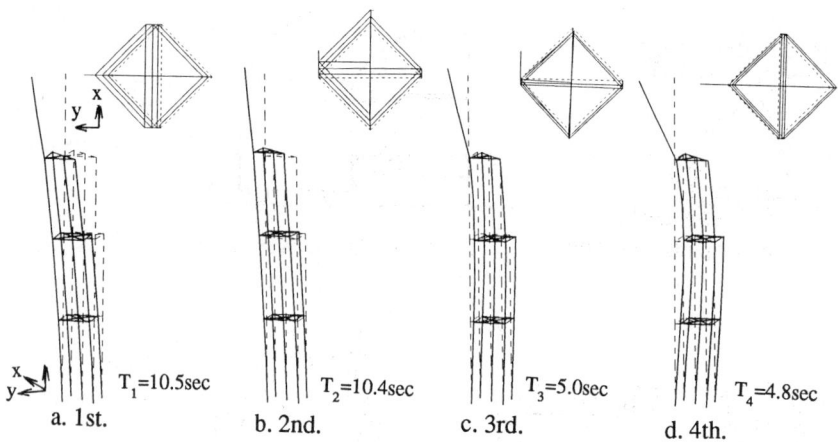

a. 1st. T_1=10.5sec   b. 2nd. T_2=10.4sec   c. 3rd. T_3=5.0sec   d. 4th. T_4=4.8sec

Fig.7 Mode shapes

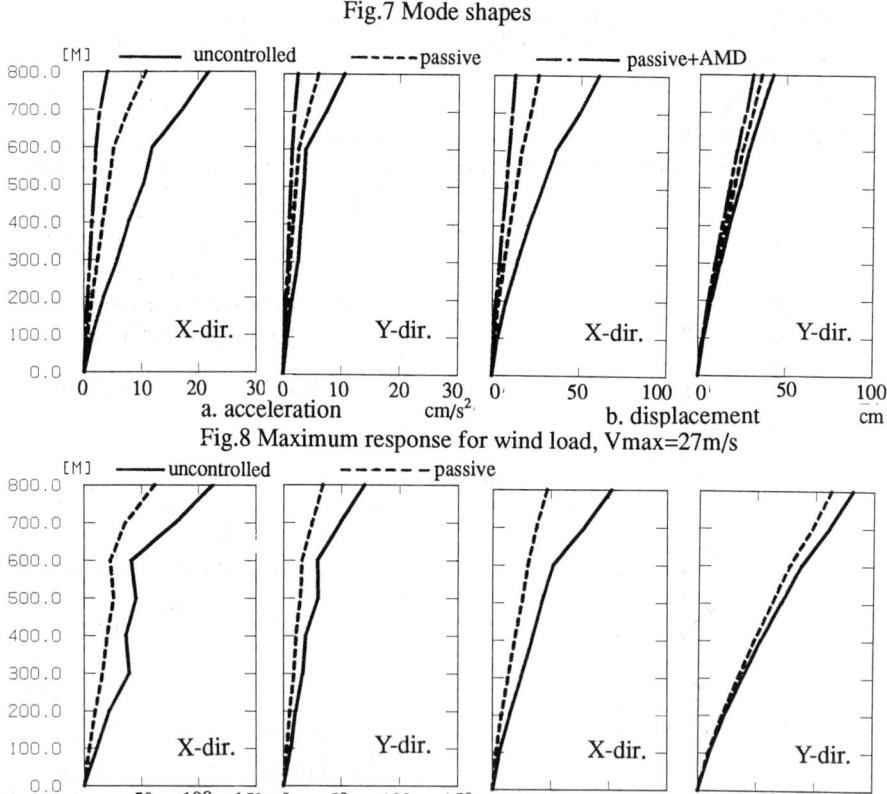

Fig.8 Maximum response for wind load, Vmax=27m/s

Fig.9 Maximum response for wind load, Vmax=70m/s

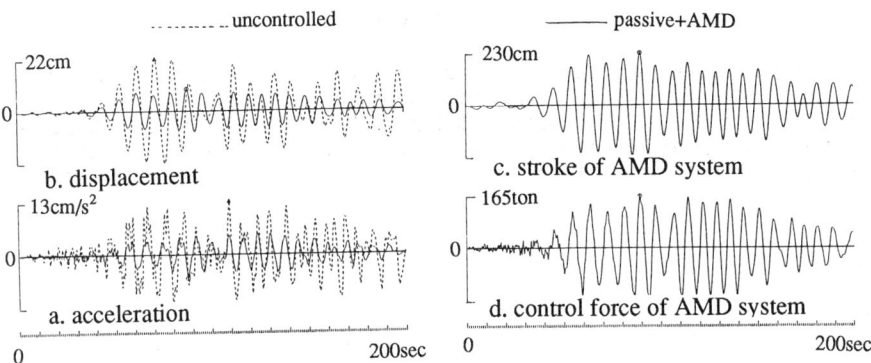

Fig.10 Response time history at 800m level for small earthquake, Koutou

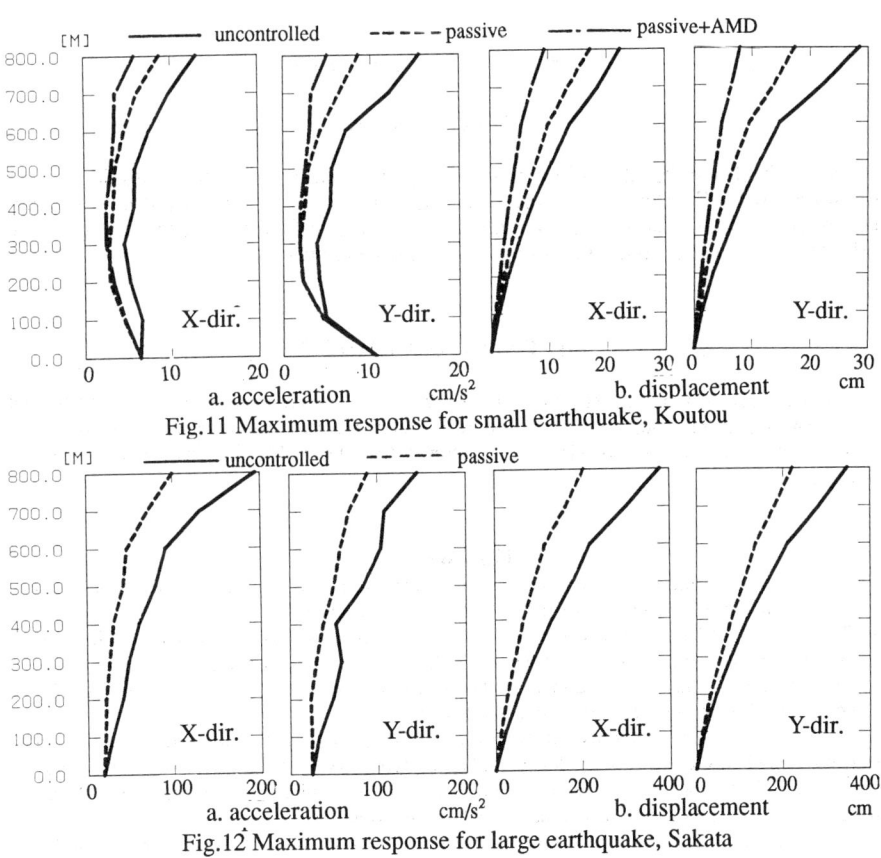

Fig.11 Maximum response for small earthquake, Koutou

Fig.12 Maximum response for large earthquake, Sakata

5. VARIATION

Our investigation was basically conducted for a 200-story building. However it can be easily modified to meet the requirements for different combinations of plural units. Fig 13. shows variations, both expanded and reduced. In addition, each 50-story unit can be made available after its completion, even if the others are still under construction.

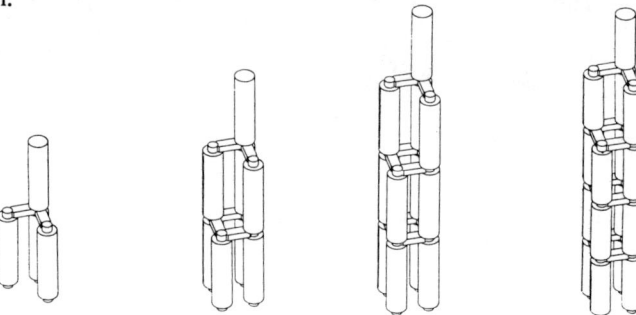

Fig.13 Variations of design

6. CONCLUSION

Super-high-rise buildings in a modern city are rational both technologically and economically. Furthermore, they have various merits: economical use of land, combination of diverse functions and provision of comfortable space. They will be effective in solving such urban problems as soaring land prices.

ACKNOWLEDGEMENT

The main design of this project, was conducted by Sadaaki Masuda and A.Scott Howe ; arrangement of equipment and fire-protection was provided by Yoshinori Kitamura, Hideo Tanaka and Hiroomi Sato; structural design and construction-planning were provided by Shigeru Ban, Takaharu Fukuda and Osamu Nohira.

REFERENCE

1) T.Kobori, et.al., 'A proposal of new anti-seismic structure with active seismic response control system - Dynamic Intelligent Building-, 9WCEE. Vol.VIII 465-470, Japan, 1988
2) T.Kobori, et.al., 'Experimental study on Active Variable Stiffness System - Active seismic response controlled structure', Tall Buildings: 2000 and beyond, 561-572, Hong Kong, 1990
3) T.Kobori, et.al., 'Study on Joint Damper system - Passive seismic response controlled structure', Tall Buildings: 2000 and beyond, 573-585, Hong Kong, 1990
4) T.Kobori, et.al., 'Study on Active Mass Driver (AMD) system (Part1) and (Part2) - Active seismic response controlled structure', Tall buildings: 2000 and beyond, 587-602, 603-616, Hong Kong, 1990
5) T.Kobori, 'Earthquake and wind induced vibration control technology towards the super-high-rise building', Tall buildings: 2000 and beyond, 917-934, Hong Kong, 1990
6) T.Kobori, et.al., 'Seismic-response-controlled structure with Active Mass Driver System. Part1: Design, Part2: Verification', Earthquake Engineering & Structural Dynamics, Vol.20, No.2 133-150, 151-166, Feb.1991

Proceedings of the Second Conference on Tall Buildings in Seismic Regions
55th Regional Conference
May 16 and 17, 1991, Los Angeles, California

ADVANCED TECHNOLOGY in RESPONSE CONTROL of HIGH-RISE BUILDINGS

Tohru KOBORI   NIKKEN SEKKEI Ltd. Tokyo JAPAN
Haruyuki KITAMURA NIKKEN SEKKEI Ltd. Tokyo JAPAN
Takayuki TERAMOTO NIKKEN SEKKEI Ltd. Tokyo JAPAN

§ 1 Introduction

In this paper, from the viewpoint of how the "input" energy of an earthquake is to be distributed and expended throughout a structure including the various associated mechanisms, an attempt is made to give a wider concept when considering the earthquake resistant design. Furthermore several examples of the responce control system are introduced.

An evaluation method from the viewpoint of energy resulting from the response of a structure, by using the response energy value integrated over time, has an advantage to simply express the response of the structure as a whole, using the energy value as a scale.

Among similar studies conducted in Japan, one of the most penetrating is the studies by Kato and Akiyama. Kato and Akiyama have stated that " if the energy input E as an aggregate value of energy that is applied to a structure throughout the duration of an earthquake, then E is the value which mainly depends upon the total mass M and the primary natural period T. The dependency of E upon the distribution of mass, rigidity distribution and the damping factor h is small. Although E is affected by the coefficient of yield shear force, the dependency thereof can be neglected from the viewpoint of seismic design. Therefore, E can be regarded as a constant regardless of the curve of restoring force characteristic. "

In the design of a high-rise steel structure subject to the earthquake motion, building height, type of structure, framing system and building weight are the basic items, which are determined in the initial stage of the design process.

As these are determined, the aforementioned total mass and the primary natural period are roughly derived and the input energy when an earthquake motion is assumed. (This is determined regardless of the detail design for each structural member.)

The earthquake input energy is divided into kinematic energy, damping energy, elastic strain energy and plastic hysteresis energy within a structure. From the aspect of how these various types of energies are distributed and expended within the structure, and to evaluate and examine the earthquake resistant design technique of the structure, becomes possible.

As the input energy can be regarded approximately constant, if any one of the response energy values is made larger, other energy values must become smaller. In seismic design, various approaches can be taken depending upon which response energy value is intended to be made larger.

To let any one mechanism other than the main structural members bear the input energy is possible and various mechanisms can be considered for the purpose of reducing the earthquake response for the structure.
Also, an important evaluation item is considered to be, to know the values of energies such as the hysteresis energy in addition to story drift, ductility factor of story and that of members.

§ 2 Evaluation of earthquake response in a structure from an energy aspect

To examine the earthquake response characteristics of a structure, the input energy and various energies of the structure must be examined.
Various energy values in the structure, subject to earthquake motions can be calculated from the equations below.

$$[M]\{\ddot{x}\} + [C]\{\dot{x}\} + \{F(x)\} = -[M]\{1\}\ddot{x}_{G(t)} \qquad \cdots \cdots \cdots \cdots (1)$$

Where, $[M]$: Mass Matrix
$\qquad$ $[C]$: Damping Mtrix
$\qquad$ $\{F(x)\}$: Restoring Force Vector
$\qquad$ $\{x\}$: Relative Displacement Vector of Mass
$\qquad$ $\{\dot{x}\}$: Relative Velocity Vector of Mass
$\qquad$ $\{\ddot{x}\}$: Relative Accerelation Vector of Mass
$\qquad\quad$ $\ddot{x}_{G(t)}$: Accerelation of Input Earthquake

By multiplying the both sides of Eq. (1) by $\{\dot{x}\}dt$ and by further integrating, the energy equation at time t is derived as follows:

$$\{\dot{x}_t\}^T[M]\{\dot{x}_t\}/2 + \int^t\{\dot{x}\}^T[C]\{\dot{x}\}dt + \int^t\{\dot{x}\}^T\{F(x)\}dt$$
$$= -\int^t\{\dot{x}\}^T[M]\{1\}\ddot{x}_{G(t)}dt \qquad \cdots \cdots \cdots \cdots (2)$$

where, $\dot{x}_t$: $\dot{x}$ at time t
$\qquad$ $\int^t$: Integration from time 0 to time t

Each term of Eq. (2) can be interpreted as follows:
$\quad$ First term of the left side : Kinematic Energy (Wk)
$\quad$ Second term of the left side : Damping Energy   (Wd)
$\quad$ Third term of the left side : Hysteresis Energy (Wh)
$\qquad\qquad\qquad\qquad\qquad\qquad\qquad$ Sum of Elastic Strain Energy (We)and Plastic
$\qquad\qquad\qquad\qquad\qquad\qquad\qquad$ Hysteresis Energy (Wp); (Wh = We + Wp)
$\quad$ Right side $\qquad\qquad\qquad\qquad$: Input Energy into System due to Earthquake (E)

The kinematic energy (Wk) is the kinematic energy belonging to the mass system at time t and the damping energy (Wd) is the energy expended due to the viscous damping of the mass system.
The elastic strain energy (We) is that elastic energy stored in the mass system at time t and the plastic hysteresis energy (Wp) is the energy expended due to yielding in the system.
The hysteresis energy (Wh) is a value which can be obtained from simple integration of the restoring force and is equivalent to the sum total of the elastic strain energy (We) and plastic hysteresis energy (Wp).
The input energy (E) is an aggregate value of energy input into the mass system due to an earthquake motion.

From the above definition, Eq. (2) can be expressed as follows:
E = Wk + Wd + We + Wp
$\quad$ = Wk + Wd + Wh $\qquad\qquad\qquad\qquad\qquad \cdots \cdots \cdots \cdots (3)$
That is to say, the input energy at a certain point in time, is the sum of kinematic energy, damping energy, elastic strain energy and plastic hysteresis energy. In other words, the energy input into the structure is divided into a kinematic energy, damping energy, elastic strain energy and plastic hysteresis energy within the structure.
At the time when an earthquake ends, as the kinematic energy and the elastic strain energy become 0, the input energy becomes equal to the sum of the damping energy and the plastic hysteresis energy. The input energy is finally expended as damping and plastic hysteresis energy.

§ 3   Design method of structures subject to earthquake motions

3 − 1   Classification of design methods for structures

A seismic design method has been established as a design method through the experience and researches. On the other hand, there are techniques in which some sort of mechanism is added to the main structure, for purpose of reducing the seismic force acting on a structure during an earthquake or for reducing deformation in a building.

In Japan, although a numerous amount of research has been conducted and proposals have been made only several of them have been implemented up until recent times. Recently, with the elucidation of the earthquake motion and seismic response analysis method, including the interaction between the building and the ground, some structural systems and mechanisms have been proposed and implemented.

Table-1 shows the classification (mainly from the viewpoint of energy) of different design methods for structures subject to earthquake motions.

A design method using any mechanism is of course not intended to resist a seismic force using the mechanism alone, but simply to reduce the earthquake input to a structure, and hence the structures are examined and designed separately. The mechanisms other than the energy absorbing mechanism are seldom used independently, and are normally used in conjunction with the energy absorbing mechanism, so as to stabilize the response characteristics.

(1)   Elastic design method

This design method is based on the concept that a structure fixed to the ground shall not be subject to failure as a result of deformation and stress resulting from seismic forces. This is the simplest type of design and all members remain within the elastic range during the earthquake.

(2)   Elasto-platic design method

Assuming a large earthquake that occurs extremely rarely, it is not considered economical to design all members to remain within the elastic domain. In the case of the elasto-plastic design method, it aims to resist seismic forces making use of the plastic hysteresis energy, by plasticizing the parts of the structural members during an earthquake.

(3)   Isolation mechanism

The isolator provides a mechanism at the base of a structure, so as to allow the structure to move relatively freely in the horizontal direction. This reduced the response acceleration during an earthquake by prolonging the natural period of the system as a whole, and also reducing the earthquake input into the structure. It is necessary that the mechanism provided at the base should safely support the weight of the structure and move freely in the horizontal direction.

The most representative example is one in which a laminated rubber isolator is placed directly under the structure. Comprehensive research has been carried out and the results thereof have been subsequently reported in recent years with some examples of actual application. (Refer to Fig.-1)

(4)   Mass effect mechanism

The mass effect mechanism aims at converting part of the input energy to kinematic energy of an added mass. This mechanism permits the mass effect by making the added mass to work in a reversed phase, thereby reducing the maximum response value.

The moment of inertia damper was proposed by Kawamata et al. and a system using the action of a lever and pendulum was proposed by Shimizu et al..

The tuned mass damper is a typical example of a mass effect mechanism.

The tuned mass damper aims at reducing the response value of the main system, by adding to the vibrating system a vibrating mass that resonates with it. By making the added mass vibrate greatly, it thereby reduces the response value of the main system. (Refer to Fig.-2)

(5) Hysteresis energy absorption mechanism

Although this method using a hysteresis absorption mechanism is basically the same concept as a elasto-plastic design method, it is the method in which a part of accessories provided as a mechanism is plasticized and not the main structural members.

The steel rod damper is the most typical example. (Refer to Fig.-3) In the case of this mechanism, it is necessary that a stable hysteresis loop can be obtained even against large deformation, so that the rupture of members will not occur. It is also desirable that rigidity of the mechanism be kept small so that the building natural period will not vary greatly as a result of providing the mechanism.

(6) Damper mechanism

This method aims at converting part of the input energy into heat using a damper mechanism in the structure and subsequently absorbing it. There are several damper mechanisms available, oil damper, viscous damper, visco-elastic damper and friction damper, etc. These dampers cause a damping force proportional to the deformation or velocity of a structure and are normally installed between floors of a building. (Refer to Fig.-4)

Many oil dampers and viscous dampers are designed to cause a damping force proportional to velocity. Visco-elastic dampers draw an elliptic loop of the force-deformation relationship and have a so-called complex rigidity. Their damping characteristics are temperature dependent.

The friction damper makes use of Coulomb's friction using the material of metallic system and therefore has advantages in that it is not temperature dependent and stable characteristics can be obtained.

Table-1 Classification of Design Methods for Structures Subject to Earthquakes

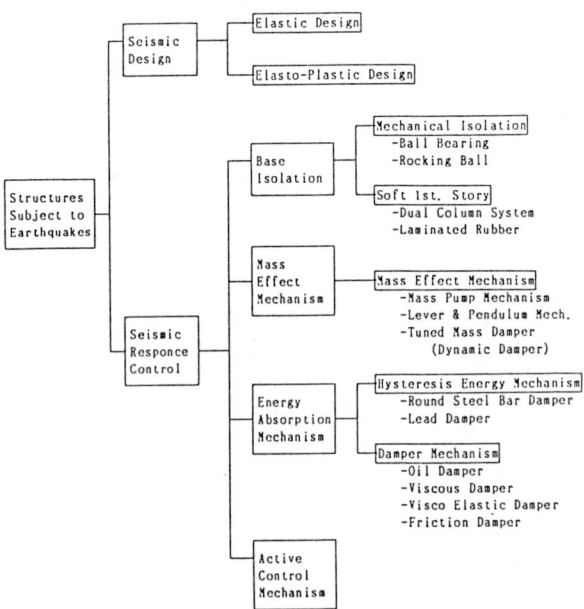

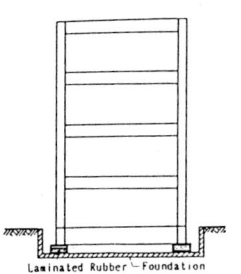

(Laminated rubber isolator)
Fig.-1 Isolation Mechanism

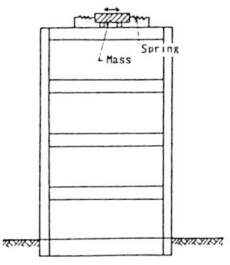

(Tuned Mass damper)
Fig.-2 Mass Effect Mechanism

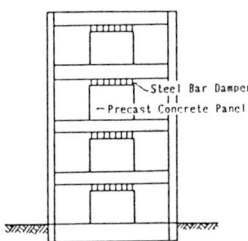

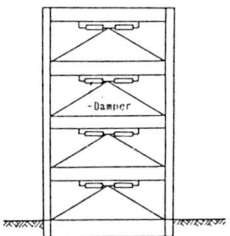

Fig.-3 Hysteresis Energy Absorption Mechanism   Fig.-4 Damper Mechanism

3 − 2   Response energy value of each structure

To verify the concept so far, several structures created from various seismic design methods are modelled and the energy value is calculated using the elasto-plastic earthquake response analysis of a 1 or 2 mass system.

3 − 2 − 1   Analytical model

The analytical model used, is a 1 or 2 mass shear-type model with a natural period of 3 seconds and a damping factor of 2%. The various factors for each analytical model are shown in Fig.-5.

Model-1 and Model-1A are structured from the elastic design method, and the shear spring is elastic. The natural period of Model-1 is 3 seconds and that of Model-1A is 1 second so as to facilitate the comparison to the isolation mechanisms (Model-3).

Model-2 is structured from the elasto-plastic method. The shear spring has an elasto-plastic restoring force characteristic and the coefficient of the yield shear force is set at 0.1.

Model-3 is a version of the structure with an isolation mechanism. The upper structure is the same as in Model-1A and a soft shear spring representing an isolator is added to it in the lower part. The natural period of the total system is set at 3 seconds.

Model-4 is the tuned mass damper system, which is a mass effect mechanism. The mass and spring that are resonant with the structure, are added to Model-1. The added mass system have a mass ratio of 1/100 with a damping factor of 10%.

Model-5 is a version of the structure with a hysteresis energy absorption mechanism, which has bi-linear restoring force characteristics. The shear spring rigidity of the hysteresis energy absorption mechanism is fixed at 1/5 of the structure and the coefficient of yield shear force is 0.01.

Model-6 is a version of the structure with a damper mechanism and has a velocity-proportional viscous damper (damping factor 3%) added to Model-1.

Although the natural periods of the analytical models were made to be equal as much as possible (so as not to vary the input energy too much), there are some where the primary natural period varied substantially.

For earthquake motion, the record of EL CENTRO 1940 NS was used with a maximum acceleration of 100 gal and the dynamic analysis was carried out for the duration of 30 seconds.

3 − 2 − 2 Maximum response value and response energy value

Table-3 shows the maximum response values from the result of the dynamic response analysis. Fig.-6 shows the calculated results of various response energy values and Table-4 shows their maximum values.
As the values shown in Table-4 are maximum response energy values, the sum of the various response energy values do not match the input energy value.

In the case of Model-1, 2, 4 and 5 where the natural period and the mass are approximately equal, the difference in input energy is small, i.e. 10,140 ∼ 11,860 t·cm, with the exception of Model-2.
This indicates that the input energy can be regarded to be about constant regardless of the system. In the case of Model-2, as the coefficient of yield shear is small, it can be seen that input energy was reduced slightly due to a prolonged period as a result of plasticization.

In the case of Model-1, as it is an elastic model, the coefficient of the response shear force shows a large response value, i.e. 0.25.
In the energy response, it can be seen that some transfer between kinematic energy and elastic strain energy resulted.

In the case of Model-1A, as it is basically an elastic model and the natural period is short, the coefficient of the response shear force became extremely large, i.e. 0.99.

In the case of Model-2, the coefficient of the response shear force, story drift, etc., were reduced by allowing the elasto-plastic deformation.
Also it can be seen that most of the response energy value was expended by the plastic hysteresis energy, enlarging its value to 58% of the input energy.
Looking at the time history of the plastic hysteresis energy, the special features are that the energy increases step-by-step every time the system enters the plastic region.

Model-3 has an isolation mechanism. Because its structural system is the same as Model-1A, the response of the upper structure only is compared with Model-1A. The coefficient of the response shear force was reduced to about 1/3 from 0.99 to 0.32. Looking at the response energy value, it can be seen that the elastic strain energy of the added spring takes as much as 52% of the input energy controlling the whole response.

In the case of Model-4, the coefficient of the response shear force for the main structure was reduced by about 20% compared with Model-1, by adding a tuned mass damper. Looking at the response energy value, it can be seen that the damping energy of the tuned mass damper accounts for 48% of the input energy. So, the maximum effect on the response seems to be the damping mechanism of the tuned mass damper.

Model-5 has a hysteresis energy absorption mechanism added to Model-1 and the coefficient of the response shear force is reduced by about 20% compared with Model-1, primarily due to the effect of the mechanism yielding relatively early. From the viewpoint of the response energy value, the plastic hysteresis energy of the added mechanism accounts for 50% of the input energy.

In the case of Model-6, as it has a velocity-proportional type damper added to Model-1, the coefficient of the response shear force was reduced by about 30% compared with Model-1. With regard to the response energy value, the damping energy of the added mechanism accounts for 55% of the input energy. The tendency of the energy response as a whole does not differ much from Model-5.

As the shearing ratio itself for each energy value can also be affected by setting up the numerical values of the analytical model, quantitative positioning of them appears to be insufficient, but it may be considered that a qualitative tendency of the response energy value in each seismic design method as positioned in this paper, has been sufficiently verified.

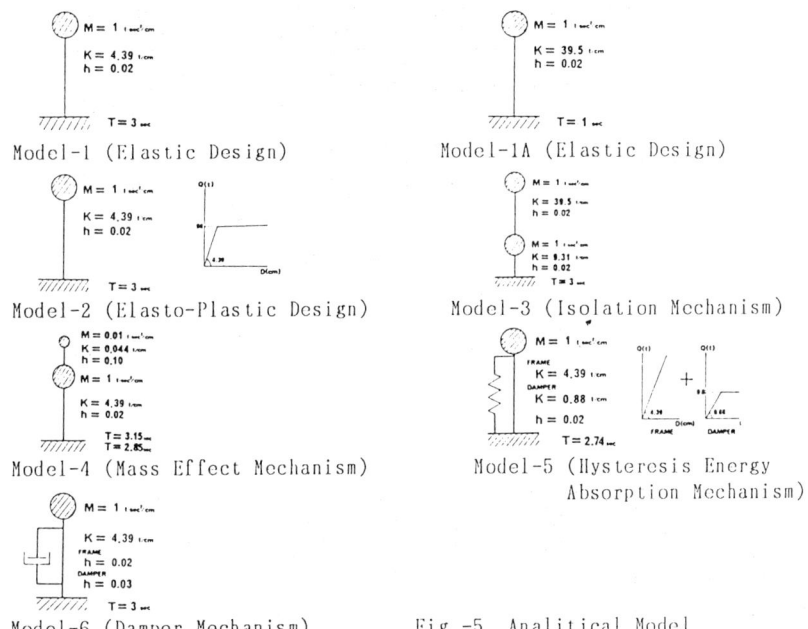

Model-1 (Elastic Design) Model-1A (Elastic Design)
Model-2 (Elasto-Plastic Design) Model-3 (Isolation Mechanism)
Model-4 (Mass Effect Mechanism) Model-5 (Hysteresis Energy Absorption Mechanism)
Model-6 (Damper Mechanism) Fig.-5 Analitical Model

Table-3 Maximum Response Value

	Structure				Added Mechanism			
	sC	sD	sV	sA	mC	mD	mV	mA
Model-1 (Elastic Design)	0.25	55	120	496	—	—	—	—
Model-1A (Elastic Design)	0.99	25	172	1287	—	—	—	—
Model-2 (Elasto-Plastic Des)	0.10	43	102	466	—	—	—	—
Model-3 (Isolation Mech.)	0.32	8	128	553	0.25	52	115	521
Model-4 (Mass Effect Mech.)	0.20	44	114	484	0.80	180	387	872
Model-5 (Hysteresis Abs.Mech)	0.19	42	119	499	—	—	—	—
Model-6 (Damper Mechanism)	0.17	37	107	476	—	—	—	—

Note) C : Maximum Coefficient of Shear Force   Suffix "s" : Structure
       D : Maximum Story Drift   cm   Suffix "m" : Added Mechanism
       V : Maximum Velocity   cm/sec
       A : Maximum Accerelation   gal

Table-4 Maximum Energy Response Value

	Input	Structure				Added Mechanism			
	E	sWk	sWd	sWe	sWp	mWk	mWd	mWe	mWp
Model-1 (Elastic Design)	1198	717	885	664	—	—	—	—	—
	1.00	0.60	0.74	0.55					
Model-1A (Elastic Design T=1.0 SEC)	1979	1485	1056	1198	—	—	—	—	—
	1.00	0.75	0.53	0.61					
Model-2 (Elasto-Plastic Design)	759	515	234	112	439	—	—	—	—
	1.00	0.68	0.31	0.15	0.58				
Model-3 (Isolation Mechanism)	2424	1476	354	121	—	—	1392	1255	—
	1.00	0.61	0.15	0.05			0.57	0.52	
Model-4 (Mass Effect Mechanism)	1053	647	434	424	—	75	503	71	—
	1.00	0.61	0.41	0.40		0.07	0.48	0.07	
Model-5 (Hysteresis Mechanism)	1188	705	478	400	—	—	—	—	592
	1.00	0.59	0.40	0.34					0.50
Model-6 (Damper Mechanism)	1014	574	373	307	—	—	560	—	—
	1.00	0.57	0.37	0.30			0.55		

Note) Upper Value: Energy Value ×10 t·cm
Lower Value: Ratio to Each Input Energy
E : Input Energy
Wk : Kinematic Energy
Wd : Damping Energy
We : Elastic Strain Energy   Suffix "s" : Structure
Wp : Plastic Hysteresis Energy   Suffix "m" : Added Mechanism

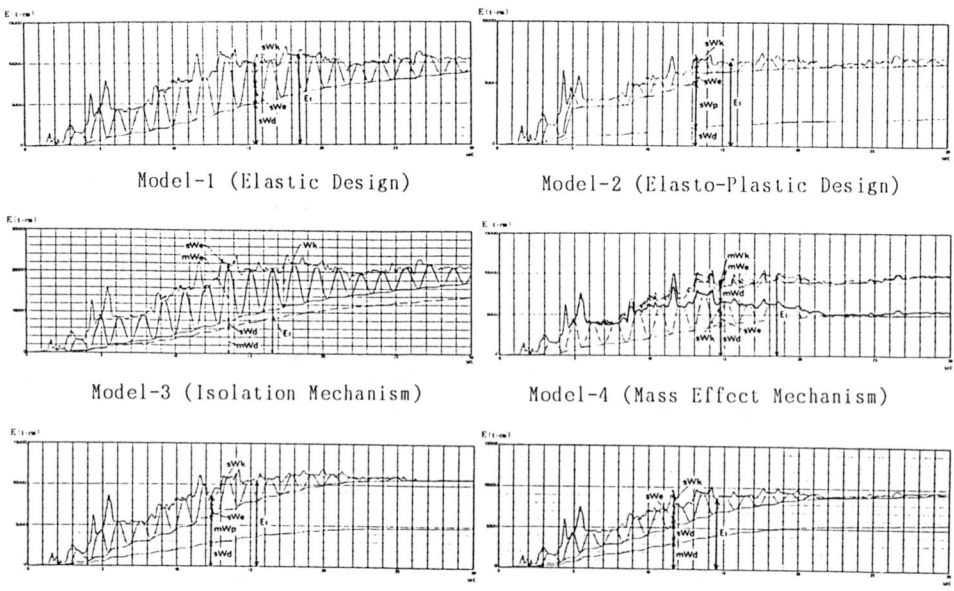

Model-1 (Elastic Design)   Model-2 (Elasto-Plastic Design)

Model-3 (Isolation Mechanism)   Model-4 (Mass Effect Mechanism)

Model-5 (Hysteresis Energy   Model-6 (Damper Mechanism)
  Absorption Mechanism)
Fig.-6 Time Hystory of Energy Value

§ 4 Examples of response control systems

4 − 1 Examples of tuned mass damper

Name of building: Chiba Port Tower
Location   : Chiba city, Japan
Constructed year: March 1986
Building height : 125 m
Number of floors: 4
Usage   : Observatory tower
Total area   : 2,308 m²
Structure   : Steel structure
Natural period   : 2.25 , 2.70 sec
Response control: Tuned mass damper
          system   (Weight:10 and 15 t)

(Chiba Port Tower)

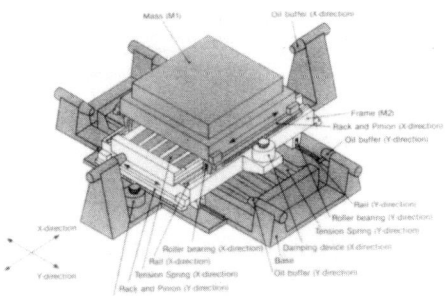

(Mechanism of Tuned Mass Damper)

(Tuned Mass Damper)

Name of building: Fukuoka Tower
Location   : Fukuoka city, Japan
Constructed year: Feb. 1989
Building height : 234 m
Number of floors: 5
Usage   : TV and Observatory tower
Total area   : 1,808 m²
Structure   : Steel structure
Natural period   : 3.08 , 3.22 sec
Response control: Tuned mass damper
          system   (Weight:25 and 30 t)

(Fukuoka Tower)

(Tuned Mass Damper)

241

4 − 2 Examples of friction damper

Name of building: Industrial Culture Center
Location   : Ohmiya city, Japan
Constructed year: March 1988
Building height : 140 m
Number of floors: 31
Usage   : Office
Total area   : 105,060 ㎡
Structure   : Steel structure
Natural period : 2.88 , 2.76 sec
Response control: Friction damper
        system (F/W=0.03
                 F:Friction force
                 W:Building weight)

(Industrial Culture Center)

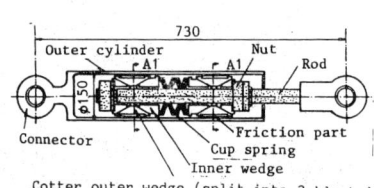

(Mechanism of Friction Damper)

(Installed Friction Damper
with Precast Concrete Wall)

Name of building: Asahi Beer Tower
Location   : Sumida-ku Tokyo, Japan
Constructed year: Nov. 1989
Building height : 95 m
Number of floors: 22
Usage   : Office
Total area   : 34,650 ㎡
Structure   : Steel structure
Natural period : 2.85 , 2.85 sec
Response control: Friction damper
        system (F/W=0.02
                 F:Friction force
                 W:Building weight)

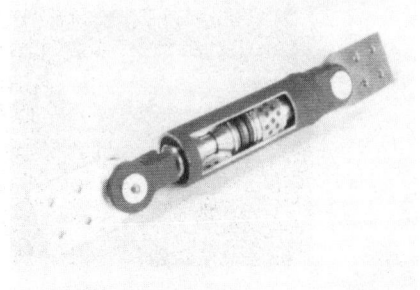

(Installed Friction Damper
with Steel Brace)

242

Proceedings of the Second Conference on Tall Buildings in Seismic Regions
55th Regional Conference
May 16 and 17, 1991, Los Angeles, California

RANDOM SEISMIC RESPONSES OF TALL BUILDINGS TO
MULTIPLE EXCITATIONS

Hongnan Li
(Dalian University of Technology, Dalian 116023, P.R.China)

Suyan Wang
(Shenyang Architectural and Civil Engineering Institute
Shenyang 110015, P.R.China)

ABSTRACT

In this paper, the analysis method of random responses and dynamic reliabilty of tall buildings is studied under the combined action of horizontal and rocking ground motions. First of all, the analytic formula of rotational power spectral density function is presented according to the statistical analysis of the curves of power spectra of the rotations derived from the translational earthquake records. Then, the vibration equation of buildings simplified as the cantilevers with continuously varied sections and a set of formulas are given to calculate the mean—square structural responses and dynamic reliability. At the end of this paper, a numerical example is given.

INTRODUCTION

In the computation of vibration, some tall buildings may be simplified as the cantilevers with variable sections. Up till now, only the effects of translational components on buildings are considered in the earthquake resistant design, while those of rotational components on ones are poorly studied. Practically, many earthquake damages have shown the fact that there exist not only translational components but rotational ones of ground motion during earthquake. The main problem, at present, lies in the shortage of records of rotational components suitble for the practical application due to the limitation of strong—motion measurement level. Therefore, it seems to be effective to estimate the rotational components in terms of the corresponding translational components recorded more extensively and to study the effects of these components on buildings.

STOCHASTIC MODELS OF GROUND MOTION

It is assumed that the medium through which the seismic waves propagate is elastic, homogeneous and isotropic, or is layered in a half—space. The incident waves near the surface of half—space consist of body wave (P wave and S wave) or surface wave (Rayleigh wave and Love wave). By some similar derivations, we have the following general form of the relationship between the rotational (torsional and rocking) and translational components of ground motion (Li,1989)

$$\ddot{\varphi}_{xj} = \frac{i\omega}{rC_x} \ddot{U}_j \tag{1}$$

where $\ddot{\varphi}_{xj}$ and $\ddot{U}_j$ are the torsional and anti-plane horizontal components (Fourier spectra) and $r=2$ for $j=y$, $\ddot{\varphi}_{xj}$ and $\ddot{U}_j$ are the rocking and vertical components (Fourier spectra) and $r=1$ for $j=z$, ω is the circular frequency, and C_x is the apparent velocity. It has been verified that the results derived from such a method are consistent with practice (Li,1989). According to the statistical analysis of the curves of power spectra for rotational components from the horizontal earthquake records by the use of Eq.(1), it has been shown that these spectra display generally double or multiple—peaks in the site I (solid soil) and site II (medium solid soil), but single—peak in the site III (soft soil). Therefore, we give the general form of rotational power spectrum by the method of curve—fitting as follows

$$S_T(\omega) = \frac{\omega^2}{\omega^2 + \mu^2} \frac{1 + 4\xi_{g1}^2(\omega/\omega_{g2})^2}{[1 - (\omega/\omega_{g1})^2]^2 + 4\xi_{g1}^2(\omega/\omega_{g1})^2} S_R \qquad (2)$$

where ω_{g1} and ξ_{g1} are the circular frequency and damping factor of soil, μ is the reduction factor of low frequency and S_R is the bedrock input given by

$$S_R = \begin{cases} \dfrac{1 + 4\xi_{g2}^2(\omega/\omega_{g2})^2}{[1 - (\omega/\omega_{g2})^2]^2 + 4\xi_{g2}^2(\omega/\omega_{g2})^2} S_1 & \text{for site I,II} \\ S_1 & \text{for site III} \end{cases} \qquad (3)$$

In which S_i is the spectral intensity, ω_{g2} and ξ_{g2} are the circular frequency and damping factor in bedrock, respectively. Eq.(2) can fairly fit most of the rotational power spectral curves in three different kinds of sites. Fig.1 shows the curve—fitting of the rocking power spectrum of the station No.153 during San Fernando earthquake in the United States.

$S_r(f)$

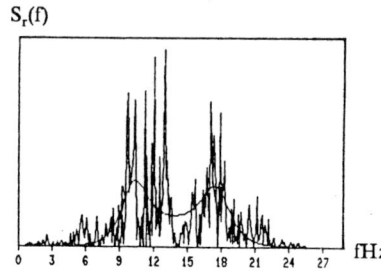

0   3   6   9   12   15   18   21   24   27   fHz

Fig.1   Rocking Power Spectrum

By the use of curve—fitting of the nonlinear least square method for Eq.(2) and the rotational power spectra derived from the three—component earthquake records of 85 stations in the western part of the United States, the statistical results of rotational power spectral parameters are obtained which may be used in engineering shown in table 1 ($\overline{X}$ is the expetation value of spectral parameters, Va is the coefficient of variation of spectral parameters, Tor. is the abbreviation of Torsion and Roc. is the abbreviation of Rocking)

The horizontal component can be described by the Kanai—Tajimi spectrum. The relationship between the horizontal acceleration peak value a_m and the parameters of Kanai—Tajimi spectrum is given by

$$a_m = R_P \left[\frac{\pi S_0 \omega_g}{2\xi_g}(1 + 4\xi_g^2) \right]^{\frac{1}{2}} \qquad (4)$$

Table 1   Statistical Values of Rotational Power Spectral Parameter

site cond.	Param.	ω_{g1}		ω_{g2}		ξ_{g1}		ξ_{g2}		μ	
		Tor.	Roc.	Tor.	Roc.	Tor.	Roc.	Tor.	Roc.	Tor.	Roc.
I	$\overline{X}$	21.84	25.86	63.90	71.31	0.677	0.614	0.263	0.310	1582.1	1523.0
	V_a	0.866	0.934	0.958	0.971	1.325	0.699	1.659	1.113	1.052	1.060
II	$\overline{X}$	20.19	24.73	52.73	67.67	0.633	0.986	0.524	0.676	4023.8	3410.5
	V_a	0.975	0.934	1.028	1.053	1.224	1.436	1.328	1.557	1.915	2.203
III	$\overline{X}$	20.07	34.16	–	–	0.861	0.412	–	–	7546.1	5601.5
	V_a	0.979	1.054	–	–	0.921	0.549	–	–	2.151	2.016

where ξ_g and ω_g are the Kanai–Tajimi spectral parameters related to local site condition, respectively and S_0 is the Kanai–Tajimi spectral intensity.

The relationship between the rotational power spectrum $S_T(\omega)$ and translational power spectrum $S_H(\omega)$ is given by (Li,1989)

$$S_T(\omega) = \frac{\omega^2}{A^2} S_H(\omega)$$
(5)

where $A = rC_x$. Hance, the relationship between the horizontal ground acceleration peak α_m and spectral intensities S_0, S_1 can be gotten by the use of Eq.(4) and by the integration for Eq.(5).

VIBRATION EQUATION OF BUILDINGS AND MEAN–SQUARE RESPONSE

The general form of vibration of tall buildings modelled as a cantilever(Fig.2) subjected to horizontal and rocking ground motion is

$$m(x)\frac{\partial^2 u(x,t)}{\partial^2 t} + C(x)\frac{\partial u(x,t)}{\partial t} + L(x)[u(x,t)] = -m(x)[\ddot{u}_{gx}(t) + x\ddot{u}_{g\theta}(t)]$$
(6)

where $u(x,t)$ is the relative displacement, L(x) a self–ajoint differential operater, $m(x)$ the mass density, $C(x)$ the damping, $\ddot{u}_{gx}$(t) the horizontal ground acceleration and $\ddot{u}_{g\theta}$(t) the rocking ground acceleration. For flexural–type strutures, the third term in Eq.(6) is

$$L(x)[u(x,t)] = \frac{\partial^2}{\partial x^2}[EI(x)\frac{\partial^2 u(x,t)}{\partial x^2} + C_s I(x)\frac{\partial^3 u(x,t)}{\partial x^2 \partial t} \quad]$$
(7)

where E is the Young's modulus, I(x) the moment of inertia, C_s the damping of strain velocity. For shear–type structures, the third term in Eq.(6) is

$$L(x)[u(x,t)] = -\frac{\partial}{\partial x}[KGA(x)\frac{\partial u(x,t)}{\partial x}]$$
(8)

where G is the shear modulus of elasticity, K the factor depending upon the geometry of cross section and $A(x)$ the cross-section area of the cantilever.

Applying the modal superposition method to Eq.(6), we have

$$\ddot{q}_n(t) + 2\xi_n \omega_n \dot{q}_n(t) + \omega_n^2 q_n(t) = -[\gamma_n(x)\ddot{u}_{gx}(t) + \gamma_n(\theta)\ddot{u}_{g\theta}(t)] \tag{9}$$

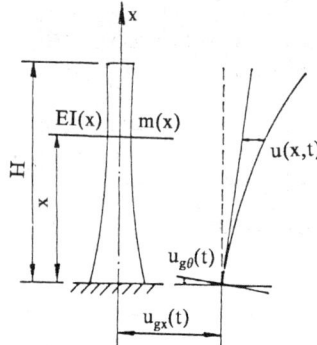

where $q_n(t)$ is the generalized coordinate, ξ_n and ω_n are the modal damping and circular frequency, and $\gamma_n(x)$ and $\gamma_n(\theta)$ are the horizontal and rocking modal participation factors defined respectively as

$$\gamma_n(x) = \frac{\int_0^H m(x)\varphi_n(x)dx}{\int_0^H m(x)\varphi_n^2(x)dx} \tag{10}$$

and

$$\gamma_n(\theta) = \frac{\int_0^H m(x)x\varphi_n(x)dx}{\int_0^H m(x)\varphi_n^2(x)dx} \tag{11}$$

in which $\varphi_n(x)$ is the nth mode.

Fig.2   Calculating sketch of Building

If x and θ in $\gamma_n(x)$, $\gamma_n(\theta)$, $\ddot{u}_{gx}(t)$ and $\ddot{u}_{g\theta}(t)$ are simply expressed as 1 and 2, Eq.(9) may be rewritten as

$$\ddot{q}_n(t) + 2\xi_n \omega_n \dot{q}_n(t) + \omega_n^2 q_n(t) = -\sum_{i=1}^{2} \gamma_n(i)\ddot{u}_{gi}(t) \tag{12}$$

From Eq.(12), the displacement response can be solved as

$$u(x,t) = -\sum_{n=1}^{\infty}\sum_{i=1}^{2} \varphi_n(x)\gamma_n(i)\int_0^t h_n(\tau)\ddot{u}_{gi}(t-\tau)d\tau \tag{13}$$

The power spectral density of displacement response is determined by

$$S_u(x,\omega) = \sum_{n=1}^{\infty}\sum_{m=1}^{\infty}\sum_{i=1}^{2}\sum_{j=1}^{2} \varphi_n(x)\varphi_m(x)\gamma_n(i)\gamma_m(j)H_n(\omega)H_m^*(\omega)S_{\ddot{u}_{gij}}(\omega) \tag{14}$$

where $S_{\ddot{u}_{gij}}(\omega)$ are auto-spectrum and cross-spectrum of excitations respectively, $H_n(\omega)$ the frequency response function.

The mean-square responses of buildings can be given in the general form

$$\sigma^2(x) = \sum_{n,m} A_n A_m I_{nm} = \sum_{n=1}^{\infty} A_n^2 I_{nm} + \sum_{n \neq m} A_n A_m I_{nm} \tag{15}$$

where

$$I_{nm} = \sum_{i=1}^{2} \sum_{j=1}^{2} \gamma_n(i)\gamma_m(j)\int_{-\infty}^{+\infty} H_n(\omega)H_m^*(\omega)S_{\ddot{u}_g ij}\quad(\omega)d\omega \tag{16}$$

$$A_n = \begin{cases} \varphi_n(x) & \text{for displacement} \\ EI(x)\varphi_n'(x) & \text{for bending moment} \\ EI(x)\varphi_n''(x) & \text{for shear} \end{cases} \tag{17}$$

For common tall buildings, the natural frequencies are well separated. The second term in Eq.(15) can be neglected.

DYNAMIC RELIABILITY OF BUILDINGS

In order to analyze the dynamic reliability of buildings under the acfion of earthquake, we should calculate the probabilistic function of which a certain control value B(x,t) of building respone will not exceed the allowable bounds during the given time interval[0,T]:

$$P_s(x,T) = P\{maxB(x,t) \leqslant \lambda_1 \cap minB(x,t) \geqslant -\lambda_2;\quad x\in[0,H]\cap\quad t\in[0,T]\} \tag{18}$$

where λ_1 and λ_2 are the upper and lower bounds of B(x,t), respectively.

Assuming now that the number $n_\lambda(T)$ of the barrier crossing of the dynamic response of building during the time interval [0,T] obays the Poisson distribution, we have

$$P\{n_\lambda(T) = i\} = \frac{1}{i!}[\int_0^T v_\lambda(t)dt]^i exp[-\int_0^T v_\lambda(t)dt] \tag{19}$$

where $v_\lambda(t)$ is the expected value of the number of barrier crossing of dynamic response during unit time. According to Eqs.(18) and (19), the conditional reliability of buildings can be obtained as

$$P_s(x,T) = P\{n_\lambda^+(T) = 0\} + P\{n_\lambda^-(T) = 0\} = exp\{-\int_0^T[v_{\lambda_1}^+(t) + v_{\lambda_2}^-(t)]dt\} \tag{20}$$

For a zero—mean Gauss stationary response, we have

$$P_s(x,T) = exp\{-\frac{T}{2\pi}\cdot\frac{\sigma_{\dot{B}}}{\sigma_B}[exp(-\frac{\lambda_1^2}{2\sigma_B^2}) + exp(-\frac{\lambda_2^2}{2\sigma_B^2})]\} \tag{21}$$

where σ_B^2 and $\sigma_{\dot{B}}^2$ are the variances of B(x,t) and $\dot{B}$(x,t).

If the curves of the risk analysis of the existing building site are given, the failure probability of the building is easily gotten from the formula of the total probability of Reference[2].

247

NUMERICAL EXAMPLE

In this section, the example of a shear–wall tall building (Li,1985) modelled as a vertical cantilever elastic beam with uniform section subjected to the horizontal and rocking ground motions is considered. The computing parameters of the equivalent cantilever is:

$$\left.\begin{array}{l} E = 2.6 \times 10^{6} t / m^{2} \\ I = 69.03 m^{4} \\ m = 10.3 t \cdot s^{2} / m^{2} \\ H = 35.5 m \end{array}\right\} \tag{22}$$

where

    E——the Young's modulus,
    I——the moment of inertia,
    m——the constant mass density per unit,
    H——the height of the building.

The vibration equation of the cantilever to horizontal and rocking support motion according to the Euler–Bernoulli theory is given by

$$m\frac{\partial^{2} u(x,t)}{\partial t^{2}} + C\frac{\partial u(x,t)}{\partial t} + EI\frac{\partial^{4} u(x,t)}{\partial x^{4}} = -m[\ddot{u}_{gx}(t) + x\ddot{u}_{g\theta}(t)] \tag{23}$$

Boundary conditions in this case are given as follows

$$u(0,t) = \frac{\partial u(0,t)}{\partial x} = \frac{\partial^{2} u(H,t)}{\partial x^{2}} = \frac{\partial^{3} u(H,t)}{\partial x^{3}} = 0 \tag{24}$$

The normal modes of vibration for cantilever are derived as follows

$$\varphi_{n}(x) = (\frac{2}{mH})^{\frac{1}{2}}[(ch\frac{\lambda_{n}}{H}x - cos\frac{\lambda_{n}}{H}x) - \eta_{n}(sh\frac{\lambda_{n}}{H}x - sin\frac{\lambda_{n}}{H}x)] \tag{25}$$

where

$$\eta_{n} = \frac{ch\lambda_{n} + cos\lambda_{n}}{sh\lambda_{n} + sin\lambda_{n}} \tag{26}$$

and λ_{n} is determined by the following frequency equation

$$ch\lambda_{n}cos\lambda_{n} + 1 = 0, \quad n = 1,2,\cdots \tag{27}$$

The first three values of λ_{n} are 1.875, 4.694, 7.855 and for higher values of n:

$$\lambda_{n} = \frac{(2n-1)\pi}{2} \tag{28}$$

The natural circular frequencies of the beam are

$$\omega_n = \frac{\lambda_n^2}{H^2} \sqrt{\frac{EI}{m}} \tag{29}$$

By the use of the integration for Eqs.(10) and (11), the horizontal and rocking modal participation factors are expressed as

$$\gamma_n(x) = \frac{(2mH)^{\frac{1}{2}}}{\lambda_n} [sh\lambda_n - sin\lambda_n - \eta_n(ch\lambda_n + cos\lambda_n - 2)] \tag{30}$$

and

$$\gamma_n(\theta) = \frac{(2mH^3)^{\frac{1}{2}}}{\lambda_n^2} \{\lambda_n(sh\lambda_n - sin\lambda_n) - (ch\lambda_n + cos\lambda_n) + 2 - \eta_n[\lambda_n(ch\lambda_n + cos\lambda_n)$$
$$- (sh\lambda_n + sin\lambda_n)]\} \tag{31}$$

It is assumed that the building is located in the site II and it is subjected to horizontal ground acceleration peak 250gal. Eq. (15) is used to calculate the root−mean−square response of the building. The results are expressed as follows

$$\frac{\sigma_{uR}}{\sigma_{uH}} = 35.32\% \tag{32}$$

and

$$\frac{\sigma_{MR}}{\sigma_{MH}} = 7.58\% \tag{33}$$

in which σ_{uR} and σ_{uH} are the root−mean−square displacement responses in the top of the building respectively to the rocking and horizontal inputs, σ_{MR} and σ_{MH} are the root−mean−square moment responses in the bottom of the building respectively to the rocking and horizontal inputs.

The calculating results of the condition reliability of the building are listed in Table 2. It is observed that the reliability Ps decreases as the duration τ_0 of ground motion increases. The effects of the rocking component on the reliability of building is important.

CONCLUSIONS

The computational method of the mean−square response and reliability of tall buildings simplified as the vertical cantilevers to horizontal and rocking seismic ground motion is given. A numerical example is provided to analyze the effects of the rocking component on the mean−square response and dynamic reliability of tall building. Based on the presented results, it has shown that the rocking of seismic ground motion influences seriously the responses of tall buildings.

Table 2   Conditional Reliability of Building

Duration / Bounds Inputs	$\Delta_1 = \dfrac{H}{600}$		$\Delta_2 = \dfrac{H}{800}$		$\Delta_3 = \dfrac{H}{1000}$	
	Trans.	Trans.+Torsion	Trans.	Trans.+Torsion	Trans.	Trans.+Torsion
2	1.0000	0.9956	0.0079	0.9312	0.9784	0.9722
4	1.0000	0.9912	0.9959	0.8672	0.9572	0.5964
6	1.0000	0.9868	0.9938	0.8875	0.9365	0.4605
8	0.9999	0.9825	0.9918	0.7520	0.9162	0.3556
10	0.9999	0.9782	0.9897	0.7003	0.8964	0.2746
12	0.9999	0.9738	0.9877	0.6521	0.8770	0.2121
14	0.9999	0.9696	0.9856	0.6072	0.8580	0.1638
16	0.9999	0.9653	0.9836	0.5655	0.8395	0.1265
18	0.9999	0.9610	0.9816	0.5266	0.8213	0.0977
20	0.9999	0.9568	0.9795	0.4904	0.8035	0.0754

Note: Trans.denotes trsnslational input,
Trans. +Torsion denotes translational and torsional inputs

REFERENCES

1. Hongnan Li and Shijun Sun, Rotational Components of Earthquake Motions and Their Stochastic Models, Proc. of Sino–Japan Conference on Seismological Research, Beijing, China, May28–30, 1989.
2. Jinren Jiang and Feng Hong, Aseismic Reliability Analysis of Multi–Story Masonty Buildings, Earthquake Engineering and Engineering Vibration, Vol.5, No.4, 1985.
3. Guiqing Li, Computational Theory and Method of Aseismic Structures, Seismological Press, 1985.

Proceedings of the Second Conference on Tall Buildings in Seismic Regions
55th Regional Conference
May 16 and 17, 1991, Los Angeles, California

SEISMIC ISOLATION TECHNIQUES FOR HIGH-RISE BUILDINGS

Lee Li *   Hai-Yan Li **

Abstract

A post supported on round balls will be unaffected by the horizontal ground shakings. Even the diameter of the round balls are very large, they are good isolators also. In practice, only the central portion of the ball is useful, then it makes a column with spherical upper surface and spherical lower surface. During earthquake, this column may rock as a round ball rolling. So it is called the rocking column. Many rocking columns make a soft-first-story for a high-rise building. It may isolate effectively the earthquake horizontal ground motion.

Introduction

Recent years, many high-rise buildings have been built in some cities in China. Even today, the problem with these high-rise buildings is how to protect them from earthquake shaking. Therefore there is an ugent need for finding a solution to this problem.

All of the seismic isolation measures, such as steel-plate-laminated rubber pad, frictional sliding joint etc, are not suitable for high-rise buildings. Because the inertia force is acting at the center of gravity of the high-rise building which is too high, and it produces a high overturning moment to the building. Then it may cause the collapse of the whole building during earthquake. So that the rubber pad and frictional sliding joint seismic isolation techniques can not be applied to high-rise buildings.

Since 1929, Martel [1] suggested that the flexible-first-story may isolate earthquakes for high-rise buildings. In 60 years, many researchers investigated this suggestion but without arriving to any definite conclusion. The main disadvantage of this suggestion is the soft columns can not support the heavy vertical loads. Especially, in San Fernando earthquake, 1971, the whole building of the Olive View Hospital suffer so large deformation that it has to be demolished, and in Tangshan earthquake 1976, the weak first story of the library of the Tangshan Mining College collapsed and the whole upper building sat down on the ground. Therefore this application of the concept of the soft-first-story seems not feasible.

In China, there were many buildings did not collapse during big earthquakes. Usually these buildings are wooden frames. Wooden beams and wooden columns are used for carrying the whole weight of the heavy roof system. Brick walls do not carry the vertical load. During earthquake, wooden frame does not collapse while the brick wall fell down (Photo 1). So that, this type of building seems to be only a partial aseismic construction. After earthquake, it is found that there are many compressive indentations at the connections of beam and column (Fig. 1). It presented that these columns were swang back and for. Finally, it did not collapsed. On the point of view of mechanical analysis, the upper part of the building might be unaffected by the horizontal ground motion due to the

---

* Professor, Central Research Institute of Building & Construction, MMI. Beijing, P.R.China.
** Assistant, Central Research Institute of Building & Construction, MMI. Beijing, P.R.China.

swang of columns. That is to say, the columns not only can carry the heavy vertical load but also can isolate the horizontal earthquake force by swinging. Another picture, in Yunan Province, a wooden pavillion (Photo 2) did not collapse during the Tonghai earthquake but it tilted after the earthquake. It means that the wooden pavillion had been swung during earthquake. This is a successful example of seismic isolation building which was supported on the rocking and swinging columns.

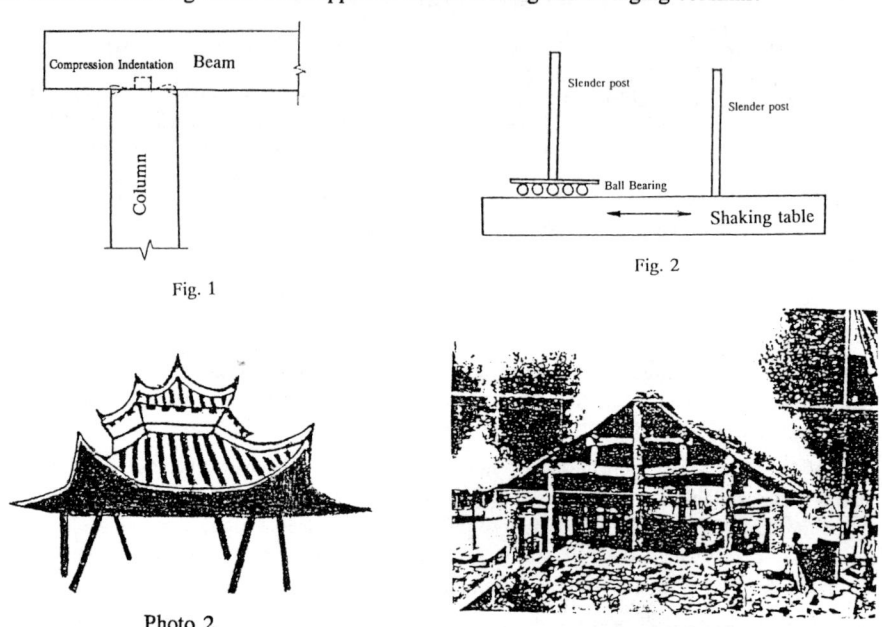

Fig. 1

Fig. 2

Photo 2

Photo 1

Today many researchers are interested in the seismic isolation techniques and some have designed structures using some of these techniques. Among these techniques, round balls are the best isolators for isolating the horizontal ground motion. Once a time, the author made an experiment, three round balls placed between two glass plates, the lower plate was fixed on the shaking table, and a slender post was supported on the upper plate (without any adhesive). (Fig. 2). When the table was shaken, the lower plate vibrated with the shaking table, but the upper plate without moving and the narrow post did not fall down. This example indicates that the round balls can isolate the horizontal ground motion well. Later, the narrow post was supported on the shaking table directly and it fell down very quickly when shaking table was forced to vibrate.

However, the ball bearing is not a practible technique for building seismic isolation. Even though there is, in Mexico, a five story high school building constructed on many small balls, [7] it had isolated already horizontal ground motions, but it is difficult to construct, and hard to maintain the rolling character during many decades, because the steel balls might be rusted or the oil might be hardened to protect the rolling of balls. All of these will defeat the utility of the isolation result.

Japanese researchers, Prof. Y. Kuo designed a big ball constructed under the floor (Fig. 3) [5] for seismic isolation. Prof. M. Izumi designed many types of deformed balls (Fig. 4) [6] . All of these method might induce an additional vertical vibration when building moving in horizontal directions.

And an additional story should be constructed for these isolators, it might rise the cost of building and makes much difficults in construction. If there were a little difference between the diameters of balls, then the vertical loads could not be distributed evenly on each ball. As a results these techniques are hard to applied to building construction.

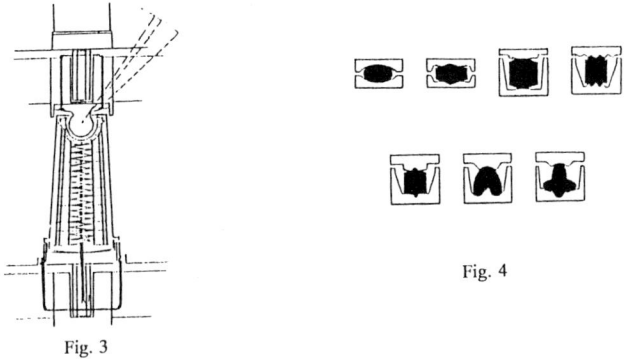

Fig. 3

Fig. 4

(2) Rocking Column

Fig. 5 shows a ball type seismic isolation technique. The diameter of the ball is just equal to the height of the first story of the building. For supporting the weight of the whole building, only the central portion of the big ball is utilized, stresses in the side portions are very little, so that the side portion may be cut out, and only the central portion remained, resulting a column as shown in Fig. 6. The top surface and the bottom surface of the column are spherical surfaces as two areas of the big ball surfaces. The column may rock as a big ball rolling. So it is called the rocking column. After many analysis, it was found that the maximum relative displacement of the first floor with respect to the ground will never exceed twenty inches, i.e. the maximum rolling distance of the column will never exceed ten inches. Then the radius of the cap (boot) of the rocking column will be more than ten inches. (Fig. 6) Usually the radii of the cap and the boot are larger than the radius of the column of the first story.

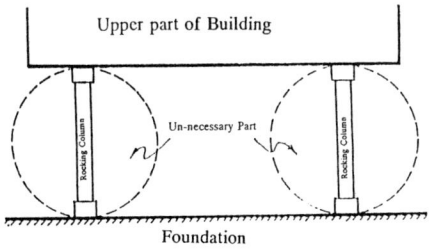

Fig. 5

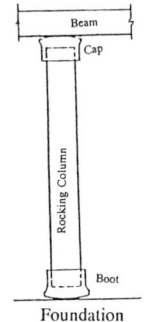

Fig. 6

During earthquake, there is not additional vertical vibration caused by this kind of rocking column. Therefore, the rocking column system is a good seismic isolation technique. Furthermore, it is easy to construct and therefore is an in-expensive isolator. It may reduce the total cost of the whole building, because the seismic force developed in the super-structure is reduced by these isolators. Otherwise this kind of seismic isolator can be applied not only to high-rise buildings, but also to the frames of industrial facilities etc.

(3) Energy Absorber

To reduce the large swinging displacement and disspate the free vibration after earthquake, energy absorbers should be adopted for isolated buildings while the rocking column system is used. There are many kinds of energy absorbers, such as oil damper, lead damper and soft-iron damper etc. They are expensive and difficult to construct. Therefore, a frictional damper is suggested.

In the first story of a high-rise building, the precast R.C. walls (outer walls and inner partition walls) are placed in steel channels. The upper edge confined horizontally to the first floor (Fig. 7),

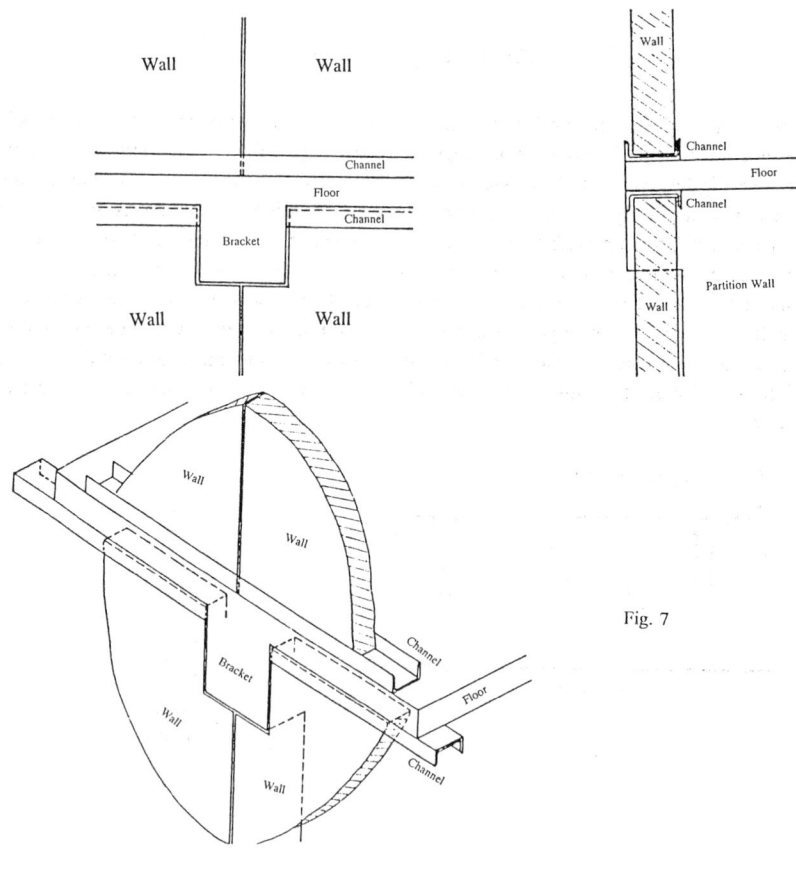

Fig. 7

254

and the lower edge is inserted in the steel channel on the ground. When there is a relative displacement between the first floor and ground, the R.C. walls will slide in the steel channels and produce frictional force F. It may absorb many vibration energy in each cycle.

$$F = W \cdot \mu \cdot (\, sign \, (\dot{Y}_1 - \dot{Y}_o))$$

Where : W-----Total weight of the R.C. walls,

μ------Frictional coefficient between the wall and the steel channel,

$\dot{Y}_1$-----Velocity of the first floor,

$\dot{Y}_o$-----Velocity of the ground.

The weight of these walls and the frictional coefficient between walls and steel channels can be adjusted to obtain a suitable damping force to control the super-structure to develop a minimum earthquake force and to minimize the free vibration times after earthquake.

(4) Locking Device

Big earthquakes are very seldom. In usual time, the vibration of buildings subjected by winds and some other horizontal forces should be protected. There are locking devices for fixing the building when the rocking column system is adopted.

There are two kinds of locking devices :

1. Electo-magnetic locker (Fig. 8). A V-type bracing under the first floor is connected to the super-structure, and an electrical magnet (Π shape) is placed at the lower end of the V-type bracing. On the ground, a steel plate just opposited to the electrical magnet, and the steel plate is fixed on the ground by four cables, so that the steel plate can not move in horizontal direction but can move a little in the up and down direction. Usually, the steel plate is hold by the magnetic force of the Π-type electric-magnet when there is not earthquake. When a strong earthquake occurs, a seismological instrument is triggered and this command a relay to shut off the electricity power supply, then the magnet will loss the suction force and the steel plate will fall down by its weight. Then the super-structure will be supported by the rocking columns only, isolating the building from earthquake and so protectives the whole building during earthquake. When the earthquake stops, the time relay will connect to the electric power supply again and the magnet will suck the steel plate , then the building will be locked on the ground again.

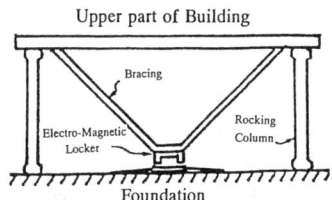

Fig. 8

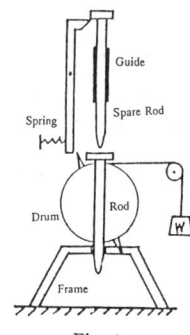

Fig. 9

2. Mechanical locker (Fig. 9). A V-type bracing under the first floor. A brittle bar (ceramic or cast iron bar) is installed at the lower end of the V-type bracing, that is to say, the brittle bar is connected to the super-structure. On the opposite side, there is a rotation drum with a hole and fixed on the ground. The brittle bar can resist the wind load and some other horizontal forces. If a strong earthquake attacks, When the inertia force of the super-structure exceed a limit, the bar will be broken, then the building will be supported by the rocking column system only. It is a good isolation technique, and the super-structure will be safe during the strong earthquake. At the same time, the drum rotates slowly. After a while (say one minute) when the earthquake stop and the drum rotates just half cycle (180°), then a new brittle bar will fall down into the hole, and the building will be fixed to the ground again.

Comparing these two types of locker :

Electro-magnetic Locker	Mechanical Locker
Controlled by the seismological instrument starters of a presetted acceleration	Controlled by a definite shearing force in the first floor.
Need electric power supply.	No power supply needed.
Suitable for high-rise buildings.	Suitable for low-rise building and Industrial facilities.

(5) Model Experiments

To verify the isolation effect of the rocking column system, three couples of high-rise building models were tested on the shaking table. (Photos 3,4,5,6,7,8).

Photo 3

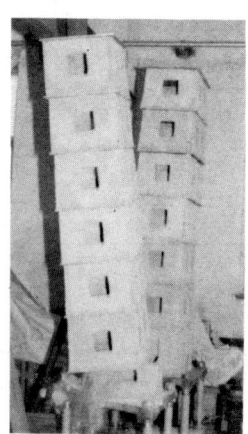

Just Broken

Photo 4

256

In each couple, one model is ordinary constructed (beams and columns are connected rigidly), another model is supported on rocking columns. The upper part of the building models are the same. During vibration tests, the responses of ordinarily constructed models were very severe and these models collapsed in a short time.* On the other hand, the responses of isolated models were very small. After vibration, there is no any crack in the isolated models. Compare these results, it proves the effect of the rocking column is good enough for isolating big earthquakes. In Photo 3, Photo 5, Photo 7, three couples of models before shaking are shown. In Photo 4, Photo 6, Photo 8 show the scenes of the models after shaking. Isolated models still stand, but un-isolated models are all collapsed.

Photo 5

Photo 6

Photo 7

Photo 8

* A video of these experiments is available.

(6) Analysis

The earthquake responses of two 16-story buildings with and without soft-first-story were analyzed. Their plan and elevation are shown in Fig. 10.

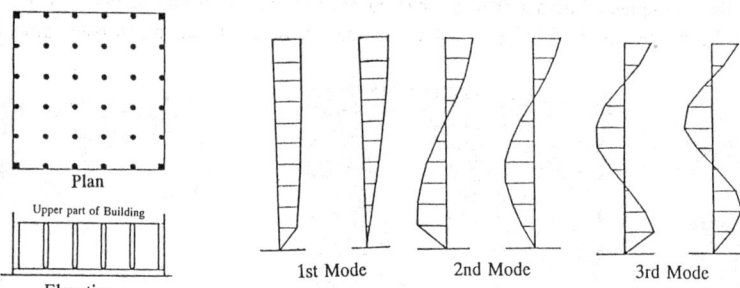

Plan

Upper part of Building

Elevation

Fig. 10

1st Mode   2nd Mode   3rd Mode

Fig. 11

There are 36 columns in each story. The horizontal stiffness of each column (these columns are rigidly connected to beams except the first story) amount 30 ton/cm, then the total horizontal stiffness of each ordinary story was $K = 36 \cdot 30 = 1080 ton/cm$. In the first story, the four corner columns are rigid connected to beams and other 32 columns are rocking columns without any horizontal stiffness. So that the horizontal stiffness of the first story was $K = 4 \cdot 30 = 120\ ton/cm$. The weight of each story was $W = 450\ tons$. The mode shapes of these two buildings are shown in Fig. 11.

The relative displacement-time curves of a same story (between 8th and 9th floors) in these two buildings when subjected to the ground motion recorded during the El Centro, 1940 (N-S direction) are shown in Fig. 12. This figure shows that the responses of the soft-first-story building are much small than the other one. In Table 1, the maximum relative displacement (mm) of each story of these two buildings are given and can be compared.

Tab. 1

Number of Story	Ordinary Construct Building	Flexible First-story Building
16	4	2
15	9	3
14	12	5
13	16	6
12	19	7
11	21	8
10	22	8
9	23	8
8	24	8
7	25	8
6	25	8
5	26	9
4	26	8
3	27	8
2	28	8
1	39	168

(7) Conclusion

The soft-first-story is certainly a good seismic isolation technique when properly used. The main problem is the engineering practice.

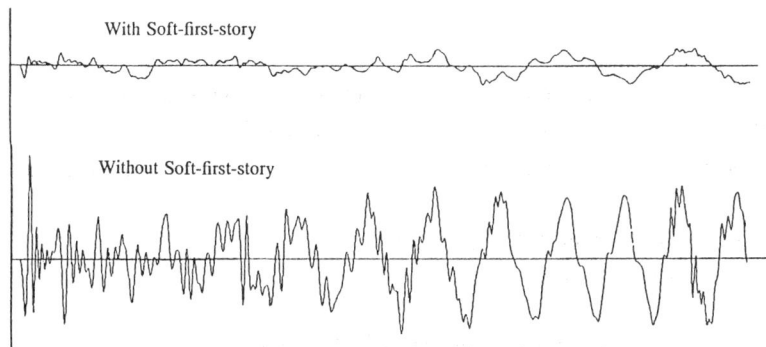

With Soft-first-story

Without Soft-first-story

Fig. 12   The Relative Displacements between the 8th floor and the 9th floor.

Rocking column is developed from the round ball bearing concept. Of course, it has the advantage of round ball bearing, i.e. it can carry the big vertical load but without stiffness in horizontal direction. On the other hand, it is easy to construct, durable and in-expensive. So that, it is an ideal seismic isolation technique.

In the example presented, a design combining rocking columns and fixed end columns has been used. It is a soft-first-story design, but it has very strong column for carrying the vertical load. It does not only isolate the horizontal ground motion, but it can resist a little amount of horizontal force also. Combined with the damping system, the side sway of the building will be limited. By the way, there has not any additional vertical vibration when the building vibrates horizontally, because the rocking columns are round balls, (central part of big round ball). So that the super-structurte will swing horizontally only on the "round balls" without vertical motion. Therefore it is easy to analyze. If there were additional vertical motion produced when the deformed balls (such as ellipsoid or other types) are used, the analysis would be more complicated, and would got some trouble in the design.

Rocking column is a column with spherical cap and spherical boot only. In construction, it may be precasted and then placed on the foundation easily, such as the construction of the ordinary precasted column. So that, it is in-expensive, especially, the anti-seismic fares for the super-structure may be cut down when this isolation technique is used, then the total cost of the whole building may be less than that of a non-isolated building.

Rocking columns can be used not only for high-rise buildings but also can be used for frames of industrial facilities. In the high-rise building, the rocking columns can be placed not only in the first story, but also in any desired stories to reduce the horizontal forces at that stories.

This work was done in China by the support of Chinese National Science Foundation.

Thanks to Prof. Vitelmo V. Bertero, he help me to translate this paper from Chinese into English.

259

References

(1) Martel, R.R.: "The Effect of Earthquake on Buildings with Flexible First Story." Bulletin of Seismological society of America.   Vol.19, No.3, pp.167-178. 1929.

(2) "Report on the San Fernando Earthquake, Feb. 9, 1971." Bulletin of Seismological Society of America. 1971.

(3) "Harzards of Tangshan Earthquake." (In Chinese)

   "唐山大地震震害"，地震出版社，1986

(4) "The report on Harzards in Tonghai Earthquake." (In Chinese)

   "云南省通海地震震灾调查报告"，国家地震局工程力学研究所资料

(5) Kuo,L.: "On the tests of seismic isolation & wind protection." (In Japanese)

   冈隆一： "免震耐风构造に关すゐ诸实验"，日本建筑学会论文集, 1939. 4. P. 204

(6) Izumi, M.: "Researchs on the reduction seismic forces acted on the buildings." (In Japanese)

   松下清夫等："建筑物に加わゐ震力の低减方法に关すゐ研究" 日本建筑学会论文

   报告集 No. 122. pp. 15-22. 1966.

(7) Flores, M.C.: "Building will roll on bearing during earthquake." Engineering News Records. Sept. 24, 1970. p.14.

(8) Lee Li : "Advances in Base Isolation in China." Proceedings of 3rd International Conference on Soil Dynamics & Earthquake Engineering. pp.297-309.   Princeton Univ. 1987.

Proceedings of the Second Conference on Tall Buildings in Seismic Regions
55th Regional Conference
May 16 and 17, 1991, Los Angeles, California

The Wind Response of Elevator Systems in a Tall Building

J. Jong Lou

Weidlinger Associates, Inc., New York, NY 10001
Polytechnic University, Brooklyn, NY 11201

ABSTRACT

The elevator outages of a tall building have been associated with wind induced building motion. It is observed that under most wind conditions, only the first pair of translational orthogonal modes of vibration play a significant part in the response. However, on a short-term basis, it has been possible to see a short period of the torsional mode being dominant and elevator malfunctions during this same time period are reported. Simplified physical models are constructed to study the possibilities of elevator cabling swinging resonantly with either the translational or torsional mode of building oscillation in the light of certain elevator operating policies and car parking procedures. Attempts have also been made to examine the possibilities of building undergoing strong torsional motion.

INTRODUCTION

Human perception to wind-induced motion of tall buildings has received extensive studies. However, it should be recognized that mechanical and/or electronic facilities housed in modern high-rise buildings, elevator systems, for instance, would also "perceive" the motion and can respond in an adverse manner.

The vulnerability of elevator system subjected to random lateral loads has been demonstrated in a number of earthquakes. The primary failure mechanism has been related to motion in the plane of counterweight frame due to building oscillation (1,2). With sufficient energy, the counterweight frame pounds against the guide rails which results in the deformation of one rail to allow the guides on the opposite rail to disengage. The counterweight system can thus swing free of the undeformed guide rail.

Modern tall buildings tend to be relatively flexible and lightly damped. Fluctuations in wind forces acting on these buildings can cause excessive motion. The wind induced building motion will impart amplitudes to the elevator's rope and cable systems. These amplitudes may build up to self-damaging proportions due to the small amount of inherent damping of the hoist, compensating, and traveling cables and to the near resonance of these ropes and cables to the fundamental frequencies of the building (3).

261

VIBRATING STRING MODELS

The basic elements of an elevator system are: a sheave, counterweight, an elevator car, a compensating cable, a hoist rope and a traveling cable (Fig. 1)(4). In most instances, the compensating cable is left to hang free without being threaded over a sheave. In high speed high rise installation elevator systems usually employ a bottom sheave or similar guidance system (Fig. 2).

The rope and cable systems can be modeled as vibrating strings, considering cable physical layout and configuration together with relevant boundary conditions (Fig. 3). The fundamental frequency f for the system shown on the left in Fig. 3 can be expressed as follows:

$$f = \frac{1}{2L} \sqrt{\frac{T + \frac{W}{2}}{m}} \tag{1}$$

where T = Applied tension on the cable,
      W = Total weight of the cable, and
      m = Mass of cable per unit length.

For the system shown on the right in Fig. 3 the fundamental frequency f is given by

$$f = \frac{1}{FL} \sqrt{\frac{g\left(\frac{W}{2}\right)}{w}} \tag{2}$$

where w = Weight of cable per unit length,

$$F = 2\pi \sqrt{\frac{4\pi}{(\pi^3 + 4\pi + 64Ea)}} \tag{3}$$

E = Young's modulus of elasticity.

Simple algebraic manipulation of these equations with appropriate parametric values confirms that the elevator cabling systems can assume a wide spectrum of natural frequencies.

Present elevator operation policies for high-rise buildings generally employ an accelerometer monitoring system which more often than not is located close to the core. During high winds the elevators can be programmed to park at positions so that elevator cabling will not swing sympathetic to the fundamental lateral frequencies, i.e., detuned for those frequencies. Another corrective action is to reduce the elevator speed. This is thought to prevent unfavorable phase relation of guide rails to guide roller positions.

Wind induced motion on the elevator systems may have similar effect as a minor earthquake. While details of elevator malfunctions attributable to wind induced building motion have not been well documented, Kabelac (5) reported that earthquake damage to elevators, especially in high-rise buildings, is primarily cable related.

It has become clear why certain buildings may be vulnerable to second mode of vibration, be it translational or torsional, so long as it contains sufficient energy since higher modes of vibration are not detuned and the monitoring system, located near the geometric center of the structure, is insensitive to pick up torsional activity.

TORSIONAL MOTION OF HIGH-RISE BUILDINGS

It is observed that the second mode of vibration of the tower in frequency content is actually a torsional one. Its contribution in terms of displacement is no greater than several percent of the total under strong wind conditions. In this case only the first translational orthogonal pair play a significant part in the response. However, on a short-term basis, it has been possible to see a short period of the second, i.e., torsional, mode being dominant (in terms of acceleration response). It is interesting to note that elevator malfunctions during this same time period were reported.

A direct evidence can be seen from Fig. 5 where the acceleration power spectra for NE corner accelerometers exhibit a higher energy content in the first torsional mode than do the first East-West lateral mode. This is significant because in most cases the energy level associated with higher modes of vibration is of order of magnitude lower than that of the fundamental modes. This type of dominant torsional activity may suggest itself to the heretofore unexplained elevator malfunctions during this same time period.

On another occasion, traces of x, y accelerations at core with a wind speed of approximately 30 mph and ranges of angles of attack between $\alpha = 10^{\circ}$ to $\alpha = 30^{\circ}$ (Fig. 4) showed a clear across-wind response with acceleration level reached 25 milli g's peak-to-peak (1 - Fig. 6). A lapse of 10 minutes saw the x, y accelerations diminished to about 5 milli g's peak-to-peak under ever increasing wind (2 - Fig. 6). The building then resumed a significant across-wind motion with acceleration level exceeded 30 milli g's peak-to-peak (3 - Fig. 6). With the significant energy level associated with the response and a lightly damped building under increasing wind forces, it is inferred that the building is undergoing strong torsional motion. An elevator outage occurred during this same time period.

In an experimental study of the vibration of a square cylinder in a steady flow, Huh (6) found that for angles of attack of 15°- 25°, almost pure torsional motion resulted. This torsional motion is excited by the nonuniform pressure distributions on the faces of the cylinder due to seperation and reattachment of the flow. Isyumov et al. (7) studied wind induced torque on square and rectangular building shapes in suburban terrain. Their findings confirm that the principal contribution to the overall dynamic torque comes from asymmetries in the pressure distributions on the side faces. The problem is clearly associated with across-wind pressures. That the tower has a strong across-wind response, and that the angle of attack lies approximately within 15°- 25°render some plausible support that the tower underwent strong torsional motion (2 - Fig. 6).

CONCLUSIONS

1. The elevator systems housed in high-rise buildings can respond to wind induced building motion in an adverse manner.

2. Evidence suggests that, from time to time, on a short-term basis, certain high-rise buildings can undergo "pure" torsional motion. Elevator malfunctions have been associated with torsional motion of the building if its energy level is comparable to that of fundamental modes of vibration.

3. Details of elevator malfunctions attributable to wind induced building motion and their frequency of occurrence need to be documented.

4. Elevator accelerometer monitoring systems should be placed at corner of the building to reflect the possibilities of strong torsional motion. Elevator systems may need to be detuned for higher modes of vibration.

5. Constraint of hoist and compensating cables for high speed high rise elevator arrangement may curtail unfavorable response to wind induced building motion.

6. Much research needs to be done on the identification of mechanisms for torsional excitation.

ACKNOWLEDGEMENTS

The author wishes to express his indebtedness to Leslie E. Robertson Associates for portion of the supporting materials reported in this study. Many thanks are due to Dr. Thomas C. Jan of DRC Consultants, Inc. for his valuable assistance in obtaining the equations in the manuscript.

REFERENCES

1. Yang, T.Y. et al., "Dynamic Response Analysis of Elevator Model," Journal of Structural Engineering, ASCE, Vol. 109, No.5, May, 1983.

2. Tzou, H.S. and Schiff, A.J., "Structural Dynamics of Elevator Counterweight Systems and Evaluation of Passive Constraint," Journal of Structural Engineering, ASCE, Vol. 114, No. 4, April, 1988.

3. Beedle, L.S., "Tall Building Systems and Concepts," Council on Tall Buildings & Urban Habitat, Van Nostrand Reinhold, 1981.

4. Barney, G.C., "Elevator Technology," Ellis Horwood Limited, 1986.

5. Kabelac, J.L., "What should an elevator do when an earthquake occurs?," Proceedings of the 50th Regional Conference on Tall Buildings in Seismic Regions, Los Angeles, CA, 1988.

6. Huh, C.K., "The Behavior of Lift Fluctuations on the Square Cylinders in the Wind Tunnel Test," Proc. 3rd Int. Conf. on Wind Effects on Buildings and Structures, Tokyo, 1971.

7. Isyumov, N. and Poole, M., "Wind Induced Torque on Square and Rectangular Building Shapes," Journal of Wind Engineering and Industrial Aerodynamics, 13, 1983.

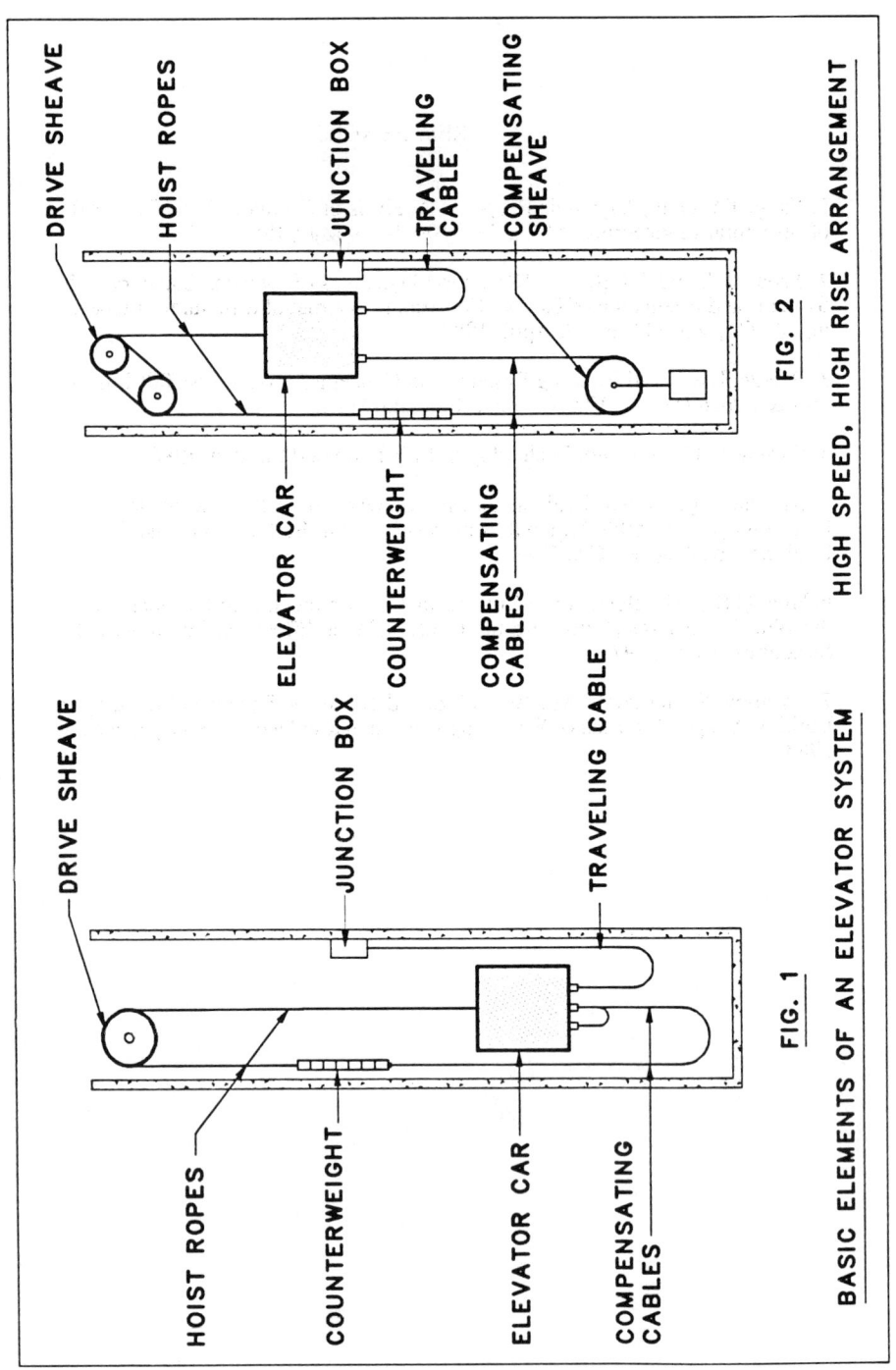

DRIVE SHEAVE

HOIST ROPES

JUNCTION BOX

TRAVELING CABLE

COMPENSATING SHEAVE

ELEVATOR CAR

COUNTERWEIGHT

COMPENSATING CABLES

TRAVELING CABLE

FIG. 2

HIGH SPEED, HIGH RISE ARRANGEMENT

DRIVE SHEAVE

JUNCTION BOX

HOIST ROPES

COUNTERWEIGHT

ELEVATOR CAR

COMPENSATING CABLES

FIG. 1

BASIC ELEMENTS OF AN ELEVATOR SYSTEM

266

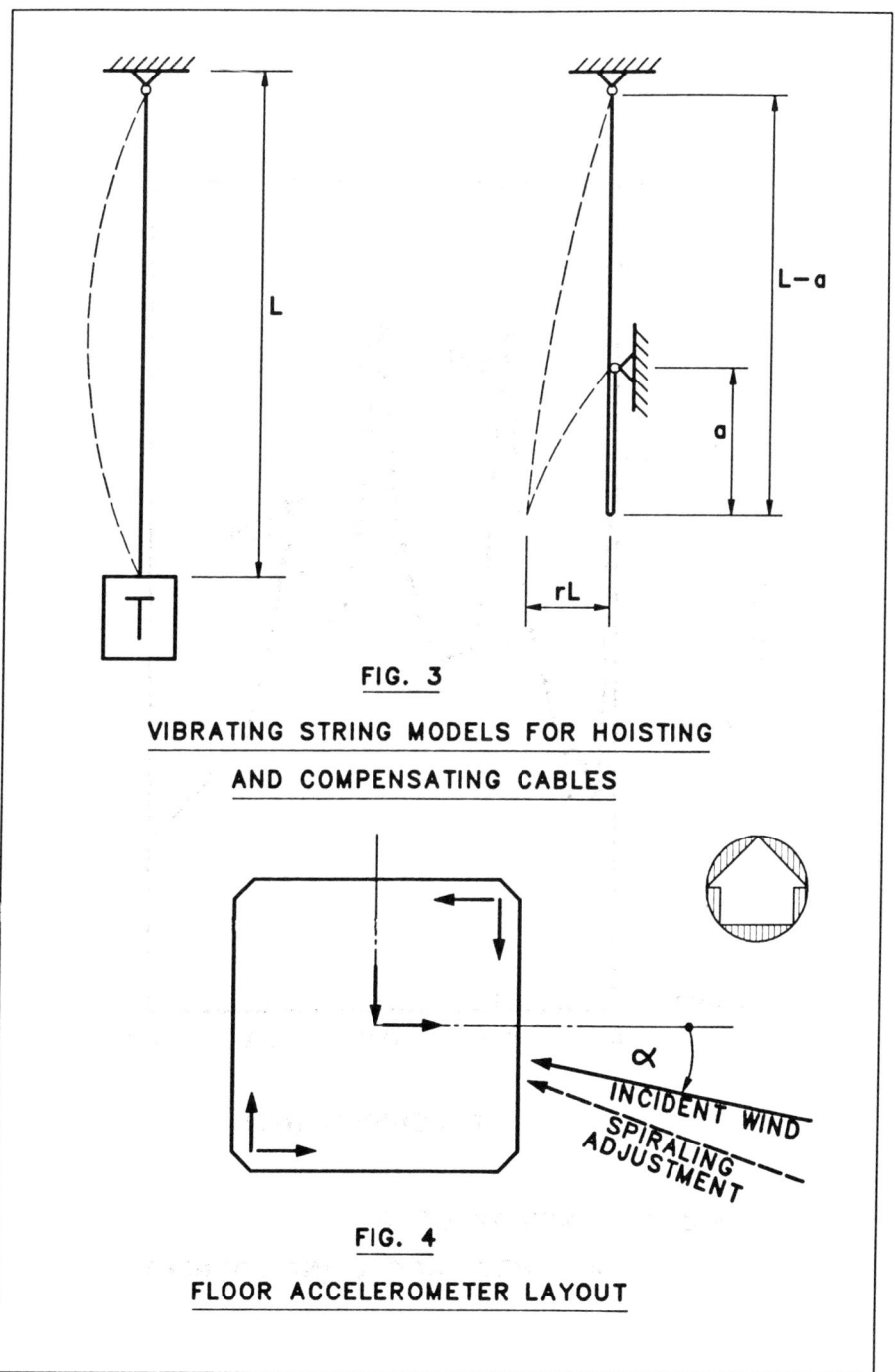

FIG. 3

VIBRATING STRING MODELS FOR HOISTING

AND COMPENSATING CABLES

FIG. 4

FLOOR ACCELEROMETER LAYOUT

267

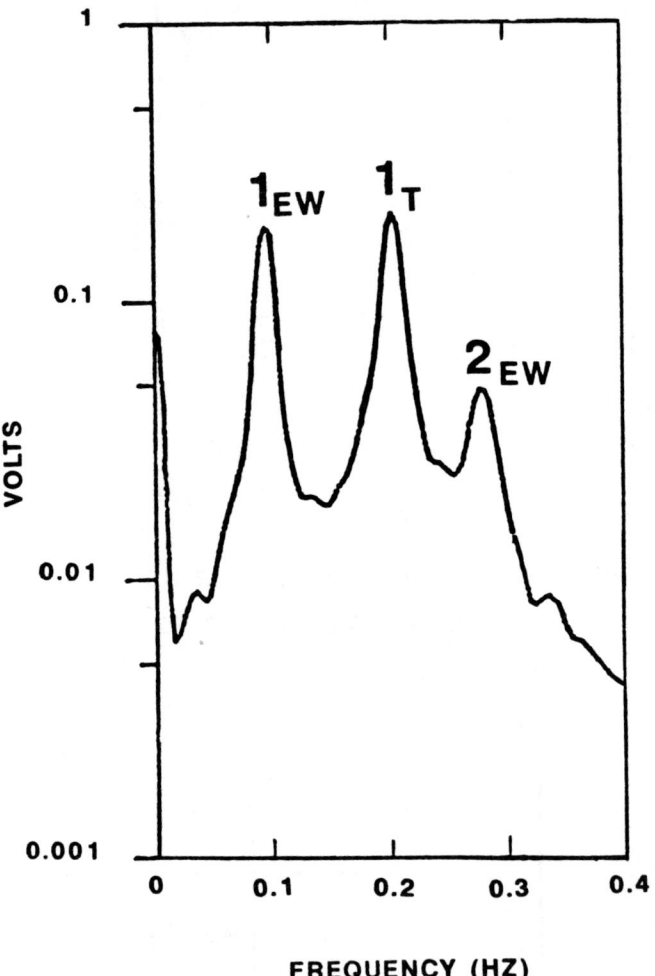

FIG. 5 POWER SPECTRA
EAST-WEST ACCEL. (NE CORNER)

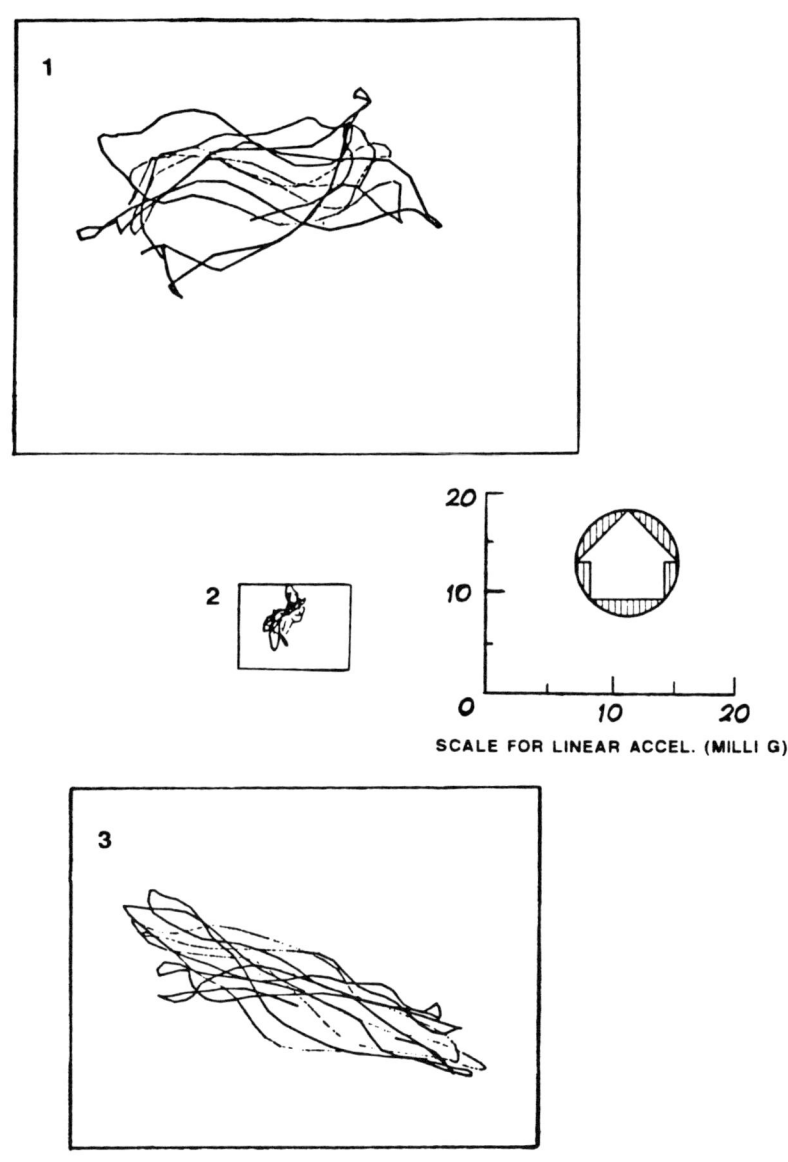

1

2

20

10

O

10   20

SCALE FOR LINEAR ACCEL. (MILLI G)

3

**FIG. 6   TRACES FROM NS AND EW
ACCELEROMETER RECORDS AT CORE**

Proceedings of the Second Conference on Tall Buildings in Seismic Regions
55th Regional Conference
May 16 and 17, 1991, Los Angeles, California

A CASE STUDY ON 'FIGUEROA AT WILSHIRE'

Mr. Oki Komada
Chief Executive Officer & President
Mitsui Fudosan, USA, Inc.

Mr. James C. Buie, Jr.
Senior Vice President
Hines Interests Limited Partnership

David C. Martin, AIA
Partner in Charge of Design
Albert C. Martin and Associates
811 West Seventh Street
Los Angeles, Ca. 90017

The fifty two story Figueroa at Wilshire Office Tower will just be completed as the Council convenes in May. It represents one of the exciting "new generation" of world class towers just being completed in Los Angeles. It also makes significant contributions to the City at a number of levels. This presentation will deal with the relevant issues that owner, development manager, and architect dealt with over the 10 year history of the site development. But first, the players.

Mitsui Fudosan, Project Owner, is a 300-year old Japanese real estate company, one of Japan's largest and most reputable. They became owners of the property in 1981. Development Manager, Hines Interests Limited Partnership, perhaps the nation's most prestigious developer of large scale urban projects. They joined with Mitsui Fudosan to jointly develop Figueroa at Wilshire in 1985. Albert C. Martin and Associates, Architect. A major architectural and engineering firm established in Los Angeles in 1906. Responsible for many Southern California landmarks, both historic and contemporary.

Mitsui Fudosan cautiously moved into the International development market in 1973, as a result of the liberalization of Japanese banking laws. Their activities over the last 18 years have been highlighted by the acquisition of the Los Angeles A.T.& T. Center, The Exxon Building in New York and now by the completion of the Figueroa at Wilshire project. The Figueroa at Wilshire site took an unusually long period to develop (10 years) because of several hurdles the team has successfully overcome. The purchase of the original site by Mitsui, was in the early 1980's from St. Paul's Church. Two adjacent parcels were acquired with much difficulty in 1983 to form the entire site 69,195 square feet. At this point, Mitsui asked Albert C. Martin and Associates to plan an approximate 1,000,000 square feet office tower. Between 1983 and 1986 two significant proposals were made to pre-lease the project (300,000 sq.ft.) each, but perhaps because of cultural differences, or simply misunderstanding neither pre-lease conditions succeeded. Finally, in 1986 at the bottoming of a real estate market, Mr. Tsuboi, Chairman of Mitsui Fudosan, made the decision to start construction without any major pre-leasing. This is when Mitsui brought on the American development manager, Hines Interests Limited Partnership.

In Los Angeles, Mitsui was the first foreign company to develop a major project, but Hines' involvement in this project was equally unique. Two points are worth mentioning. At any particular time the Hines Group is constructing five or more similar buildings across the U.S., that is 5 buildings in the 750,000 to 1,000,000 square feet range. As you can imagine, this kind of experience brought specific attitudes about the design and purchase of such major systems as elevators, structural concepts and purchase of exterior skin, etc. The second point deals with marketing. Hines Interests Limited Partnership has dealt with many of the major Architects in the country, realizing the value of design in the image of the project. They have also participated successfully in some of the most competitive dynamic markets in the County. This experience has helped them develop marketing tools that at every level imaginable, helps the potential tenant visualize the excitement of the design, while simultaneously understanding the depth and commitment to quality. Hines created for Figueroa at Wilshire, brochures, graphic identity, videos, fantastic, large scale models (15 x 15 x 12) with the hydraulically operated, moving walls, video back drops and real water in the fountains. Room window mock-ups, photographic view simulation, extensive material boards, all to help one visualize and to some extent, fantasize, how the environment will be built - and it worked.

Albert C. Martin and Associates worked for Mitsui Fudosan since 1980, but the real activity started when the project team was formed and development in detail began in 1986. The mildly provincial firm was confronted with two team mates with different cultures. The aggressive Texans with their specific technique, at times seems as foreign at the Japanese. However, after an initial "getting to know you" period, the team was united by a quest for the quality both in material and design. The Martins sought to find a solution which reflected the desire of the team, that is quality, function, uniqueness and yet respond to a number of issues which helped make the project compatible with Los Angeles. The Martins were interested in making a design that had regional roots rather than "International Style" architecture. Although contemporary in design, the building profile has a terraced top and strong vertical emphasis which recalls many significant Los Angeles structures of the Art Deco period. The building, like several of the Los Angeles towers, has a 45 degree rotation, but to square off with the skyline an octagon is formed and to square off to Figueroa two large atriums form a street wall. At the pedestrian level, a number of "moves" were made to make the project friendly to the city. The tower was shifted north to create a small sunlit plaza on the south of the site. This plaza is activated with food service, umbrellas and chairs, landscape, recirculating fountains to mask street noise, and the edge of almost all fountains and planters are designed to 14" to allow the community to stop, sit, rest and appreciate city life.

The bronze obelisk art work by artist Eric Orr has a 35' water cascade, but offers a wonderful urban surprise of catching on fire every 30 minutes, for 2 minutes - I mean a 35' flame at the corner of Wilshire and Figueroa. Because the buildings along Figueroa do not form a definitive street edge, the Figueroa facade is set back just far enough to create a pedestrian promenade formed by a canopy of street trees and diamond pattern granite pavers. The atriums are large skylit spaces but are unique by the addition of lattice-like sunscreens. They form two transition spaces which connect the outdoors to the interior elevator lobbies. Brazilian granite, Italian marble and back light Portugese marble add a richness often missing in contemporary structure.

Summary

The results of 10 years of patient planning and 5 years of active design and construction has created a structure which apparently is extremely well accepted by the community. It is financially successful with 85% lease up at opening and both Mitsui Fudosan and Gerald Hines Interests are considered vital, active and permanent players in the Los Angeles community.

Proceedings of the Second Conference on Tall Buildings in Seismic Regions
55th Regional Conference
May 16 and 17, 1991, Los Angeles, California

THE CALIFORNIA PLAZA COMPLEX

John A. "Trailer" Martin, Jr., and Farzad Naeim

John A. Martin and Associates, Inc., Los Angeles, California

ABSTRACT

California Plaza Complex is the largest urban revitalization project in a zone of high seismicity in North America. Covering five blocks in downtown Los Angeles, the project is being developed in accordance with a master plan approved by the Los Angeles City Council and Community Redevelopment Agency. The estimated total construction cost of the complex is $1.2 billion.

INTRODUCTION

California Plaza Complex is the largest urban revitalization project in a zone of high seismicity in North America. The 11.2 acre site of the project is located at the crest of Bunker Hill in downtown Los Angeles. The site is bounded by Grand Avenue, Fourth Street, Hill Street, Olive Street and Second Street. The Metro-Rail system will bring commuters directly to California Plaza's station at Fourth and Hill Streets.

When completed, the project will include three office towers totaling 3.5 million square feet, the Museum of Contemporary Art or MOCA, 750 residential units, a 469 room hotel, 160,000 square feet of retail space, on-site parking facilities for 4,650, the recreation of Angel's Flight Funicular Railway, A Bunker Hill history museum, the Dance Gallery, and a 5.5 acre park system of gardens, terraces and water features, including a major performing arts plaza.

The first phase of the project which was completed in 1985 includes: One California Plaza, a 42-story, one million square feet office tower; the Museum of Contemporary Art; the Spiral Court Amphitheater; 27,000 square feet of retail space; and eight levels of parking structures (four above and four below the ground level) accommodating 1,170 parking spaces.

The Second phase of the project which is currently under construction (to be complete in 1992) includes: Two California Plaza, a 52-story, 1.3 million square feet office tower; The Watercourt, a 1.5 acre central performance plaza; the California Plaza Hotel; the Museum Tower, composed of 217 condominium residences; 52,000 square feet of retail space; and 1,263 parking spaces.

The third and the final phase of the project which is to start in 1994 includes: Three California Plaza a 50+ story office tower; Angel's Flight Funicular Railway and Bunker Hill History Museum; 533 condominium residences; The Dance Gallery, a 100, 000 square feet dance theatre and teaching institute; and about 32,000 square feet of retail space.

The California Plaza Complex is a development of Bunker Hill Associates which is a joint venture of Metropolitan Structures and Cadillac Fairview as general partners and Shapell industries and Goldrich & Kest as limited partners.

The overall California Plaza Master Plan is the work of Arthur Erickson Architects who are also design architects for the California Plaza I and II towers.

CALIFORNIA PLAZA TOWERS

Erection of structural steel for the Two California Plaza which is the second of three towers to be built within the complex, topped out in December 1990. The design of this tower is reminiscent of the completed One California Plaza with curving walls of glass anchored by a textured granite base. However, at 52 stories high, it is ten stories taller than its predecessor. The lateral system of both towers is composed of a curved framed-tube system with typical column spacing of 15 feet 6 inches. The third tower is believed to reflect on the in place architecture of the first two.

THE MUSEUM OF CONTEMPORARY ART

MOCA opened to public on December 10, 1986. The architectural design of the MOCA is the work of the famous Japanese architect Arata Isozaki. This 98,000 square foot building lies at the heart of California Plaza.

Seven years of planning and construction went into the new museum, Isozaki's first major commission in the United States. MOCA's design has praised for its originality and beauty. When plans for the Museum were unveiled to the press in 1983, the received acclaim form Time Magazine, Art in America, House & Gardens and several other major publications. MOCA is composed of two interconnected, relatively low buildings bracketing a sunken courtyard. The outside of the structure presents a profile of simple geometric shapes: barrel vault, cubes and rooftop pyramids. The inside spreads over seven levels and comprises seven interconnecting galleries, an auditorium, bookstore, cafe, sculpture court, staff offices, library, support service areas and storage spaces for MOCA's permanent collection. As Isozaki asserts: "The symbolism of the building is not related to its volume. It is surrounded by gigantic buildings and is, in a sense in the middle of the valley of the skyscrapers .... That's why I broke the building into fragments -- little pyramids, the vault, small cubes. These elements are facing each other and, in a way, look like a small village inside the valley created by the skyscrapers." The structural system of the MOCA which is set on the top of the California Plaza One parking structure, is composed of a set of very complex and inter-related shear wall systems.

THE MUSEUM TOWER

Museum Tower, scheduled to open in late 1991, is a 22 story residential condominium complex. The structural system for this tower consists of a perimeter ductile concrete moment frame. lateral columns are spaced at 13 ft distances around the building perimeter and are connected with upturn spandrel girders. The exterior frame is composed of painted, exposed, architectural concrete.

The gravity system for the typical floor framing consists of 8 inch post-tensioned flat plate system with banded tendons running in one direction and uniform tendons running in the other direction.

The three base levels of the project are to be clad in red, clefted, finished sandstone and polished white travertine marble. The sandstone recalls the cladding of MOCA, and connects Museum Tower to the entire complex.

THE CALIFORNIA PLAZA HOTEL

The California Plaza Inter-Continental Hotel is scheduled to open in the fall of 1992. The project is currently under construction. The hotel consists of 17 floors above grade and two levels below grade. The lateral resistance is provided by perimeter reinforced concrete ductile moment frames. Typical hotel floor system is composed of 8 1/2 inches of normal-weight, post-tensioned concrete flat plates. One special structural feature of the project is the existence of four 75 ft. long, 14 ft high steel plate girders which carry the weight of the concrete hotel structure across the Third Street tunnel which crosses the site below the basement parking.

The hotel's granite base with sandstone detailing is reminiscent of the bases of the adjoining One and Two California Plaza Office towers and the texture of MOCA's sandstone. The hotel tower's residential appearance is a function of its distinctive sandstone finish in conjunction with a cementitious coating which interplays with its steel and glass details.

To distinguish it from surrounding towers of California Plaza curving painted metal panels will conceal roof-mounted mechanical systems and form a base for the emergency helipad.

ACKNOWLEDGEMENTS

The authors wish to express their gratitude to Ms. Susan Roberts of Metropolitan life Structures West, Inc. who provided them with some very important background information. Thanks is also due to all our office staff who were involved in structural design of California Plaza.

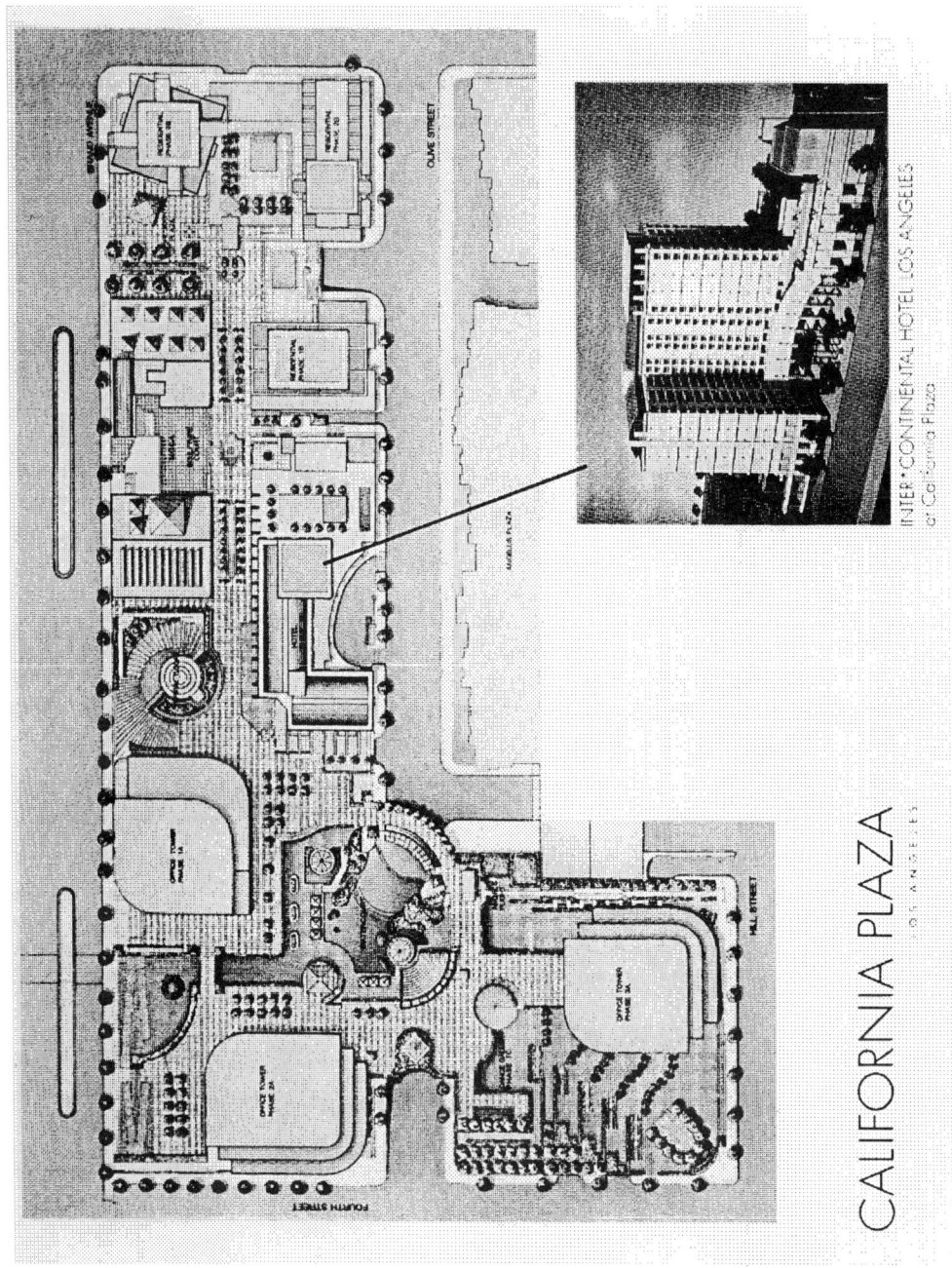

Figure 1. CALIFORNIA PLAZA MASTER PLAN

Figure 2. California Plaza Tower I

Figure 3. California Plaza Tower II under construction

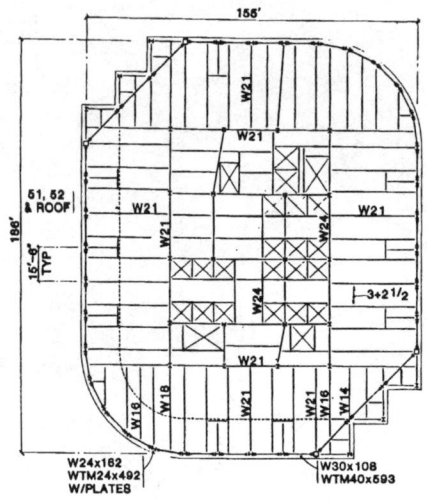

Figure 4. Typical Floor Framing Plan of California Plaza Tower II

Figure 5. A View of MOCA

Figure 6. Views of MOCA

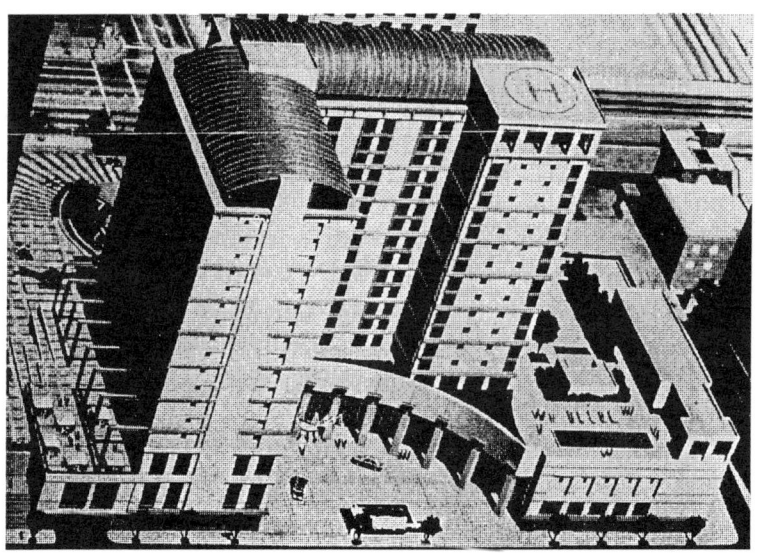

Figure 7. The California Plaza Hotel

Figure 7. An Artist's vision of the Watercourt

Figure 8. SpiralCourt Amphitheater

Proceedings of the Second Conference on Tall Buildings in Seismic Regions
55th Regional Conference
May 16 and 17, 1991, Los Angeles, California

SEISMIC RESPONSE OF A THIRTY-STORY BUILDING DURING THE LOMA PRIETA EARTHQUAKE

Eduardo Miranda, James C. Anderson and Vitelmo V. Bertero

ABSTRACT

This paper describes the seismic response of a thirty-story ductile moment-resistant reinforced concrete building during the October 17, 1989 Loma Prieta earthquake. A brief description of the building and its instrumentation is presented. System identification techniques were used to identify the dynamic characteristics of the building from the records obtained during the earthquake. These results are compared with those previously obtained by using ambient and small-amplitude forced vibration, showing that the periods of vibration of dominant modes obtained from the earthquake records are significantly longer than those obtained in small amplitude measurements. Moving window Fourier analyses were conducted to identify possible changes in dynamic characteristics during the earthquake. No changes in dynamic characteristics were identified in one translational direction, and only minimal changes in the other direction. Higher mode contribution was found to be a particularly important part of the response, indicating the importance of dynamic analysis in the design of this kind of structures. Results from simplified analyses are presented to evaluate the effectiveness of simplified (linear elastic and time-invariant) time history analyses to capture the response of tall buildings to moderate earthquake shaking. The absence of damage (structural and non-structural) is evaluated and discussed.

INTRODUCTION

On October 17, 1989 a 7.1 surface wave magnitude earthquake struck northern California. This earthquake, the largest magnitude earthquake in California since 1906, caused 62 deaths, approximately 3750 injuries, and more than $8 billion in damages in the San Francisco Bay area [1].

The Pacific Park Plaza is a 30-story symmetrical three-winged Y-shaped structure built in 1983, located in the city of Emeryville next to San Francisco Bay. A typical plan of the building is shown in Fig. 1. The structural system is ductile moment-resistant reinforced concrete frames. The site is underlain by a layer of soft silty clay known locally as Bay Mud. The foundation of the building consists of a 5 ft. thick concrete mat and 900 prestressed concrete piles, 60 to 70 feet long. Non-structural elements consist of precast lightweight concrete elements in all facades of the building with interior partions located at beam lines.

(EM,VVB) Dept. of Civil Engineering, University of California at Berkeley, Berkeley CA, 94720.
(JCA) Dept. of Civil Engineering, University of Southern California, Los Angeles, CA 90089.

The building forms part of the strong-motion network operated by the United States Geological Survey (USGS). The instrumentation consists of 21 CRA-1 analog acceleration sensors distributed over the three wings and central core on the 13th, 21st, and 31st level (roof), and at the ground level. Additionally, there is a 3-component CRA-1 free-field analog accelerometer 40 meters (131 ft) north of the building, and a 3-component SMA-1 free-field analog accelerometer 100 meters (328 ft) south of the building. Fig. 2 shows the location of instruments within the building.

The Pacific Park Plaza Building is located approximately 97 km (60 miles) north of the epicenter of the Loma Prieta earthquake. Major damage occurred within 5 km (3 miles) of the building, including the collapse of the Cypress Street Viaduct, the collapse of one segment of the San Francisco-Oakland Bay Bridge, and damage to facilities at the Port of Oakland. The building suffered no damage during the earthquake (neither non-structural nor structural). The parking structure next to the building experienced flexural cracks in the floor system due to north-south motion of the structure, as well as shear failure of two columns in the first floor.

A total of 27 accelerograms were obtained in the Pacific Park Plaza Building in the Loma Prieta earthquake [2]. In general, recorded motions are characterized by a strong phase lasting approximately 9 seconds. Table 1 lists peak values of acceleration, relative velocity and relative displacement for each instrument location.

The objectives of these paper are first, to describe analytical studies conducted to obtain the dynamic characteristics of the building from the records obtained during the earthquake using system identification techniques; and second, to evaluate the effectiveness of simplified, linear elastic and time-invariant, time history analyses to capture the response of tall buildings under moderate earthquake shaking.

SYSTEM IDENTIFICATION

The occurrence of an earthquake can be viewed as a full-scale, large-amplitude experiment on a structure, and if the structural motion is recorded, it offers the opportunity to make a quantitative study of the structure at dynamic force and deflection levels directly relevant to earthquake-resistant design. In this study, three different frequency domain system identification techniques were used. These were: i) Non-parametric, time-invariant; ii) Non-parametric, time-variant (moving window Fourier analysis); and iii) Parametric time-invariant.

In the first technique the structure is idealized by a non-parametric (black box), time-invariant linear model in which the dynamic properties are determined from the transfer function, $H(i\omega)$, defined as the ratio of the Fourier transform of the input and output signals. The second system identification technique (non-parametric, time-variant) is essentially the same as the approach described above except that in order to identify the variation of structural parameters in time, a window smaller than the total duration of the record, is "moved" in time. In this study, a window with a duration of approximately four times the fundamental period of the building is employed.

In the third identification technique the structure is idealized by a simple mathematical model which defines the input-output relation of the building. The parameters of the model are adjusted through least square procedures to minimize the difference between the smoothed Fourier transform of the recorded response (output) and the Fourier transform of the computed response [3].

Fig. 3a shows the Fourier amplitude spectra of input and output signals, for the 260° component recorded motions. The input signal (dotted line) corresponds to the motion recorded at the ground level, and the output signals (solid lines) correspond to the motions recorded in the central core at the 31st, 21st, and 13th levels. It can be seen that the ground motion has its strongest input in a band between 0.6 and 1.2 Hz and around 2.0 Hz. Transfer functions corresponding to motions recorded in the central core in the 260° component are shown in Fig. 3b. From this figure, 1st, 2nd, and 3rd modes were identified at 0.37 Hz, 0.94, and 1.81 Hz, respectively. Identified translational periods and mode shapes for both directions of the building are summarized in Table 2.

Estimation of the damping ratio in the first two translational modes in each direction was found to vary depending on the resolution, filter and smoothing used in the signal processing. Their values range between 2.4 and 3.0%.

Ambient and forced-vibration measurements of the building were made in 1983, details of these measurements are given in Ref. 4. Table 3 compares the periods of vibration obtained through ambient vibrations, small-amplitude forced vibrations, and those obtained in the present study. As shown in this table there is a very good agreement between periods obtained through ambient and forced vibrations, however, there exists very large differences between the small-amplitude measurements and the results from the records obtained in the Loma Prieta earthquake. For 350° component the ratio of fundamental period measured during the earthquake to that measured through ambient vibrations is 1.51. For 260° component this ratio is 1.57. A comparison of mode shapes obtained with the three methods is shown in Fig. 4, where it can be seen that, unlike periods of vibration, mode shapes identified from earthquake records are very similar to those obtained in small-amplitude vibrations.

Results from moving window Fourier analyses showed no change in dynamic characteristics in the 350° component and only a small change for 260° component.

SIMPLIFIED ANALYSES

Earthquake time history analyses were conducted using a simplified model of the building consisting of two-dimensional models (one for each orthogonal direction) with only one degree of freedom per floor. These models are an extension of those used in the parametric system identification. The mechanical characteristics of these two-dimensional models were prescribed such that their dynamical characteristics matched those identified from the earthquake records. The purpose of these simplified models was to have a small model in which time history analyses could be performed relatively quickly in order to calibrate a detailed three-dimensional finite element model of the building, and to evaluate the effectiveness of linear elastic, time-invariant models to capture the response of the building.

A time domain comparison of the recorded and calculated response is shown in Fig. 5. This comparison corresponds to absolute acceleration, relative velocity, and relative displacement time histories of the west wing at the 31st level (350° component). Relative motions correspond to the difference of the motion at the roof and the recorded motion at the ground floor of the building (for this component). As shown in this figure, with the exception of the first 10 seconds, correlation between the recorded and calculated response is very good.

It was found that the relative importance of each mode in the total response depends on the response function (acceleration, relative velocity, or relative displacement). For relative displacements, the first mode response dominated the total response, for relative displacements the response was dominated by the first two modes, whereas for absolute accelerations, at least three modes were needed to adequately capture the recorded response. Fig. 6 shows how the calculated acceleration time history is improved with the incorporation of higher modes.

In addition to the time history analyses described above, response spectrum analyses were conducted using the 1988 UBC design spectrum based on a response modification factor (R_w) of 12 and a soil factor (S) of 2 (soil type 4).

Fig. 7 shows an envelope of displacements of the core of the building (260° component), resulting from the UBC response spectrum analysis and the time-history analysis unsing the recorded motions in the Loma Prieta earthquake. The maximum interstory drift computed in the time history analyses is 0.004 which correlates well with the absence of structural and non-structural damage. The displacements shown in this figure computed with UBC forces are multiplied by $3R_w/8$. It is important to note that the 1988 UBC requires that for irregular buildings (like the one studied here), element forces and displacements be scaled up by the ratio of the maximum base shear resulting from an equivalent static analysis to the maximum base shear resulting from the response spectrum analysis. The profile of scaled-up displacements is also shown in Fig. 6. The maximum roof displacement according to the 1988 UBC is 32% smaller than that observed in the Loma Prieta earthquake. Scaled 1988 UBC displacements underestimate observed displacements by 17%. The disparity in displacements is expected to be larger for closer and/or larger magnitude earthquakes.

CONCLUSIONS

- Periods of vibration identified from earthquake records are significantly longer than those previously measured through ambient and small-amplitude forced vibrations. Mode shapes inferred from small-amplitude vibrations are similar to those identified in the records from the Loma Prieta.

- Simplified time-invariant, linear elastic two-dimensional models of the building capture the recorded response relatively well.

- Maximum interstory drifts computed for the building explain the absence of damage during this earthquake.

- Displacements computed according to the 1988 UBC underestimate the recorded response of the building by 17%.

REFERENCES

1. Governor's Board of Inquiry on the 1989 Loma Prieta Earthquake, "Competing Against Time," State of California, May 1990.

2. Maley, R., et. al., "U.S. Geological Survey Strong-Motion Records From the Northern California (Loma Prieta) Earthquake of October 17, 1989," *Open-File Report 89-568*, Department of the Interior, United States Geological Survey, October 1989.

3. McVerry, G.H., "Frequency Domain Identification of Structural Models from Earthquake Records," *Report EERL 79-02*, Earthquake Engineering Research Laboratory, California Institute of Technology, Pasadena, CA, 1979.

4. Stephen, R.M., Wilson, E.L., Stander N., "Dynamic Properties of a Thirty-Story Condominium Tower Building," *Report No. UCB/EERC-85/03*, Earthquake Engineering Research Center, University of California, Berkeley, April 1985.

5. Building Officials Conference, "Uniform Building Code," *1988 Edition*, Whittier, California, 1988.

LEVEL	LOCATION	COMP.	FILENAME	PEAK ACC. [cm/sec^2]	PEAK VEL.[I] [cm/sec]	PEAK DISP.[II] [cm]
31st Level	West Wing	350	PLAZA4	257.4	30.69	5.83
31st Level	South Wing	050	PLAZA5	298.7	62.77	17.16
31st Level	North Wing	290	PLAZA6	466.7	70.08	17.32
31st Level	Central Core	350	PLAZA7	240.3	26.05	5.45
31st Level	Central Core	260	PLAZA8	359.1	77.42	19.45
21st Floor	West Wing	350	PLAZA10	185.6	13.02	3.77
21st Floor	South Wing	050	PLAZA11	165.4	26.05	8.46
21st Floor	North Wing	290	PLAZA12	235.9	30.08	6.60
21st Floor	Central Core	350	PLAZA9	179.4	15.00	5.98
21st Floor	Central Core	260	PLAZA15	239.2	32.1	8.76
13th Floor	West Wing	350	PLAZA16	206.3	19.21	2.77
13th Floor	South Wing	050	PLAZA17	218.4	34.42	7.80
13th Floor	North Wing	290	PLAZA18	303.0	40.84	8.11
13th Floor	Central Core	350	PLAZA13	265.6	25.37	4.24
13th Floor	Central Core	260	PLAZA14	253.7	40.13	9.10
Ground Floor	North Wing	350	PLAZA24	173.4	15.81	2.90
Ground Floor	North Wing	260	PLAZA22	208.3	37.40	6.66
Ground Floor	North Wing	UP	PLAZA23	46.8	4.17	0.94
Ground Floor	West Wing	UP	PLAZA19	55.9	4.36	0.90
Ground Floor	South Wing	UP	PLAZA20	55.4	4.34	0.87
Ground Floor	Central Core	UP	PLAZA21	37.5	4.24	0.79
Ground South	Free Field	350	PLAZA1	210.3	21.93	3.79
Ground South	Free Field	UP	PLAZA2	58.5	4.54	0.74
Ground South	Free Field	260	PLAZA3	252.8	40.84	8.15
Ground North	Free Field	350	PLAZA25	178.7	15.74	2.57
Ground North	Free Field	UP	PLAZA26	82.2	5.32	1.01
Ground North	Free Field	260	PLAZA27	225.1	37.94	6.67

I.- Velocities obtained from highpassed, baseline corrected integral with respect to time of recorded accelerations

II.- Displacements obtained from highpassed, baseline corrected integral with respect to time of computed velocities

Table 1. Peak responses in the building during the Loma Prieta earthquake.

PARAMETER	COMPONENT 350° (NS)			COMPONENT 260° (EW)		
	1st MODE	2nd MODE	3rd MODE	1st MODE	2nd MODE	3rd MODE
Period [sec]	2.59	0.89	0.46	2.69	1.07	0.55
Frequency [Hz]	0.39	1.12	2.15	0.37	0.94	1.81
Damping Ratio [%]	2.4-2.9	2.5-3.0	-	2.5-2.9	2.5-3.0	-
Mode Shape 31st	1.00	1.00	1.00	1.00	1.00	1.00
21st	0.69	-0.36	-1.02	0.63	-0.29	-0.79
13th	0.38	-0.82	0.60	0.34	-0.84	0.34

Table 2.   Translational dynamic characteristics identified from earthquake records.

COMPONENT	MODE	FORCED VIBRATION	AMBIENT VIBRATION	EARTHQUAKE RESPONSE
350° (EW)	1st	1.68	1.71	2.59
	2nd	0.60	0.59	0.89
	3rd	0.32	0.32	0.46
260° (NS)	1st	1.69	1.71	2.69
	2nd	0.60	0.59	1.07
	3rd	0.32	0.32	0.55

Table 3.   Comparison of translational periods of vibration

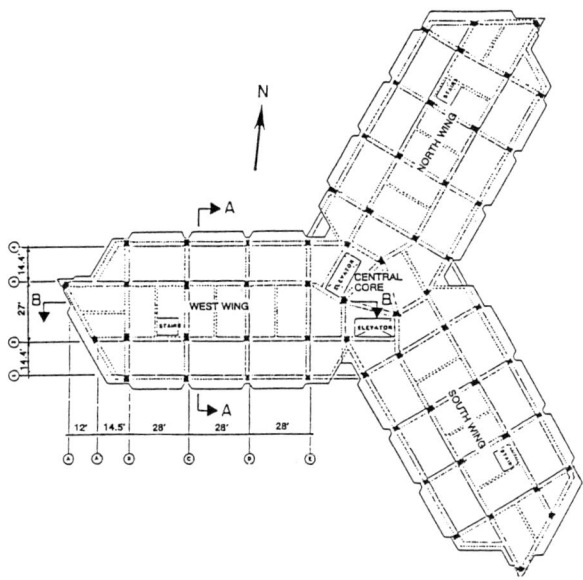

Figure 1. Typical floor plan of the Pacific Park Plaza Building.

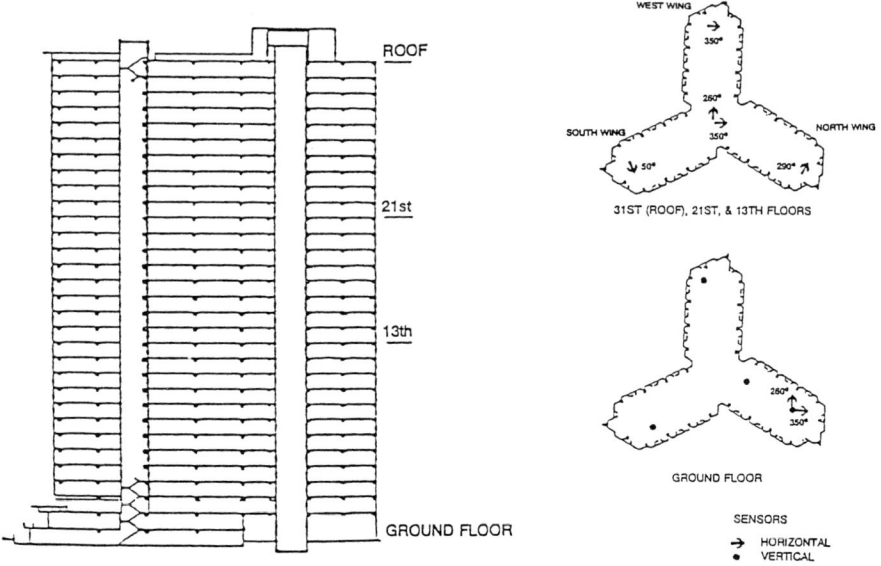

Figure 2. Location of instruments within the building.

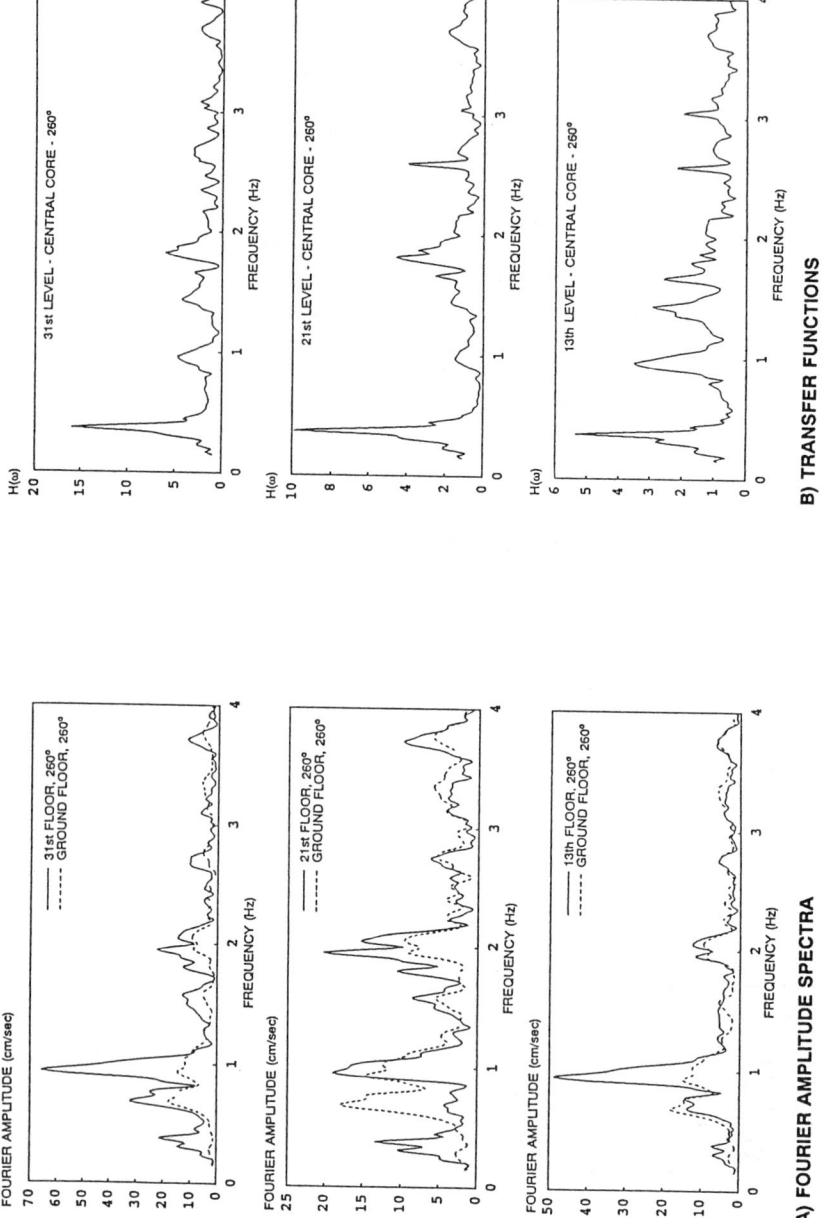

Figure 3. Fourier amplitude spectra and transfer functions in the central core of the building for the 260° component.

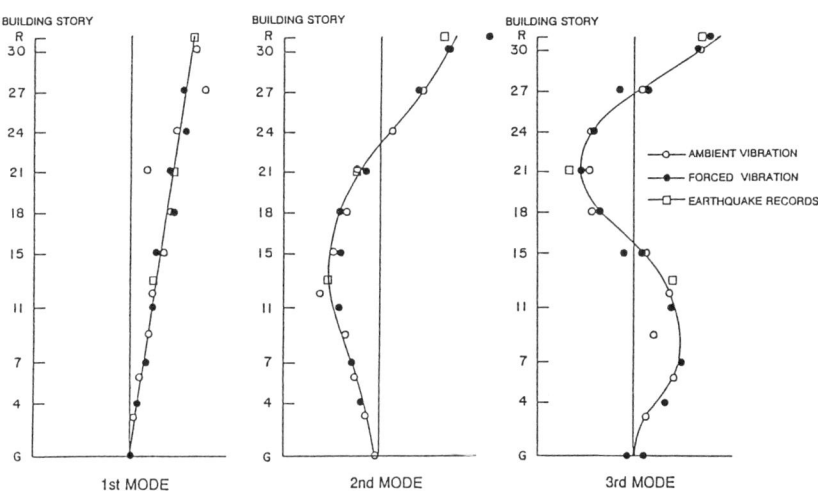

Figure 4. Comparison of translational mode shapes of the building (350°) obtained through three different procedures.

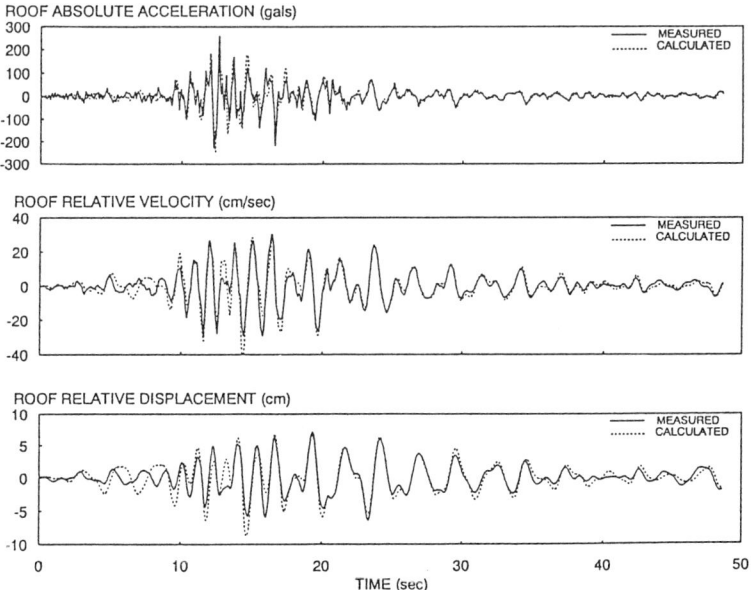

Figure 5. Comparison of measured and calculated response at the roof level of central core of the building (component 260°).

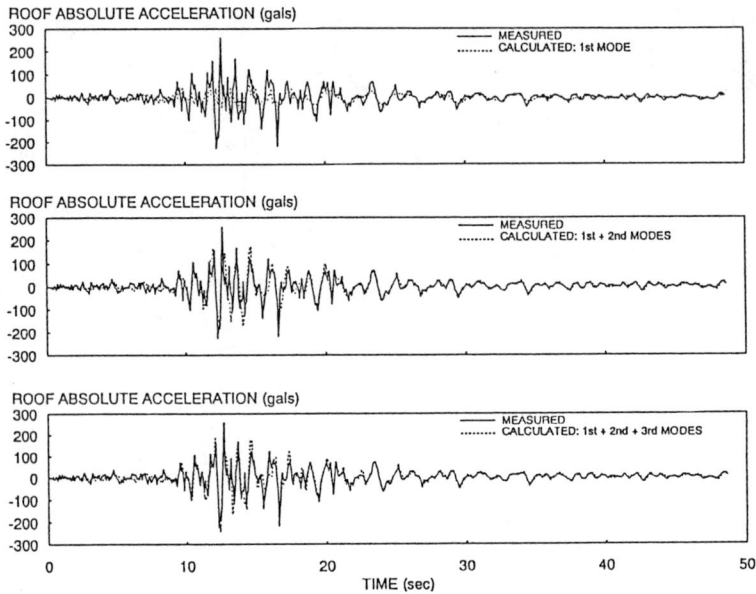

Figure 6. Influence of higher modes in calculated acceleration time histories at the roof of the building (component 350°).

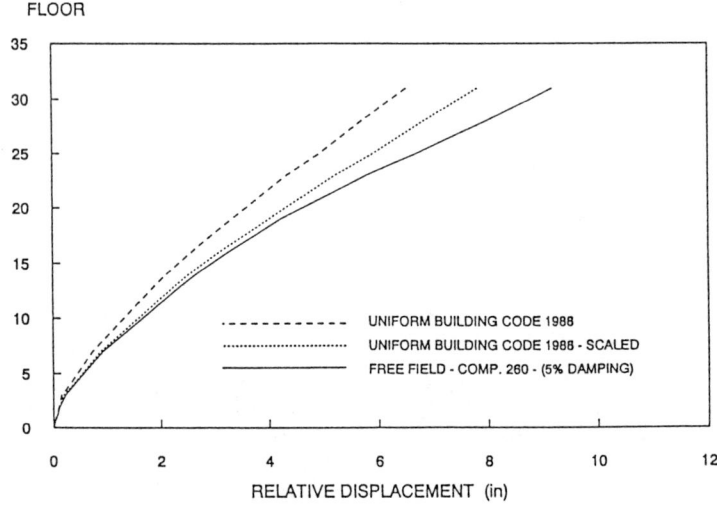

Figure 7. Comparison of calculated displacements in the Loma Prieta earthquake (260° component) and those obtained using the 1988 UBC.

Abstract of Presentation

NCNB Corporate Center
Charlotte, North Carolina

This presentation will discuss the development and design of an 870-foot-tall tower currently under construction in Charlotte, NC. When completed in 1992, this building will be the third tallest concrete frame building in the world. The tower was designed for an 80 MPH Basic Wind Speed, as well as Zone 2 seismic forces.

It will highlight the team process used to develop an economical, practical concrete frame structural system which responds to the leasing needs of the developer and prime tenant, as well as key architectural design criteria.

The tower utilizes a perimeter concrete "tube" frame with eight- and four-level-tall vierendeel truss frames to frame major building offsets in two locations. The discussion will include how the dimensions of the concrete frame were refined to integrate with the desired lease depths and open exterior area, as well as the intricate granite cladding.

Value engineering led to a change to 75 KSI reinforcing steel, saving $500,000. Designers also took advantage of a change in ACI 318-89 to reduce seismic column ties and save another $489,000.

The presentation will address how the project was evaluated for wind using state-of-the-art wind tunnel techniques. Although wind directionality is not addressed by most building codes, the project designers took advantage of the local wind climate to develop lower, but more realistic, frame forces to realize significant project savings without sacrificing safety. Designers found that the controlling design wind forces were essentially equivalent to the seismic forces as required by ANSI Zone 2.

The project team includes developer Lincoln Property Company; design architect Cesar Pelli & Associates; production architect HKS, Inc.; structural engineer Walter P. Moore and Associates, Inc.;[1] and general contractor McDevitt & Street Company.

---

[1] Walter P. Moore and Associates, Inc., 3131 Eastside, Second Floor, Houston, Texas 77098.

Proceedings of the Second Conference on Tall Buildings in Seismic Regions
55th Regional Conference
May 16 and 17, 1991, Los Angeles, California

CONTROL OF THE BEHAVIOR OF THE SYP BUILDING UNDER WIND ENVIRONMENT

Yoshinori Murai, Yuuichi Takase, Hiroshi Sugimoto and Hideo Nakashima
Design Division, Shimizu Corporation, Tokyo

Even in an earthquake-prone region like Japan, wind environment is one of the principal design considerations from the serviceability point of view as well as structural safety. However, optimum structural systems for earthquakes are not always the optimum structure for wind. For example, longer periods of building sway may reduce seismic force, but, increase wind-induced motion considerably. On the other hand, adding mass to a building decreases acceleration response to wind, but, increases seismic force which is proportional to mass. Therefore, control of mass, if possible, stiffness and damping of the building is necessary to design buildings to be safe, comfortable and economical.

The Shin Yokohama Prince Hotel Tower (SYP) is a cylindrical building with a height of 149.4 meters and a diameter of 38.2 meters (Fig. 1). Because of its cylindrical shape, wind engineering studies which include studies on wind-induced motion and studies on large negative pressure on curtain walls were done from the early stage of the design process. To improve behavior of this cylindrical structure under wind environment, as a result of the studies, following are necessary:

1. Cylindrical framed tube structure to obtain enough stiffness of the building to reduce wind-induced motion, but not to increase seismic force too much.
2. Tuned Liquid Dampers on top of the tower to increase damping for better serviceability.
3. Choice of appropriate surface texture to reduce negative wind pressure on curtain walls.

295

STRUCTURAL SYSTEM

The wind resisting system for this tower is a double framed tube structure with steel box columns and steel H-shape girders (Fig. 2). The outside tube has 30 columns, about 4 meters on center, and resists 80 to 90 percent of total wind shear. The inside tube has 15 columns. By locating columns as close to the building perimeter as possible, large moment of inertia of the total building is obtained. As a result, the flexural deformation (lateral displacement due to axial deformation of columns) is reduced to only 25 percent of top lateral displacement under wind.

Fig. 1   Model of The Shin Yokohama Prince Hotel, Yokohama, Japan

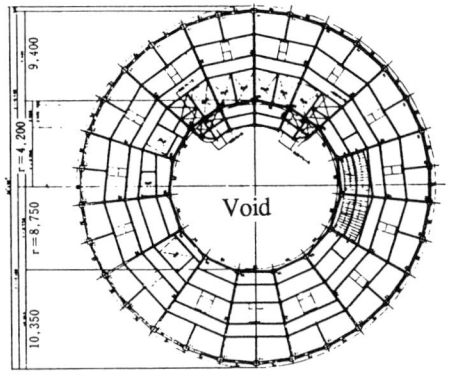

Fig. 2   Framing Plan of The SYP Tower

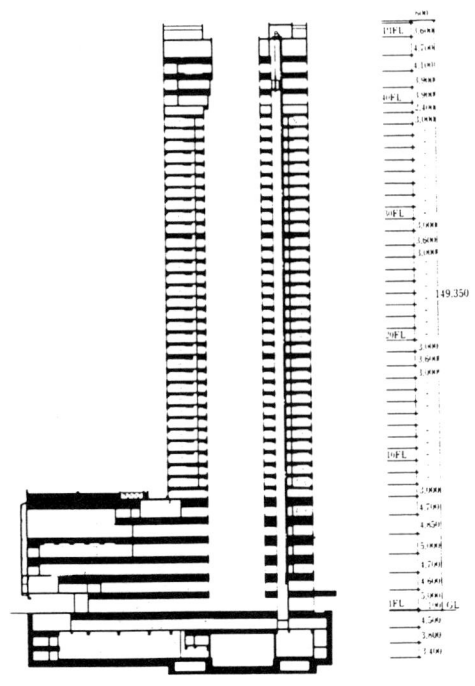

Fig. 3   Section of The SYP Tower

TUNED LIQUID DAMPERS

The acceleration response due to ordinary wind should be limited for serviceability. There are three ways to reduce acceleration response to wind: increasing mass, increasing stiffness and increasing damping of the building. In many cases, increasing damping by adding devices to buildings is the most economical way for modern buildings with very small damping capacity.

Tuned Liquid Damper(TLD) is chosen as a damping device for SYP, based on our experiences with TLDs installed on several towers. On-site measurements of the wind induced motions of those towers show that wind induced motions are reduced to less than that of the motions without TLD. The way to predict wind-induced motion by numerical analysis was established and checked by on-site measurement data of those towers.

TLD is a tuned mass damper, except for its nonlinearity dependent upon vibration amplitude, and the mass is the containing water itself. Optimum TLD can be designed by using linear theory for tuned mass damper. In the case of the SYP building, mass ratio (ratio of liquid mass in TLD container to structural generalized mass of fundamental mode) of 1 percent and damping constant of liquid water in TLD of 5 percent are selected as a result of preliminary studies. To tune the frequency of sloshing to the building natural frequency, a water depth of 124 mili-meters (4-7/8 in.) is required for a circular container with a diameter of 2 meters (6 ft. 6 3/4 in.) (Fig. 4). Ribs are attached inside the water container to add damping characteristics of the motion of water (Fig. 5). 30 units of TLDs each containing 9 layers of water will be installed on the top of the SYP building.

Numerical simulation for the wind-induced response were conducted using time-history wind pressure records obtained by wind tunnel tests. Wind pressure measurements were conducted in the boundary-layer wind tunnel testing facility of Shimizu Corporation. Wind pressures were measured at more than 300 points on a 1/300 scaled model of the tower. Electronically Scanned Pressure Sensors made it possible to measure and record wind pressures simultaneously (Fig. 6).

The TLD-Structure Interactive Model is used to simulate motion of the tower under wind environment. Motion and the interactive force of liquid water in the TLD container is simulated using Boundary Element Method (BEM). The motion of the tower is modeled as Multi-degrees of Freedom (MDOF) Model (Fig. 7). Numerical calculations were made in a step-by-step mannor using time-history wind pressure records from wind tunnel measurement.

Some of the simulation results are shown on Fig. 8 and Table 1. Accerelations on the top floor due to wind with TLDs are predicted to be less than half of those without TLDs.

Fig. 4   Tuned Liquid Damper to be Installed on The SYP Tower

Fig. 5   One Layer of TLD

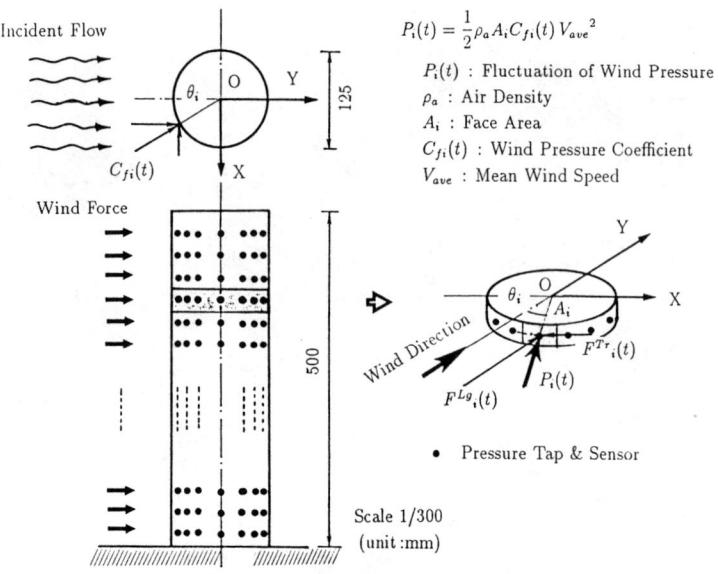

Incident Flow

$$P_i(t) = \frac{1}{2}\rho_a A_i C_{f_i}(t) V_{ave}{}^2$$

$P_i(t)$: Fluctuation of Wind Pressure
ρ_a : Air Density
A_i : Face Area
$C_{f_i}(t)$: Wind Pressure Coefficient
V_{ave} : Mean Wind Speed

Wind Force

● Pressure Tap & Sensor

Scale 1/300
(unit :mm)

Fig. 6   The Scaled Model of SYP Tower for Wind Tunnel Tests

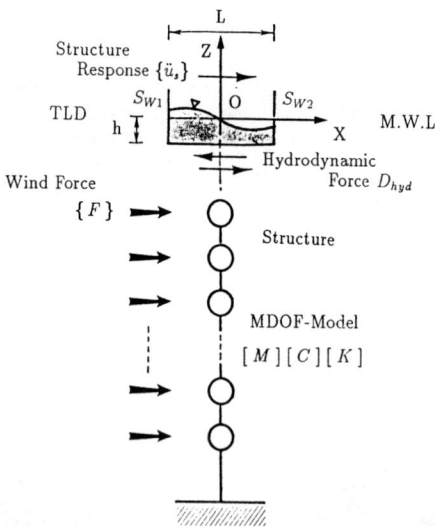

Fig. 7   The TLD-Structure Interactive Model

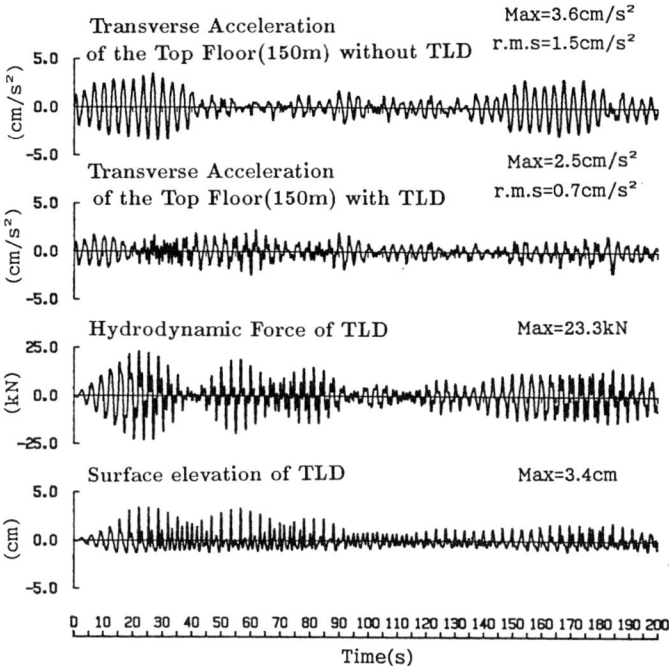

Max=3.6cm/s²
r.m.s=1.5cm/s²

Max=2.5cm/s²
r.m.s=0.7cm/s²

Max=23.3kN

Max=3.4cm

Fig. 8 Calculated Time History of Wind-induced Response
(Mean Wind speed of 25 meter/second at top of the tower)

Table 1 Calculated Accelerations on The Top of The SYP Tower

Wind Speed	without TLD		with TLD	
V_{ave} (m/s)	Maximum (cm/s²)	R.M.S. (cm/s²)	Maximum (cm/s²)	R.M.S. (cm/s²)
20.0	1.688	0.671	1.212	0.395
25.0	3.614	1.544	2.479	0.721
30.0	7.541	3.539	6.250	1.792

(V_{ave} : Mean Wind Speed at 150m above the Ground)

APPROPRIATE SURFACE TEXTURE

Because of the cylindrical shape of the tower, large negative wind pressure (wind pressure coefficient C of less than -2.0) for the cladding is anticipated locally, if the surface is flat. We have conducted a series of wind pressure tests, using models with different surface textures. Models for preliminary tests have a reduced scale of 1/127 of the real building, and have a diameter of 300 milimeters (Fig. 9). Five different textures are:

SF-1   flat surface
SF-2   with one milimeter ribs
SF-3   with 0.5 milimeter ribs
SF-4   with rough surface having average height of 0.5 milimeters
SF-5   with 0.5 milimeter ribs only on spandrels (on one third of each floor height)

Fig. 10 shows that negative wind pressures on the rough surfaces are smaller than that on the flat surface. Ribs only on one third of each floor height works to reduce negative pressure for curtain wall (including flat part of the curtain wall) (SF-5). Fig. 11 shows ribs on spandrels of SYP building.

Fig. 9   Wind Tunnel Test for Appropriate Surface Texture

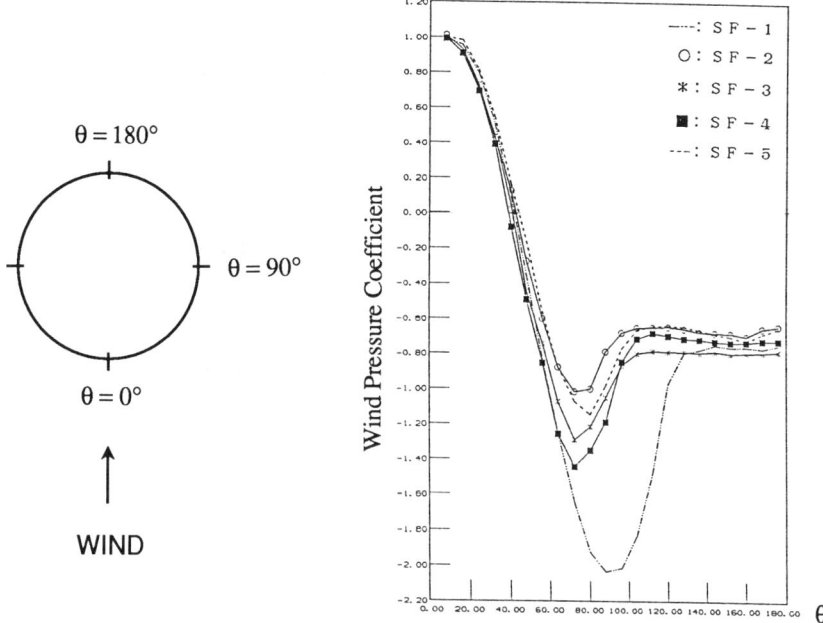

θ = 180°

θ = 90°

θ = 0°

↑

WIND

Wind Pressure Coefficient

‒‒‒: S F ‒ 1
O : S F ‒ 2
* : S F ‒ 3
■ : S F ‒ 4
‒‒‒: S F ‒ 5

θ

Fig. 10   Pressure Distribution on Cylindrical Building with Different Surface Textures

Fig. 11   Texture of Spandrels used on The SYP Tower

CONCLUSION

Three structural and mechanical proposals are made for the SYP building.

1. Cylindrical framed tube structure to obtain enough stiffness of the building to reduce wind-induced motion, but not to increase seismic force too much.
2. Tuned Liquid Dampers on top of the tower to increase damping for better serviceability.
3. Choise of appropriate surface texture to reduce negative wind pressure on curtain walls.

Wind tunnel studies and numerical analysis have predicted the effectiveness of these proposals. On-site measurements will follow to evaluate these ideas as well as the studies done for the building.

REFERENCES

1. TAMURA, Y et al.: "Wind-Induced Vibration of Tall Towers and Practical Application of TSD", Proceeding of Workshop on Serviceability of buildings, 1988.
2. HIBI, K. et al.: "Visualization of Fluctuating Surface Pressure Distribution on Bluff Body Using Electronically Scanning Pressure Sensors", Proceedings of 5th International Symposium on Flow Visualization, 1989.

SCALING DESIGN SPECTRA AND SEISMIC SAFETY
OF TALL BUILDINGS IN GREATER LOS ANGELES

Farzad Naeim[*]   and Marshall Lew[**]

ABSTRACT

For over a decade the Department of Building and Safety of the City of Los Angeles has required the application of site-specific design spectra for design of buildings taller than 160 feet. In direct contrast, the seismic analysis provisions of the 1988 and 1991 editions of the Uniform Building Code permit using predetermined spectral shapes and scaling the resulting base shear to match that obtained from equivalent static lateral force analysis. This paper examines the relative wisdom of these two approaches for Los Angeles through site-specific seismic hazard evaluations for several sites across the City of Los Angeles and a building design case study.

INTRODUCTION

The 1988 edition of the Uniform Building Code (UBC-88) seismic design requirements have been adopted by city after city in the western United States. In general,
this is a sound move. Adoption of the latest seismic design and detailing recommendations of the seismological committee of the Structural Engineers Association
of California (SEAOC) has made the new UBC a much better code than its predecessor, the 1985 edition (UBC-85). For the first time, specific requirements for dynamic analysis have appeared in the Uniform Building Code. Among other things,
UBC embodies a standard dynamic analysis procedure based on using code suggested design spectral shapes scaled to match the code specified static base shear.
UBC-88 and 91 treat the design spectrum only as a vehicle for better distribution of
lateral forces. Hence, it ignores several important and site-specific ground motion
parameters that each design spectrum represents. For cities where dynamic analysis of buildings has not been enforced by building officials (to our knowledge, every
major city except Los Angeles and San Francisco), this is a step forward. It serves
to familiarize engineers with essentials of dynamic analysis, provides them with an
opportunity to consider the effects of higher modes on seismic response of their
buildings, and at the same time furnishes a tool for a more sensible distribution of
seismic lateral forces. The wisdom of such a scaling, however, for a city like Los
Angeles where dynamic analysis of major buildings is the *modus operandi* and
where significant historical and geological seismicity information is readily available, is the subject of investigation of this paper.

CURRENT PRACTICE

The current seismic design practice for major buildings in Los Angeles is explicitly
formulated around the following well-known, two-level seismic design philosophy
(Harder and Martin, 1989):

   1-   Buildings should resist moderate earthquakes with essentially no

---

* Director of Research & Development, John A. Martin and Associates, Inc., Los Angeles, California.
** Vice President and Dir. of Earthquake Eng., LeRoy Crandall and Associates, Glendale, California.

structural damage.

2- Buildings should resist catastrophic earthquakes with some structural damage but without collapse and major injuries or loss of life

Conforming to the above levels of performance, two design earthquakes are used: a Maximum Probable design earthquake to model the moderate events and a Maximum Credible design earthquake to model the catastrophic events.

The Maximum Probable event is defined as the anticipated ground motion at the site due to an earthquake with a moderate (about 50%) probability of occurrence during the anticipated life of the structure considered as 60 years or more (roughly corresponds to the once per 100 year event). In generation of a design spectrum corresponding to the Maximum Probable event, the following factors should be considered:

* Past seismic history of the site

* Faults within a radius of 100 km of the site that might be active within next 100 years

* Mechanism of fault rupture for each of the faults considered

* Statistical seismic recurrence analysis to establish the design earthquake magnitude

In addition, the postulated Maximum Probable design magnitude should not be less than that of historical earthquakes that have occurred within the 100 km distance of the site.

The Maximum Credible event is defined as the maximum ground motion at the site that the presently known tectonic and geological framework appears to be capable of generating. In development of a design spectrum corresponding to the Maximum Credible event, the following factors are considered:

* Tectonic and/or structural history of the entire region

* Past seismic history of the entire geological province

* Known faults within a radius of 100 km of the site, their potential rupture lengths, and type of faulting

It is clear that the various time factors and recurrence relations are not considered as deciding parameters in establishing the Maximum Credible design earthquake.

Factors such as soil conditions at the site, design earthquake magnitude, duration of strong ground motion, travel path from source to site, and corresponding attenuations, should be considered in construction of these two site-specific design spectra.

In contrast with the UBC seismic lateral loads which are *working stress* loads, the forces generated by these design spectra are considered to represent *ultimate loads*. In steel design, for example, the City permits the use of plastic design equations as enumerated in *Part II* of the *AISC Specifications* in conjunction with the site specific design spectrum. Many engineers, however, prefer to use modified working stress design equations (where the elastic safety factors are stripped out to permit reaching yield or elastic stability limit) when designing for the Maximum Probable event. Clearly the latter approach is more conservative and more consistent with the City of Los Angeles design philosophy which requires structures to remain basically elastic during the Maximum Probable event.

Between the two, the Maximum Probable design spectrum is the basic vehicle of proportioning structural members. The Maximum Credible spectrum is used in design for more brittle types of failure such as shear in structural walls and braces in ordinary braced frames. The maximum permitted inter-story drift ratio is 0.0075 for the Maximum Probable event and 0.0150 for the Maximum Credible event.

SEISMIC HAZARD EVALUATIONS

Scaling spectral base shear to the level of code specified equivalent static base shear, as UBC recommends, implies the same level of design lateral forces for a given building independent of its location within a code specified *Seismic Zone*. UBC's seismic zone 4 covers most of the state of California, including the entire metropolitan Los Angeles area.

In this section we explore seismic hazard exposure at various sites within the City of Los Angeles to show that based on current seismological knowledge, seismic risk varies significantly, from location to location, within this city. As a result, normalization of design spectrum subjects similar buildings (and their occupants) in different parts of Los Angeles to various levels of seismic risk. This will be shown to be clearly unjustified.

Selected Sites and Procedures

Eight sites representing various localities within the City of Los Angeles were selected and subjected to detailed historical and geological seismic hazard evaluations. Locations of these sites are summarized in Table 1 and shown on Figure 1.

For each site, two soil conditions were considered: shallow alluvium and deep alluvium. The unconstrained mean attenuation relationships developed by Campbell (1987) were used to estimate the peak ground acceleration at the sites. The same principal conclusions, however, could be reached using any of the several other available attenuation relationships. The peak ground accelerations were modified where applicable to represent the *"repeatable high ground acceleration (RHGA)"* peaks as outlined by Ploessel and Slosson (1974). The concept of a lower-amplitude *"repeatable acceleration"* has also been used by ATC 3-06 and NEHRP-85 where it has been referred to as *"Effective Peak Acceleration"* or EPA. The same concept has been implicitly used by SEAOC and UBC in the development of their seismic zone maps, which in turn are based on ATC 3-06 EPA maps. The estimated *RHGA* values reported in this paper are intended for this comparative analysis only. Hence, they should not be perceived as suggested design values for these or similar sites.

Historical Data

For each of the eight selected sites, a search was made for historical earthquakes of local Richter magnitude (M_L) of 4.0 and greater that have occurred within 100 km of each site between the years 1800 and 1988 AD. The EQSEARCH computer program (Blake, 1989a) and its data base containing earthquake catalogs of California Division of Mines and Geology, data from Towenley and Allen (1939) for the 19th century earthquakes, and preliminary determination of epicenters and magnitudes by USGS (for more recent earthquakes) were used for this search. For each site a probability analysis was performed to establish the cumulative Gutenberg-Richter (Gutenberg and Richter, 1954) seismicity relationship needed to estimate the magnitude of the Maximum Probable earthquake.

The cumulative Gutenberg-Richter relationship is expressed as:

$$\log_{10}(N) = A - bM$$

where N is the number of events of magnitude M or greater per year in the search area, and A and b are constants derived from linear regression analysis of the data.

This probability analysis ignores the provincial geology and is based on the assumptions that earthquakes are independent, random events, and that the historical distribution of epicenters is representative of the future distribution of epicenters. Generally, these assumptions are not valid. However, such a probability analysis of the brief historical past (189 years) provides a lower-bound estimate of what can be expected in the future. Results of this analysis are summarized in Table 2 where a Maximum Probable earthquake magnitude between 6.5 and 6.7 is forecasted for the sites assuming a return period of 100 years.

Geological Data

Several known active and semi-active faults pass through the Los Angeles basin. A data base of 108 digitized, late-Quaternary faults, including their dominant type of faulting, estimated Maximum Credible and Maximum Probable (once per 100 year) magnitudes (Blake, 1989b) was used in this study. Principal reference sources used in selection of the faults were Jennings (1975), Anderson (1984), and Wesnousky (1986). For each site, all the faults passing within a distance of 100 km (62 miles) of the site were considered. An alphabetic list of such faults, along with their estimated Maximum Credible and Maximum Probable magnitudes, and respective distance to each of the eight selected sites is presented in Table 3.

There are two distinct methods of estimating *RHGA* at each site:

1- *The Deterministic Approach* which assumes fault ruptures to occur on each fault at its closest distance to the site and uses this distance, the given design earthquake magnitude, and a set of attenuation relationships to estimate the largest expected *RHGA* at the site.

2- *The Probabilistic Approach* [McGuire (1979); Idriss (1985); Blake(1989c)] which assumes that there is an equal chance of ruptures occurring anywhere along the entire length of each fault. In this approach, recurrence relationships developed for each fault, a set of attenuation relationships, and a relatively complex probability analysis are used to estimate the exceedence probabilities of various levels of *RHGA* at a given site.

Among the two methods, the deterministic approach is clearly more conservative and results in larger estimates of *RHGA*. In this study, the Maximum Probable *RHGA* values are estimated using both methods. To be consistent with the current Los Angeles City guidelines, however, the Maximum Credible *RHGA* values are based on the deterministic approach only.

The estimated deterministic Maximum Probable and Maximum Credible *RHGA* values for the selected sites are shown in Table 4. As this table clearly indicates, there is a maximum difference of 106% for *RHGA* values at these sites.

Probabilistic estimates of the Maximum Probable *RHGA* (once per 100 years) values are summarized in Table 5. Here, it can be seen that while the *RHGA* estimates are , as expected, lower than the corresponding deterministic values, the maximum difference is again high: 94% for shallow soil and 100% for deep soil sites. Clearly, to retain the same level of seismic hazard potential, a building located at site 7, should be designed for much higher level of lateral forces than a similar building located at site 4. An objective that can not be achieved by the UBC design spectrum scaling procedure.

BUILDING DESIGN CASE STUDY

To further illustrate the undesired effects of adoption of a spectrum scaling procedure, let us consider a typical 20-story building. The lateral system for this building consists of a peripheral steel moment resistant frame as sketched.

The frame was proportioned to satisfy the current Los Angeles City requirements for gravity, wind, and seismic forces. Site-specific design spectra developed for shallow alluvium soils at Site 1 were used to represent design earthquake forces. Fifteen modes were considered in the dynamic analysis of the building and the modal responses were combined using the CQC method. The computed fundamental period of the building was 3.05 seconds. The computer model of the building was then subjected to site-specific design spectra for shallow and deep alluvium soil developed for other selected sites.

The design spectra for the sites were developed using the techniques suggested by Mohraz and Elghadamsi (1989) and the forementioned deterministic estimates of *RHGA* for each site . Critical damping ratios of 5% and 7.5% were assumed for Maximum Probable and Maximum Credible events, respectively.

The dynamic base shears so calculated are shown in Table 6. Here again, the wide difference among the expected base shears at various sites may be observed. It is interesting to note that while the *RHGA* estimates for deep soils were significantly less than the corresponding values for shallow soils (see Table 4), the reverse is true for the resulting base shears.

Using UBC static lateral force procedure, two sets of base shears are obtained: one for stress design and the other for drift design. UBC recognizes four different soil types with the Site Coefficient (S) varying from 1.0 for rock sites to 2.0 for deep clay sites. We consider the two extreme cases of S=1.0 and S=2.0 for comparison with the site specific dynamic base shears shown in Table 6.

For stress design, UBC Formula 12-1 yields a base shear of 1739 kips for rock sites and 2324 kips for deep clay sites (assuming an Importance Factor of 1.0. For drift design, UBC Formula 12-1 yields reduced base shears of 1144 kips for rock sites and 2288 kips for deep clay sites.

Since UBC forces correspond to working stress state of behavior, they should be converted to yield level forces before a comparison between the code design forces and displacements and the corresponding current practice values can be made. Let us Consider the simple case of members whose design is governed by flexural stresses. For compact sections, the code allowable stress for major axis bending is $0.66F_y$. Under combined gravity and seismic forces, this value may be increased by 33% to $0.878F_y$. The current practice permits running stresses up to yield (F_y) for the Maximum Probable forces. Since a linear elastic analysis is performed in both cases, the design forces and displacements are directly proportional to the corresponding design base shear values. For example, the ratio of UBC design moments to current design moments for site 1 on rock is:

$$(1739 \text{ kips} / 3480 \text{ kips}) \times (1.000/0.878) = 0.569$$

which means that UBC design value for this situation is 76% less conservative.

This same value varies from 13% less conservative for site 4 on rock, to 141% less conservative for site 7 on deep soil. Regardless of less conservatism exhibited in this particular case, the large differences in these values clearly indicate that the UBC scaling procedure is incapable of providing a uniform level of seismic risk for similar buildings located in various parts of the city. The adaptation of this procedure, hence, is equivalent to a process of discrimination against residents based on their location of work or residence within the city.

If we compare the current 0.0075 drift index limitation for the Maximum Probable event to the UBC drift design limits, a similar pattern of non-uniformity is witnessed. For example, UBC drift limitations are equivalent to limiting the Maximum Probable drift index at site 4 on deep clay soil to an extremely conservative value of 0.003, while permitting a relaxed Maximum Probable drift index of 0.010 for site 7 on rock.

CONCLUSIONS

The undesirable and serious side-effects of forgoing the current two level site-specific seismic design requirements for UBC Design spectrum scaling procedure were discussed and illustrated through application to an example 20-story building.

It is strongly recommended that the City of Los Angeles retain its two-level site specific design philosophy for major buildings and adopt UBC seismic design and detailing requirements except for spectrum normalization procedure, UBC Sec. 2312 (f) 5 C, and the corresponding drift requirements, UBC Sec. 2312 (e) 8. Some other minor changes are necessary in adoption of code provisions. These changes are needed in order to retain consistency with the two-level seismic design procedure.

For example, UBCoverturning design requirement, UBC-88 Sec. 2312 (e) 7 B, specifies load combinations with respect to static lateral forces: $3(R_w/8)E_{static}$. These factors should be converted to appropriate site-specific dynamic values such as $(1.0)E_{Credible}$. In addition, lower-bound dynamic base shear values may be specified in order to provide a minimum across-the-board level of safety. These lower-bound values may be specified in terms of overall building weight (e.g. 0.03 times the total weight of the building), or in terms of the design method (e.g. if dynamic base shears are less than 1.4 times code static base shears, they should be treated as working stress forces). The exact wording and extent of these and similar changes, of course , should be worked out by appropriate building authorities.

REFERENCES

Anderson, J.G. (1984), "Synthesis of Seismicity and Geological Data in California," U.S. Geological Survey, Open File Report 84-424, 186 pp.

Applied Technology Council (1978), "Tentative Provisions for the Development of Seismic Regulations for Buildings," ATC Publication ATC 3-06.

Blake, T.F. (1989a), "EQSEARCH A Computer Program for the Estimation of Peak Horizontal Acceleration from Southern California Historical Earthquake Catalogs -- User Manual," Thomas F. Blake Computer Services and Software, Newbury Park, California.

------------ (1989b), "EQFAULT A Computer Program for the Deterministic Prediction of Peak Horizontal Acceleration from Digitized California Faults -- User Manual," Thomas F. Blake Computer Services and Software, Newbury Park, California.

------------ (1989c), "FRISK89 A Computer Program for the Probabilistic Estimation of Seismic Hazard Using Faults as Earthquake Sources -- User Manual," Thomas F. Blake Computer Services and Software, Newbury Park, California.

Campbell, K.W. (1987), "Predicting Strong Ground Motion in Utah," in Gori, P.L. and Hays, W.W., eds., *Assessment of Regional Earthquake Hazards and Risk along the Wasatch Front, Utah*, U.S. Geological Survey, Open File Report 87-585.

Federal Emergency Management Agency (1985), "NEHRP Recommended Provisions for the Development of Seismic Regulations for New Buildings," Washington, D.C.

Gutenberg, B. and Richter, C.F. (1954), "Seismicity of the Earth," Princeton University Press, 310 pp.

Harder, R.N. and Martin, J.A., Jr. (1989), "Seismic Design Requirements for Tall Buildings in Los Angeles," Proceedings of 1989 Annual Meeting, Los Angeles Tall Building Structural Design Council.

International Conference of Building Officials (1988), "Uniform Building Code," 1988 Edition, Whittier, California.

International Conference of Building Officials (1985), "Uniform Building Code," 1985 Edition, Whittier, California.

Jennings, C.W. (1975), "Fault MAp of California with Locations of Volcanoes, Thermal Springs, and Thermal Wells," California Division of Mines and Geology, Geologic Data Map No. 1.

McGuire, R.K. (1978), "FRISK: Computer Program for Seismic Risk Analysis Using Faults as Earthquake Sources," U.S. Geological Survey, Open File Report 78-1007.

Mohraz, B. and Elgahdamsi, F.E. (1989) "Earthquake Ground Motion and Response Spectra," in Naeim, F., ed., *The Seismic Design Handbook*, Van Nostrand Reinhold, New York.

Ploessel, M.R. and Slosson, J.E. (1974), "Repeatable High Ground Accelerations from Earthquakes," *California Geology*, Vol. 27, No. 9., pp. 195-199.

Seismology Committee of Structural Engineers Association of California (1988), " Recommended Lateral Force Requirements and Tentative Commentary," SEAOC, San Francisco, California.

Wesnousky, S.G. (1986), "Earthquakes, Quaternary Faults, and Seismic Hazard in California," *Journal of Geophysical Research,* Vol. 91, No. B12.

TABLE 1. SELECTED SITES AND THEIR LOCATIONS

SITE NO.	LOCATION	LATITUDE	LONGITUDE
1	Downtown Los Angeles	34.05 N	118.25 W
2	Wilshire/Westwood	34.02 N	118.44 W
3	LAX 405 Fwy/Century Blvd	33.96 N	118.38 W
4	San Pedro	33.75 N	118.29 W
5	Van Nuys	34.21 N	118.45 W
6	Venice Beach	34.00 N	118.45 W
7	Sun Valley	34.24 N	118.36 W
8	Woodland Hills	34.19 N	118.59 W

TABLE 2. SUMMARY OF HISTORICAL INVESTIGATION

SITE NO.	SEISMIC CONSTANTS		MAGNITUDE (Return Period in Yrs.)					D*	NEQ**
	A	b	10	30	50	75	100		
1	3.797	0.868	5.53	6.08	6.33	6.53	6.68	3.0	408
2	4.008	0.917	5.46	5.98	6.22	6.42	6.55	2.7	401
3	3.881	0.888	5.50	6.03	6.28	6.48	6.62	3.7	403
4	3.780	0.865	5.53	6.08	6.33	6.54	6.68	2.0	401

* Distance in miles to the closest historical event (1800-1988 AD)
** Number of events ($M_L \geq 4$) in the 100 km radius search area (1800-1988)

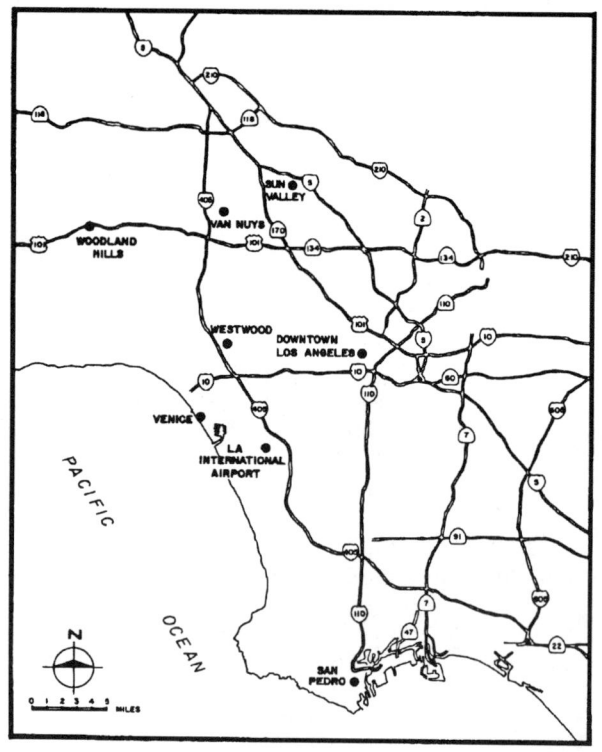

FIGURE 1. LOCATIONS OF SELECTED SITES

TABLE 3. FAULTS AND THEIR DISTANCES FROM THE SELECTED SITES*

ABBREVIATED FAULT NAME	MAX. CRED. MAG.	MAX. PROB. MAG.	APPROXIMATE DISTANCE (Miles) SITES							
			1	2	3	4	5	6	7	8
ANACAPA	7.00	6.25	31	20	22	31	25	19	31	20
ARROYO PARIDA - MORE RANCH	7.00	6.00	57	49	54	-	42	49	46	35
BIG PINE	7.50	5.75	-	-	-	-	53	-	54	49
CHINO	7.00	4.75	28	39	36	36	41	40	37	49
CLEARWATER	7.00	3.00	37	39	43	57	26	40	24	28
CLEGHORN	6.50	6.25	48	59	58	61	57	60	52	-
CUCAMONGA	7.00	6.75	25	36	34	37	36	37	32	44
ELSINORE	7.50	6.75	46	56	51	44	61	56	58	-
FRAZIER MOUNTAIN	6.50	3.00	61	56	61	-	45	57	47	42
GARLOCK (West)	7.75	6.75	-	61	-	-	50	-	50	47
GLN.HELEN-LYTLE CR-CLREMNT	7.50	7.00	40	50	49	54	47	52	42	55
HOLSER	6.75	6.25	31	28	33	48	16	29	17	15
MALIBU COAST	7.50	5.00	18	7	12	25	14	7	19	11
MOJAVE RIVER (Ord Mtn.)	7.00	6.25	58	-	-	-	-	-	62	-
NEWPORT - INGLEWOOD	7.50	6.50	8	(2)	(1)	7	11	3	13	14
NORTHRIDGE HILLS	6.50	4.00	18	15	19	35	(2)	16	6	(5)
OAK RIDGE	7.50	6.25	36	30	36	51	20	31	23	16
OFFSHORE ZONE OF DEFORM.	7.50	6.00	44	49	44	31	59	49	58	-
OZENA	7.00	3.00	-	-	-	-	61	-	-	55
PALOS VERDES HILLS	7.00	5.50	17	9	7	(0)	22	8	25	22
PINE MOUNTAIN	7.00	4.50	50	45	51	-	34	46	35	31
PLEITO	7.50	6.00	-	-	-	-	55	-	55	53
RAYMOND	7.50	5.50	4	13	13	25	14	15	11	21
RED MOUNTAIN	7.50	6.00	-	54	59	-	49	54	53	41
SAN ANDREAS (Mojave)	8.50	8.25	35	41	43	54	30	43	25	34
SAN ANDREAS (Southern)	8.00	7.25	47	58	57	59	56	59	51	-
SAN CAYETANO	7.50	6.25	39	33	39	54	23	34	26	19
SAN CLEMENTE	7.50	6.25	-	59	56	45	-	57	-	-
SAN GABRIEL	7.50	6.25	13	21	22	32	11	23	6	15
SAN GORGONIO - BANNING	8.00	7.00	58	-	-	-	-	-	-	-
SANTA CRUZ ISLAND	7.50	5.00	-	57	60	-	58	56	-	50
SANTA MONICA - HOLLYWOOD	7.50	6.00	4	2	7	22	10	(3)	10	12
SANTA SUSANA	7.00	6.50	22	20	25	40	8	21	7	7
SANTA YNEZ (East)	7.50	5.75	52	47	52	-	36	47	39	32
SIERRA MADRE-SAN FERNANDO	7.50	6.50	11	18	20	31	5	19	2	8
SIMI - SANTA ROSA	7.00	4.25	29	23	29	44	14	24	18	9
VENTURA - PITAS POINT	7.00	6.25	56	46	51	-	42	46	47	35
VERDUGO	7.00	4.50	6	14	15	26	4	15	(2)	10
WHITE WOLF	8.00	7.75	-	-	-	-	-	-	-	61
WHITTIER - NORTH ELSINORE	7.50	6.25	(4)	14	13	22	16	15	13	23

* Distance from the closest fault to each site is circled. Dashes ('-') indicate distances greater than 100 km (62 miles).

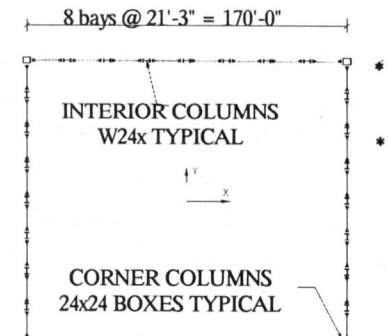

8 bays @ 21'-3" = 170'-0"

INTERIOR COLUMNS
W24x TYPICAL

CORNER COLUMNS
24x24 BOXES TYPICAL

* ALL STORY HEIGHTS
ARE 12 FEET

* ALL STORY WEIGHTS
ARE 2898 KIPS

TABLE 4. DETERMINISTIC RHGA ESTIMATES *(in g's)*

SITE NO.	MAXIMUM CREDIBLE		MAXIMUM PROBABLE	
	Shallow Soil	Deep Soil	Shallow Soil	Deep Soil
1	0.449	0.298	0.257	0.170
2	0.490	0.325	0.238	0.158
3	0.370	0.245	0.229	0.152
4	0.298	0.197	0.166	0.110
5	0.415	0.275	0.296	0.197
6	0.468	0.310	0.223	0.148
7	0.489	0.325	0.342	0.227
8	0.345	0.229	0.264	0.175
MAX. DIFF.	**64%**	**106%**	**65%**	**106%**

TABLE 5. PROBABILISTIC ONCE PER 100 YEARS RHGA

SITE NO.	MAXIMUM PROBABLE	
	Shallow Soil	Deep Soil
1	0.185	0.120
2	0.195	0.118
3	0.162	0.113
4	0.138	0.090
5	0.220	0.137
6	0.155	0.113
7	0.268	0.180
8	0.200	0.133
MAX. DIFF.	**94%**	**100%**

TABLE 6. COMPUTED DYNAMIC BASE SHEARS *(kips)* FOR THE 20 STORY BUILDING

SITE NO.	MAXIMUM CREDIBLE		MAXIMUM PROBABLE	
	Shallow Soil	Deep Soil	Shallow Soil	Deep Soil
1	6076	8371	3480	4775
2	6631	9129	3220	4438
3	5007	6882	3100	4270
4	4032	5534	2246	3090
5	5616	7725	4006	5534
6	6333	8708	3018	4157
7	6618	9129	4628	6376
8	4670	6430	3573	4916

Proceedings of the Second Conference on Tall Buildings in Seismic Regions
55th Regional Conference
May 16 and 17, 1991, Los Angeles, California

TUNED PENDULUM MASS DAMPER USING ICE THERMAL STORAGE TANK INSTALLED IN CRYSTAL TOWER

Tadashi NAGASE* and Toshiharu HISATOKU**

* Structural Eng.Sect. and ** Dr.Eng., Principal Engineer
Building Design Dept., Takenaka Corporation, Nishi-Ku,Osaka,Japan

SUMMARY

A new damper to reduce the lateral vibration of the building has been installed in the 37-story Crystal Tower completed last year in Osaka, Japan. The tower has a slender proportion of height-width ratio of 6. We have developed the tuned pendulum mass damper to decrease the wind-induced building motion by 50%. Crystal Tower is the first high-rise building with the damper in Japan and the damper weight of 540 tons (1190 kips) is of a comparatively great class in the world. This paper presents some intrinsic features of the pendulum damper integrating air conditioning facilities and demonstrates that fundamental tuning by the pendulum length and/or additional spring devices is simple and practical. Special considerations are given both to the connecting joints between the tank and the building and to the sloshing behavior of the liquid contents. The typhoon and earthquake records observed in Crystal Tower have proved that the damper works well as expected.

INTRODUCTION

A tuned mass damper (TMD) using a pendulum has been developed and installed in the 37-story Crystal Tower recently completed in Osaka, Japan. The tower is fully equipped with the latest facilities for intelligent office buildings and creates comfortable environment and fulfills the office function efficiently. As shown in Photo 1, the tower is wholly covered by the half-mirror glass walls. The tower is 157m(515ft) high and 28m(91ft) by 67m(221ft) in plan dimension as illustrated in Fig.1 and weighs 44,000 tons (97,000 kips) above the ground. The east and west elevations are very slender and symbolic while the north and south sides are large enough to afford a fine view of Osaka Castle Park.

The height and shape of the tower make it prone to lateral vibrations due to strong winds. The lateral motion of the tower in the NS (north-south) direction is more affected by winds rather than earthquakes. The designed fundamentals of the tower are 4.7 sec in the NS direction and 4.3 sec in the EW (east-west) direction. We have decided at the early phase of the structural design of this tower to apply a TMD to lessen the usual wind-induced motion of the building.

As a rule the weight of the mass damper is approximately 1% of the weight of the building above the ground. The installation of the TMD to a high-rise building needs a huge weight of the mass and a large room at the top of buildings (Refs.1,2), causing production cost and storage space problems. It is also needed that the huge mass weight should begin to slide as soon as the building sways. The newly developed Tuned Pendulum Mass Damper is our answer to these problems. Crystal Tower is the first high-rise building with a TMD in Japan and the first in the world to use a pendulum for this purpose. Figure 2 sketches a conceptual diagram of the tuned pendulum mass damper. The ice thermal storage tanks are used as the pendulum weight hung from the top roof girders. The system is of a passive type control which needs no energy supply to actuate the damper.

ICE THERMAL STORAGE TANK   Crystal Tower features an advanced system for cooling and heating the building by using the transfer of latent heat through the phase variation of evaporation and condensation of the refrigerant. In the air conditioning part the refrigerant vapor is cooled and condensed by the condenser in the penthouse machine room and transfers to the liquid. The liquid refrigerant falls by its own weight to the office rooms to be air conditioned and is evaporated by getting heat there and goes up to the top machine room naturally. In the heating part conversely the evaporator for heating at the basement machine room also enables the natural circulation of the refrigerant.

The ice thermal storage method for cooling the refrigerant through the condenser can make the best use of low-cost and off-peak night-time electric power service. As shown in Fig.3, the new air conditioning system needs vast capacity of the ice thermal storage tanks at the top of the building.

Photo 1 Crystal Tower

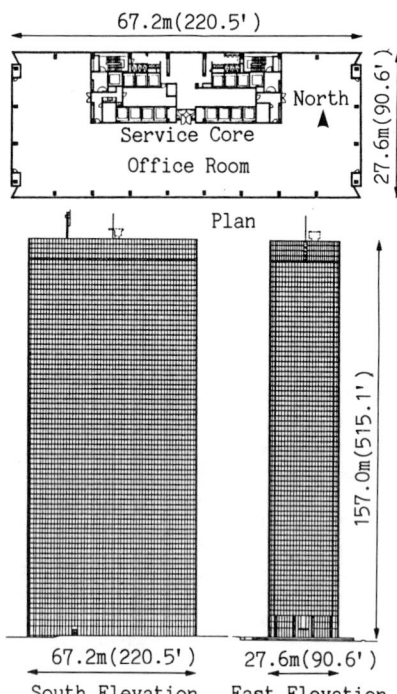

Fig.1 Dimension of Crystal Tower

Crystal Tower has the ice thermal storage of 720 m³ (25,400 ft³) divided into 9 tanks on the roof floor, and 6 tanks are also used as the mass damper. Total mass damper weighs 540 tons (1190 kips) including tanks and steel framing units. The TMD system of Crystal Tower, therefore, needs no additional mass weight nor space for installing.

PENDULUM DAMPER USING ICE THERMAL TANKS

When the strong wind blows against the building, the building swings side to side mainly by the natural period of the building, and the pendulum can also swing by its own period corresponding to the hanging length only. When the pendulum is tuned to swing by the same period as the building, the pendulum will swing in the opposite direction to decrease the building motion. The TMD of the pendulum type is very simple and can respond quickly to the building motion.

As demonstrated in Fig.4, the mass damper in Crystal Tower comprises 6 tanks suspended from the top roof girders to slide back and forth by the same periods as the building. Each tank weighs 90 tons (198 kips) and slides in one direction, NS or EW. Four tanks of 360 tons (794 kips) swing in the NS direction and two tanks of 180 tons (397 kips) in the EW direction. The weight ratio of the total mass damper to the building above the ground is 1.2 %, so the effective mass ratio is 3.7 %.

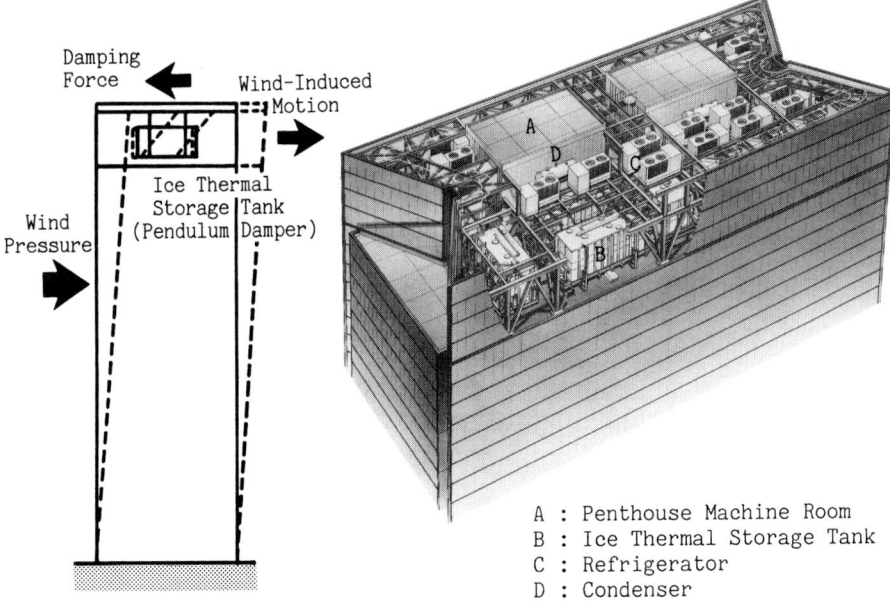

A : Penthouse Machine Room
B : Ice Thermal Storage Tank
C : Refrigerator
D : Condenser

Fig.2 Conceptual Illustration of
Tuned Pendulum Mass Damper

Fig.3 Equipment for Air Conditioning
in Crystal Tower

The natural periods of the tower are estimated to be in the range from 3.5 sec to 4.5 sec in the NS direction while 3.0 sec to 4.0 sec in the EW direction. The exact periods of the tower can be obtained by vibration tests at the final step of the construction when the TMD system is also installed in the building. So the pendulum length of the tanks to be swayed in the NS direction is designed to be able to change from 3.5 m (11.5 ft) to 4.5 m (14.8 ft) and from 3.0 m (9.8 ft) to 4.0 m (13.1 ft) in the EW direction.

Since the ice thermal storage tank containing liquid ice or sleet is used as the damper to be swayed back and forth, special considerations have been given both to connecting pipe joints between the tank and the building and to the sloshing motion of the liquid contents. The rubber flexible pipe joint is used that will follow the sway of the damper within the half amplitude of 25 cm (9.8 in). The tank dimension, 3.5 m (11.5 ft) × 10.5 m (34.4 ft) × 3 m (9.8 ft), is determined for the sloshing of the liquid content not to be stimulated. If the liquid moves up and down, the surface sleet can suppress the motion (Ref.3).

EFFECT OF MASS DAMPER   The wind tunnel test and the computational model analyses (Ref.4) have proved that the pendulum mass damper of Crystal Tower can reduce the lateral vibration due to winds by 50 % at least. Since $\sqrt{2}\sigma$ (standard deviation of acceleration) at the top floor of the tower is lessened from 10 cm/sec^2 (3.9 in/sec^2) to 5 cm/sec^2 (2.0 in/sec^2) when the wind blows with maximum 10-minute average speed of 16 m/sec (52 ft/sec) at 10 m (33 ft) above the ground, and office workers even at the top floor will not feel the building vibration. The wind of this class might be expected within a 5-year interval in Osaka.

The damper is effective until the stopper works when the 10-minute average wind velocity exceeds 22 m/sec (72 ft/sec) which is expected during a 20-year interval. The wind speed at the top of the tower will be twice as large as that near the ground. Crystal Tower is designed to resist the severe winds expected during a 100-year interval keeping the structure wholly within the elastic range without the damper. The damper is installed to improve the usual serviceability to the strong wind during up to a 20-year interval in Osaka.

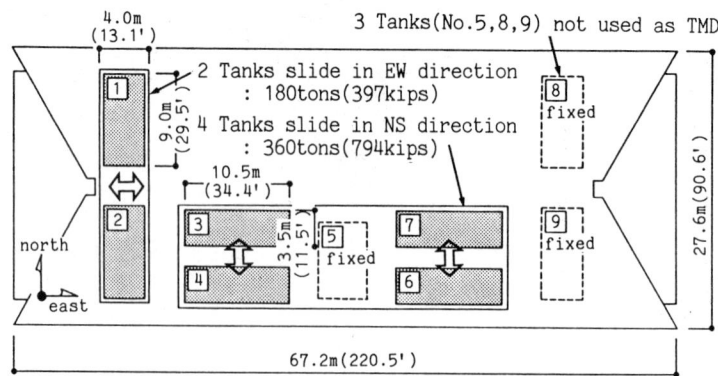

Fig.4 Layout of Ice Thermal Tanks on the Roof Floor

<u>DETAIL OF MASS DAMPER</u>   The damper system of Crystal Tower shown in
Fig.5 consists mainly of 7 devices as follows. Details and functions
are designed and examined by a full-scale model test (Ref.5).

Storage Tank : sliding mass weight
Hanging Rod : to suspend the tank as a pendulum and swing very
               smoothly with roller bearings at both ends
Oil Damper   : to consume the motion energy of the pendulum tank and
               to damp the building sway motion
Guide Roller : to constrain the sway direction of the tank in one
               direction
Stopper   : to limit the peak amplitude of the tank and also to
               fix the damper not to move at regular check-up and
               maintenance of the TMD or the tank
Coil Spring : to serve the fine tuning of the pendulum swing period
Rubber Joint : to absorb the sway gap of the connecting pipes
               between the tank and the building

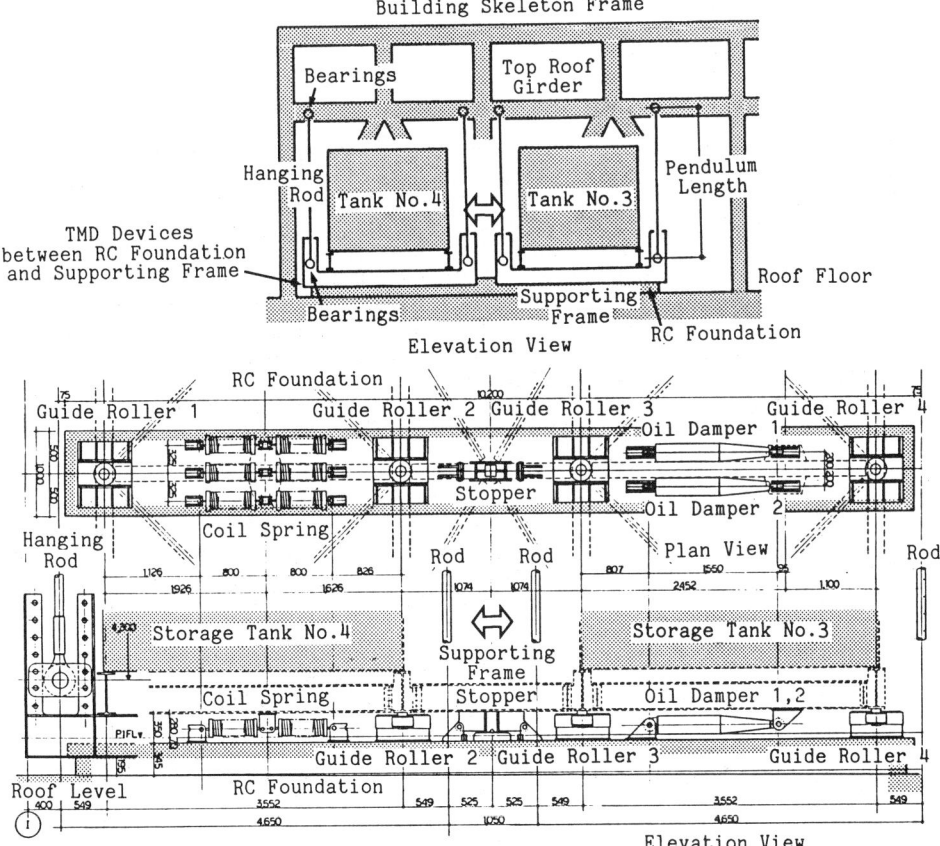

Fig.5 Devices of Mass Damper

319

DESIGN CONSIDERATIONS ON TUNED PENDULUM MASS DAMPER

COUPLED FUNDAMENTALS   The weight of pendulum is the ice thermal tank which contains liquid of water, sleet and ice. It has been proved by the sloshing model test (Ref.3) that the sloshing frequency of sleet or ice liquid is equal to that of water as long as the contents can move. The sloshing liquid is modeled by one mass system (Ref.6) and the coupled fundamentals of pendulum-sloshing-spring system shown in Fig.6 are expressed as follows.

$$\omega^2 = [(\omega_p{}^2 + \omega_s{}^2) \pm \sqrt{(\omega_p{}^2 + \omega_s{}^2)^2 - 4\mu \, \omega_p{}^2 \, \omega_s{}^2} \,] / 2\mu \tag{1}$$

where μ indicates the mass ratio $(m_p + m_f)/(m_p + m_w)$ and $m_w = m_s + m_f$. ω_p is the circular frequency of the pendulum with spring and is given by

$$\omega_p{}^2 = g / l_p + k_a / (m_p + m_w) \tag{2}$$

where g is the gravity constant. The sloshing frequency ω_s and mass m_s can be written by

$$\omega_s{}^2 = (\pi g / L) \tanh (\pi H / L) \tag{3}$$

$$m_s = \tanh (\pi H / L) \, m_w / \{ (\pi / 2)^3 (H / L) \} \tag{4}$$

where H and L indicate the depth of the liquid and the tank length in the vibration direction. The pendulum length of the TMD in Crystal Tower can be changed by every 20 cm (7.9 ft) interval and is to be tuned finally by additional coil springs if necessary. Combining three pairs of several springs, each pendulum can be stiffened up to 90 kgf/cm (504 lbf/in) by every 10 kgf/cm (56 lbf/in). Figure 7 demonstrates the fundamental tuning chart in the NS direction. The tanks have the dimensions for the sloshing behavior not to be stimulated by the swing of the TMD. It should be noted that the pendulums with length from 3.5 m (11.5 ft) to 4.5 m (14.8 ft) can be tuned from 3.5 sec to 4.5 sec by using additional springs together. And the period of this TMD can be shortened by up to about 15 % by the additional springs only.

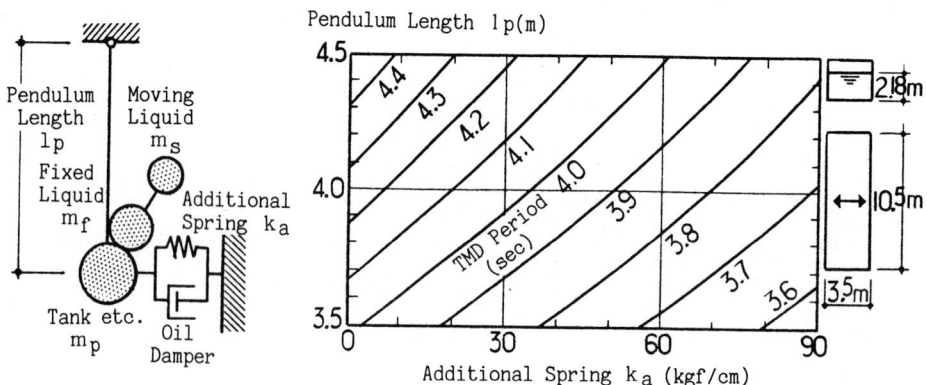

Fig.6 TMD Model   Fig.7 Design Chart of Fundamental Period of TMD

<u>RESPONSE OF TMD</u>   As the peak amplitude of the pendulum damper is limited for protecting pipe joints between the tank and the building, it is very important to study the relations among the responses of the building and the TMD and the damping of TMD itself. Let the building be modeled by SDOF as sketched in Fig.8, mass m_1, circular frequency ω and damping h_0. The building response of dynamic displacement due to winds is denoted by X_0 without TMD, while the building response with TMD, mass $m_2 (\equiv \beta m_1)$ and damping $h_2 (\equiv \xi h_0)$, is reduced to $X_1 (\equiv \eta X_0)$ and the response of TMD is $X_2 (\equiv \zeta X_0)$. The equivalent damping corresponding to X_1 is represented by $h_1 (\equiv \gamma h_0)$. Assuming that the external works done by winds are the same in these models leads to

$$\eta = 1 / \sqrt{\gamma} , \quad \zeta = \sqrt{(1 - 1 / \gamma) / (\beta \xi)} \qquad (5,6)$$

The normalized responses of the building η and TMD ζ are expressed by the mass ratio β and normalized dampings ξ and γ. Though the equivalent damping γ cannot be expressed explicitly, it may be given easily by simulation analyses (Ref.5) and has the form $\gamma = 1/(a\xi + b/\xi)$ where a and b are positive constants. As shown in Fig.9, $\xi = \sqrt{b/a}$ corresponds to the optimal damping of TMD. When $\xi \geq \sqrt{b/a}$, γ decreases monotonously with respect to ξ, so η increases and ζ decreases as ξ increases. The feasible damping of TMD can be obtained considering the TMD peak response

Figures 10 and 11 summarize the relations among ξ, γ, η and ζ in the NS direction of Crystal Tower. It should be noticed that TMD with damping from 20% to 40% critical can reduce TMD response greatly while the building response is still lessened by 50%. The damping of TMD is set to be 20% critical larger than the optical damping 10% critical.

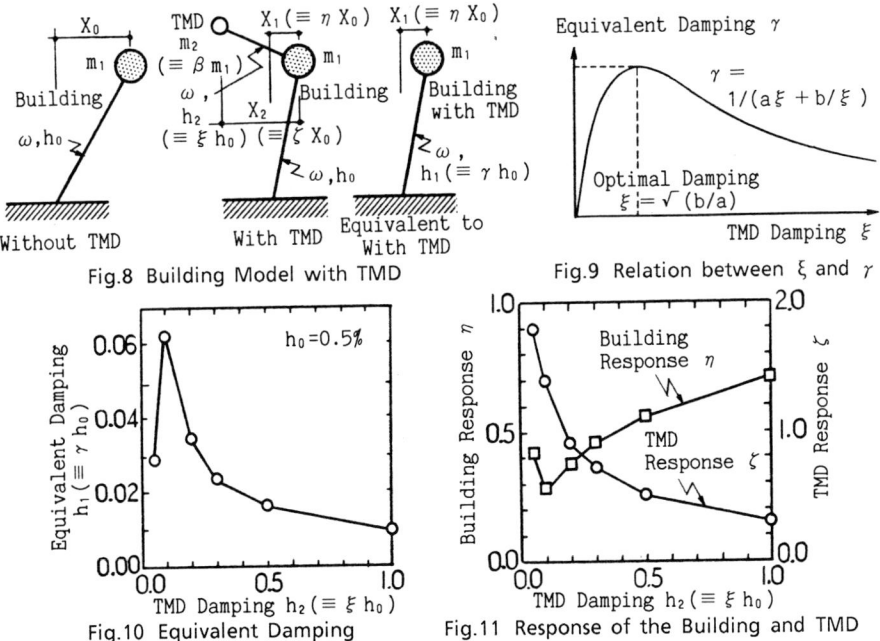

Fig.8 Building Model with TMD

Fig.9 Relation between ξ and γ

Fig.10 Equivalent Damping

Fig.11 Response of the Building and TMD

321

TYPHOON AND EARTHQUAKE RECORDS IN CRYSTAL TOWER

TYPHOON 9019   The strong Typhoon 9019 (No.19 in 1990) hit Osaka on Sept. 19 through 20 and challenged Crystal Tower soon after it was built. As illustrated in Fig.12 the peak wind velocity observed at the top of the tower was 39 m/sec (128 ft/sec) and the peak acceleration of the rooftop was 9 cm/sec² (3.5 in/sec²) or $\sqrt{2}\,\sigma \fallingdotseq 4$ cm/sec² (1.4 in/sec²). The TMD worked for 9 hours with the peak amplitude of 3.8 cm (1.5 in). Figures 13 and 14 demonstrate that the relation of the sway direction between the building and the TMD was exactly the same as designed. The effect of the TMD is shown in Fig.15 comparing with analytical results and it is noted that the TMD lessen the response by 50 %. It should be emphasized from Figure 16 that the huge weight started moving when the rooftop vibration reached a mere 1 cm/sec² (0.4 in/sec²) or 0.5 cm (0.2 in), proving that the frictional coefficient of the system is 0.001 and the slightest sway of the building gears the device into action.

EARTHQUAKE RECORD   The tower was struck also by the small earthquake on Sept. 24 last year. The time histories both observed and simulated are compared in Fig.17. The peak displacement observed at the rooftop of the building and the relative displacement between the building and TMD are 0.9 cm (0.4 in) and 1.9 cm (0.7 in), respectively. A simulation analysis with TMD shows a good agreement with the observed records and compensates the time histories in the last parts of the records. It is concluded through the simulation analyses that the TMD can damp the building lateral motion quite quickly after earthquakes, which is particularly remarkable in the high-rise buildings.

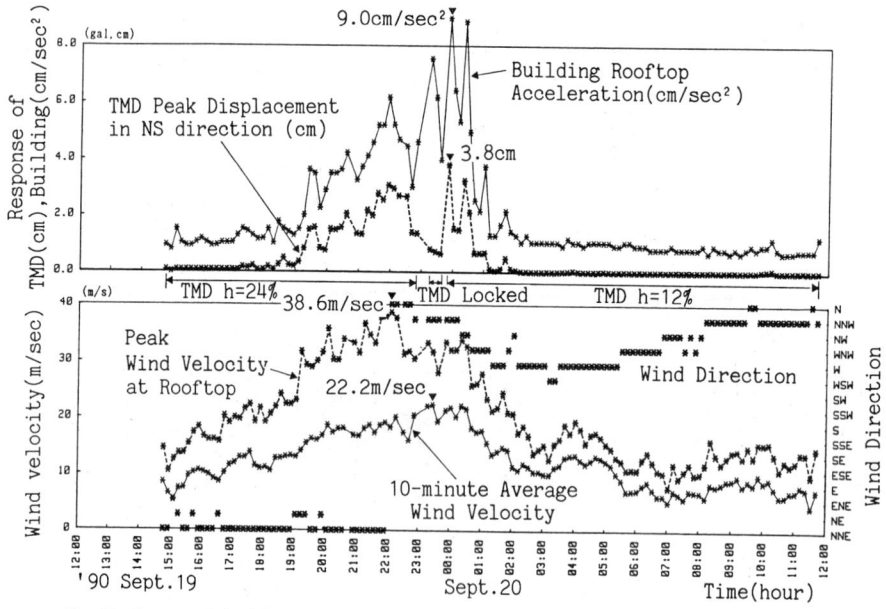

Fig.12 Strong Wind by Typhoon 9019 and TMD Observed at Crystal Tower

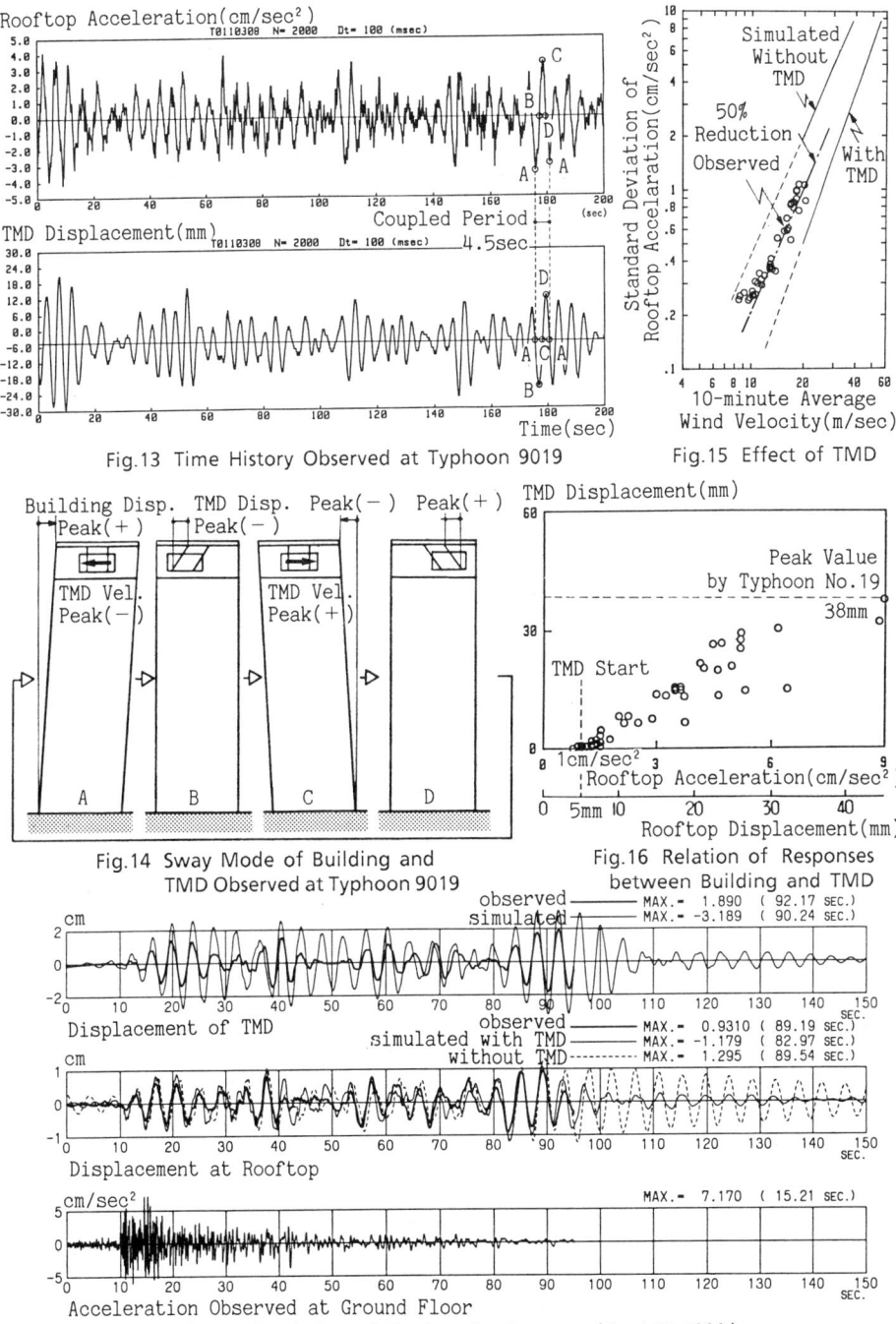

Fig.13 Time History Observed at Typhoon 9019

Fig.15 Effect of TMD

Fig.14 Sway Mode of Building and
TMD Observed at Typhoon 9019

Fig.16 Relation of Responses
between Building and TMD

Fig.17 Simulation of Earthquake Response (Sept.24,1990)

CONCLUDING REMARKS

The pendulum mass damper has been developed and installed in the 37-story office building to reduce the usual wind-induced motion by 50%. As the ice thermal storage tank at the rooftop of the building for air conditioning is integrated and used as a pendulum weight, the damper itself needs no additional mass nor installing space, so it is very economical. In fact the total cost of the damper system installed in Crystal Tower is about 50 millions yen (350 thousands US dollars) which is less than 0.2% of the construction cost of the building.

Six ice thermal tanks are suspended from the top roof girders so as to swing by the same periods as the building. Fundamental tuning by adjusting the pendulum rod length and/or additional spring devices is simple and practical. The rubber flexible pipe joints enable the damper to slide up to the amplitude of 25 cm (9.8 in) and the damper is effective until the stopper works, which is expected during a 20-year interval. The tank dimensions are determined for the sloshing behavior not to affect the damper.

The strong typhoon proved that the friction of the hanging TMD is small enough for the damper to work successfully in a wide range of the wind load. Also the earthquake record indicates that the TMD can damp quickly the building motion after earthquakes particular to the high-rise building. The tuned pendulum mass damper in Crystal Tower is our solution that satisfies the aesthetic demand for the shape and proportion of the tower and the engineering requirements thoroughly.

ACKNOWLEDGMENT

The authors would like to be grateful to T. Ohtake and K. Katayama, Takenaka Corporation, for their valuable assistance in the observation and analyses of typhoon and earthquake records at Crystal Tower.

REFERENCES

1. McNamara,R.J.,"Tuned Mass Dampers for Buildings", *Journal of the Structural Division*, ASCE, Vol.103, No.ST9, Sept., 1977, pp.1785-1798.
2. "Hancock Tower Now to Get Dampers", *Engineering News Record*, Oct.30, 1975, p.11.
3. Nagase,T. *et al.*, "Sloshing Test of Thermal Storage Tank of Liquid Ice" (in Japanese), *Summaries of Tech. Papers, 1989 Annual Meeting*, Archi. Inst. Japan, 1989, pp.603-604.
4. Mataki,Y.*et al.*, "A Theoretical Approach to Wind-Induced Response for a Tall Building with a Tuned Mass Damper" (in Japanese), *Summaries of Tech. Papers, 1989 Annual Meeting*, Archi. Inst. Japan, 1989, pp.581-582.
5. Nagase,T. *et al.*, "Application of Tuned Pendulum Mass Damper to High-rise Building using Ice Thermal Storage Tank" (in Japanese), *Journal of Structural and Construction Eng.,Trans. Kinki Branch*, Archi.Inst.Japan,1990, pp.465-468.
6. Graham,E.G. and A.M. Rodriguez, "The Characteristics of Fuel Motion which Affect Airplane Dynamics", *Journal of Applied Mechanics*, Vol.19, No.3, 1952, pp.381-388.

Proceedings of the Second Conference on Tall Buildings in Seismic Regions
55th Regional Conference
May 16 and 17, 1991, Los Angeles, California

STRUCTURAL DESIGN OF C-1 BUILDING
THE BIGGEST BASE ISOLATED BUILDING IN THE WORLD

By
Susumu Nakagawa*

C-1 building is located in Fuchu City, Tokyo and is at present under construction. It will be used as a computer center, being the biggest base isolated building in the world.
The function of this base isolated system is such that even if a large earthquake occurs the Computer Center will not stop working.

C-1 Building Outline

C-1 building consists of a 7-floor super structure, a penthouse and 1 basement floor. The total weight of the building is 62800t and the total area is 37846m². A composite structure (Steel - Reinforced Concrete) is used for the design. In figure 1 is shown the C-1 building outline.

Soil Characteristic and Foundation

The boring log is shown in figure 2. The allowable bearing capacity of the soil for permanent load is 50t/m².
A direct foundation lying on fine sand layer (Kazusa Layer) at GL-10.8m was used.

Base Isolators

68 lead-rubber isolators (LRB) are used in this project. (Figure 3)
There are four types of LRB: 1100φ, 1200φ, 1300φ, 1500φ, being 1500φ the biggest LRB in the world up to date.
Standard bolts are used to connect the flange plate of the base isolator and the lead-rubber portion of the isolator.

Tests were performed in order to check the stiffness and damping factor of all the LRB used in this project, before installing them. (See table 1 and figure 4). By setting LRB between the girders of the basement floor and the foundation, the aseismic safety of the building has been improved thus preserving the computer center equipments.

*Structural Engineering Department
 Nihon Sekkei, Inc.
 Tokyo, JAPAN

Aseismic Design Philosophy

As for the super structure, allowable stress design for seismic loads, vertical loads, and also the response analysis to check its aseismic safety were performed.

The retaining wall and the mat slab are designed to be continuous. 1.5 times the vertical load and the seismic load were used for the allowable stress design of the retaining wall.

A base shear coefficient of 0.25 was used for the allowable stress design of the LRB and building connection. This base shear coefficient is large in comparison to the Level 3 response coefficient equal to 0.148, which will be discussed later.

The LRB system connection is designed such that it will not collapse before the LRB system collapses.

To check the aseismic safety of the building, three levels of earthquakes in terms of velocity were used: 25cm/sec (level 1), 50cm/sec (level 2), 75 cm/sec (level 3). See table 2.

As the live load will change if the use of the building is different to the one that it was designed for, the design was performed for different types of live loads in order to take into account this possibility.

Design for Vertical Load

The design was done following the Japanese Building Standard Code. The possible need to change one or more LRB in the system has also been considered in the design.

Design for Seismic Load

The seismic story shear coefficient is calculated following the Japanese Building standard Code:

$$C_i = Z \; R_t \; A_i \; C_o$$

Where

C_i: Seismic story shear coefficient of the aboveground part of a building at a given height

Z: A value to be specified by the Minister of Construction within a range between 1.0 and 0.7 according to the extent of earthquake damage, seismic activity and other seismic characteristics based on the record of earthquakes in the region concerned

Rt: A value representing vibration characteristic of buildings to be obtained by the calculation method specified by the Minister of Construction according to the natural periods of buildings and kinds of ground

Ai: A value representing a vertical distribution of seismic story shear coefficients according to the vibration characteristics of buildings to be obtained by the calculation method specified by the Minister of Construction

Co: Standard shear coefficient

As there is the possibility of a large displacement of the LRB system when a large earthquake occurs, a distance of 65cm has been set between the building and the retaining wall.

LRB System Characteristics

Hysteresis curves are defined for each LRB. As the LRB stiffness changes with the maximum displacement, bilinear P-δ curves are done for the 3 levels of earthquakes mentioned before.
Level 1 (LRB maximum displacement =12cm,
        rubber maximum strain   = 50%)
Level 2 (LRB maximum displacement =24cm,
        rubber maximum strain   =100%)
Level 3 (LRB maximum displacement =36cm,
        rubber maximum strain   =150%)

As mentioned before there are 4 types of LRB, therefore there are 4 corresponding P-δ curves. By adding these 4 P-δ curves of the LRB system, it is possible to get a general curve that represents these 4 types, i.e. a total trilinear P-δ curve for the LRB system.

Seismic Response Analysis

Analysis Model

A nine mass-spring system and the equivalent shear stiffness have been used to model the 7-floor super structure, 1 basement and 1 penthouse. This is fixed at the bottom level of the LRB System.
The stiffness value for the isolation interface was calculated from the total trilinear P-δ curve mentioned above.
The P-δ curve for the Super Structure is in the elastic stage as it was set in the design.

The criteria for the earthquake level 1 is that the shear stress response for the shear wall is less or equal to Fc/10, and the shear wall stiffness reduction factor ß=0.8. The internal viscous damping is 1%. For the earthquake levels 2 and 3 ß=0.3. The internal viscous damping is 2%.

The natural period for the first mode has been calculated for two cases. When the isolator interface displacement is small (i.e. the LRB remains elastic): T1s=1.45s, T1y=1.34s. And when the isolator interface displacement is large (where an equivalent stiffness is calculated for a displacement of 30cm): T1x=3.01s, T1y=2.97s.

The earthquake waves used for the response are shown on table 3.

Response Analysis Results

There is a difference between the rubber stiffness when it is brand new and when it gets older. To take into account this, the response analysis for level 2 was checked for the case when the stiffness is 20% smaller and when it is 30% larger. On table 4 the response analysis results are presented, besides these results are shown on figure 5.

Wind Design

The story shear calculated following the Japanese Building Standard Code for wind design is less or equal than the 13% of the story shear calculated for seismic design. As the maximum base shear due to wind load is 923t and, the isolator interface yield shear strength is 2133t which is much larger, it is possible to say that the building is safe against winds.

Conclusions

• The Construction of C-1 building was started in August 1990 and it will be finished by August 1992. It is at present under construction and the LRB System set up has been finished by December 1990.

• From the tests performed to find the stiffness and damping factor of all the LRBs, it has been possible to check the LRB System design.

• Through the design of the C-1 building, it has been possible to take the maximum advantage of the use of the base isolator system and also to improve the building function itself.
Base isolator systems should be improved, by taking the maximum advantage, through experiences with other new building developments in the future.

Table 1. LRB Characteristics

		LRB1 (600TON)	LRB2 (800TON)	LRB3 (1100TON)	LRB4 (1500TON)
rubber ϕ	mm	1,100	1,200	1,300	1,500
t x n	mm	8.0x30	8.0x30	8.0x30	8.0x30
Lead ϕ	mm	180	210	240	280
LRB width	mm	1,500	1,600	1,700	1,900
LRB height	mm	530	530	530	650
K1 ε=100%	ton/cm	12.71	15.47	18.52	24.75
K2 ε=100%	ton/cm	1.96	2.38	2.85	3.80
QY ε=100%	ton	31.85	43.53	56.62	77.08
Equivalent Stiffness, δ=30cm	ton/cm	2.88	3.47	4.55	6.13
Vertical Stiffness	ton/cm	3,609	4,427	5,322	7,390
ϕ/t.n	---	4.6	5.0	5.4	6.3

n: number of rubber layers
t: rubber thickness

Table 2. Seismic Design Criteria

Seismic Level	Upper Structure	LRB System
Level 1 20 cm/s	Design Shear Force≥QR Shear Wall Stress ≤0.1Fc	
Level 2 50 cm/s	Elastic Limit Shear Force QEL≥QR	ε≤200% δmax=30cm Rubber strain safety factor≥1.5 There is no tension in the LRB
Level 3 75 cm/s	Ultimate strength QU≥QR	ε≤300% δmax=50cm rubber strain safety factor ≥1.0

Table 3. Earthquake Waves used for the Response

Wave Name	Analysis time (s)	Maximum Acceleration (cm/s²)		
		Level 1	Level 2	Level 3
		25 cm/s	50 cm/s	75 cm/s
EL CENTRO NS 1940/05/18	30.0	253.90	507.80	761.70
TAFT EW 1952/07/21	30.0	256.73	513.46	777.19
TOKYO 101 NS 1956/02/14	11.3	256.67	513.34	770.01
HACHINOHE NS 1968/05/16	30.0	178.97	357.94	563.91
HACHINOHE EW 1968/05/16	30.0	135.10	270.20	405.30

Table 4. Response Results

Level		Upper Structure		Isolator Interface
		Criteria	Response Results	
X	1	Design Shear Force QD≥QR	QR=0.997QD	
	2-1	QEL≥QR	QR=0.988QEL	δEL(=30cm) ≥δR / δR=24.0 cm δR=0.8δEL
	2-2	QEL≥QR	QR=0.787QEL	δEL(=30cm) ≥δR / δR=23.2 cm δR=0.773δEL
	2-3	QEL≥QR	QR=1.115QEL	δEL(=30cm) ≥δR / δR=23.1 cm δR=0.778δEL
	3	QU≥QR	QU≥≥QR	δU(=50cm) ≥δR / δR=38.1 cm δR=0.762δU
Y	1	Design Shear Force QD≥QR	QR=0.895QD	
	2-1	QEL≥QR	QR=0.965QEL	δEL(=30cm) ≥δR / δR=23.8 cm δR=0.793δEL
	2-2	QEL≥QR	QR=0.77QEL	δEL(=30cm) ≥δR / δR=23.7 cm δR=0.79δEL
	2-3	QEL≥QR	QR=1.1QEL	δEL(=30cm) ≥δR / δR=26.3 cm δR=0.876δEL
	3	QU≥QR	QU≥≥QR	δU(=50cm) ≥δR / δR=37.1 cm δR=0.742δU

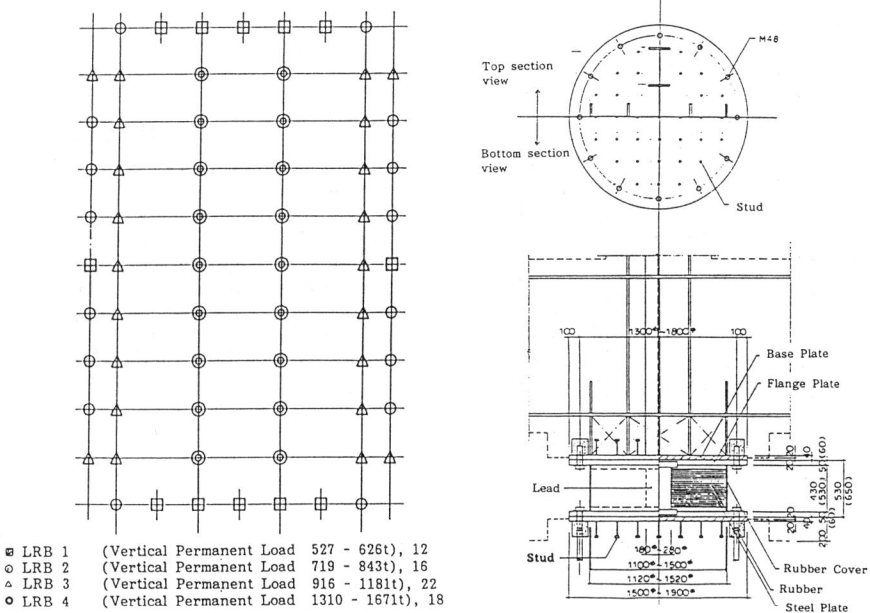

▣	LRB 1	(Vertical Permanent Load	527 - 626t),	12
⊙	LRB 2	(Vertical Permanent Load	719 - 843t),	16
△	LRB 3	(Vertical Permanent Load	916 - 1181t),	22
○	LRB 4	(Vertical Permanent Load	1310 - 1671t),	18

Figure 3 LRB System Layout   Figure 4 Lead-rubber Isolator (LRB)

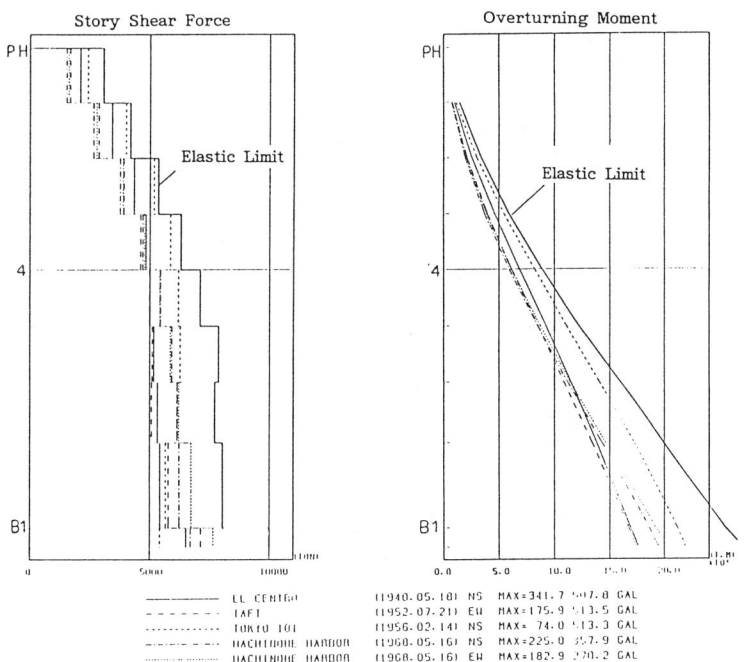

Figure 5 Response Analysis Results

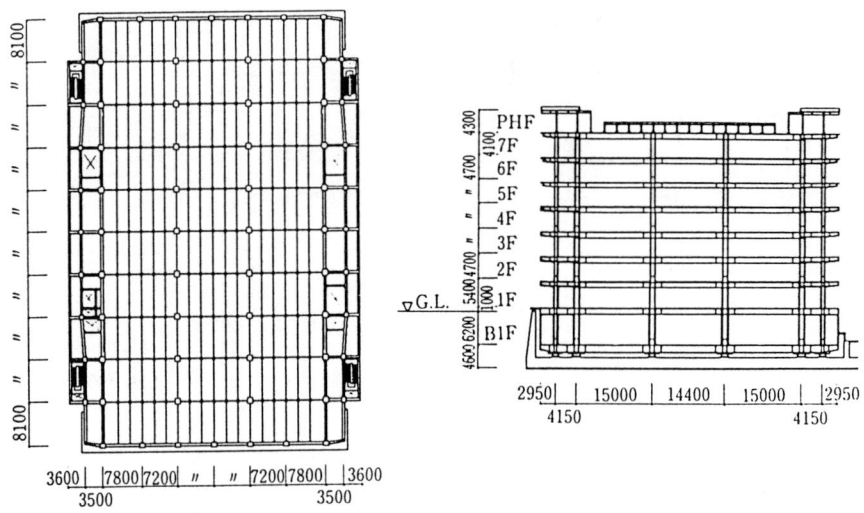

Figure 1 C-1 Building Plan and Elevation

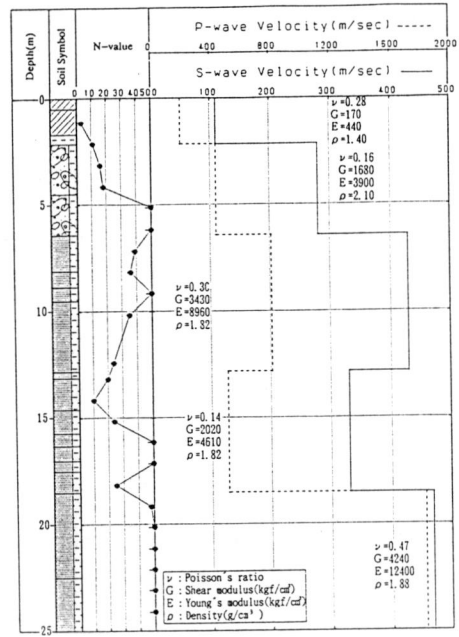

Figure 2 Boring Log

Proceedings of the Second Conference on Tall Buildings in Seismic Regions
55th Regional Conference
May 16 and 17, 1991, Los Angeles, California

STRUCTURAL ANALYSIS AND DESIGN OF THE MINAMI TOWER

John M. Nissen S.E. and Hung C. Lee P.E.

John A. Martin and Associates, Inc., Los Angeles, California

ABSTRACT

Structural analysis and design of the lateral load resisting systems in the thirty six story Minami Tower are presented. Special attention is given to the analysis and detailing of the discontinuous ductile steel moment resisting tube frame and the resulting structural design issues.

INTRODUCTION

The Minami Tower is a thirty six story, steel framed, mixed use building containing office, restaurant and retail areas. Twenty five stories of office floors are supported by a seven story steel framed base, which encloses a sixty-eight foot high open atrium lobby space with retail areas interspersed around the perimeter of the third and fifth levels. The uppermost levels of the tower are dedicated to restaurant and observation deck usage. The tower is flanked to the north and to the south by individual six story concrete parking garages and sits on a split level mat foundation with the lower half accommodating a vehicle drive-through which connects the parking structures. The partial subterranean level is braced by shear walls.

The project site is located in Downtown Las Vegas, Nevada and is contained in the block bounded by South Las Vegas Boulevard, Sixth Street, Bridger Avenue and Clark Avenue. The site area is approximately 260,000 square feet of which 178,000 square feet are being developed for this phase of the project. The tower encompasses roughly 700,000 gross square feet. The lower six floors starting at the plaza level consists of approximately 73,500 square feet of retail space. Typical office floors range from approximately 14,400 to 19,500 gross square feet.

The tower is symmetrical and has a maximum plan dimension of 137 feet at the lower levels and a maximum plan dimension of 126 feet at the upper levels. The tower rises 452 feet above the plaza level which acts as the shear base. The total overall height,including the radio mast, is 588 feet making this project the tallest structure in the state of Nevada. The critical aspect ratio, excluding the radio mast, is 3.30:1. The typical story height is 13 foot 4 inches with a mechanical level story height of 22 feet occurring at the seventh floor.

The building is essentially symmetrical about the centerline axis of the structure. There is a six story steel framed annex attached to the eastern portion of the tower base. Additionally, as a result of the horizontal setbacks dictated by the architecture, a number of different size floor plates are present. As a consequence, relatively complicated vertical and lateral load carrying system evolved which exhibited a great deal of variation throughout the structural entity (Figures 1,2,3 and 4).

CRITERIA FOR STRUCTURAL ANALYSIS AND DESIGN

Design dead and live loads were typical for this class of office structure.

Seismic recommendations for the tower were specified by the Project Geotechnical Engineer, Western Technologies, Inc. There is a noticeable lack of seismic events which have epicenters in the Southern Nevada area although underground nuclear testing at the Nevada Test Site (approximately 90 miles north of Las Vegas) has simulated numerous small earthquakes in the immediate locale. Consequently, code lateral forces were generated. For stress considerations, forces were derived by use of equations 12-1 and 12-2 of the 1988 Uniform Building Code and members were checked in combination with vertical loads against AISC allowable stresses including a one third increase. Seismic drift considerations utilized seismic forces modified in conformance with Section 2312(e)8 of the aforementioned Code. Drifts were limited to 0.005 times the story height.

The American National Standards Institute (ANSI) code was used to derive wind loads. For stress considerations, a 75 mile per hour wind velocity was used in conjunction with Exposure B for design of the tower. This analysis was supplemented by actual recorded wind speed data supplied by Englekirk and Hart Inc. For drift considerations, a 55 mile per hour wind velocity was used. Wind drifts were limited to .0025 times the story height in both of the principle directions. Wind stresses were checked, for the 75 mile per hour wind, in combination with vertical loads, against AISC allowable stresses including a one third increase.

PRELIMINARY DESIGN

The initial concept for a lateral load resisting system consisted of a continuous perimeter ductile steel tube frame utilizing WTM24 columns and wide flange beams as deep as 40 inches.

A preliminary model was developed to study this concept wherein all special architectural features such as setbacks and transfers were neglected . This simple three dimensional model of the frame was generated using the CSI-ETABS+ computer program and a single, typical vertical load was used throughout.

At this stage it was found that the perimeter tube frame appeared to be adequate to satisfy the seismic and wind stress criteria in both of the primary directions of the structure. The model demonstrated that the short diagonal bays (Lines K to H.7 and 19 to 19.7) at the buttress corners were creating a hard spot in the levels corresponding with floors 25 through 30. These bays were removed prior to completing a second, more accurate, computer model (Figure 5).

At this point in the modelling, due to the unsupported height of the atrium vertical supports, a variety of box column sizes and plated WTM24 columns were added to the member database to facilitate the stress analysis for the final analysis.

The second, more complete computer model of the frame incorporated the main transfer level at the twenty seventh floor as well as the termination of the tube frame at the twenty ninth floor. The completion of this portion of the analysis allowed a timely identification of the variety of column dimensions which then facilitated the architectural design process and further refined and delineated specific areas of concern within the lateral force resisting system.

LATERAL LOAD RESISTING SYSTEM

The final structural system is a special moment resisting space frame consisting of a perimeter frame with a full setback at the twenty seventh floor and terminating entirely at the twenty ninth floor. The lateral system then becomes four separate, stand alone perimeter frames, which begins at the twenty seventh floor and extends to the thirty third floor. The frame is then moved to the core area of the building where it terminates at the thirty sixth level. At each of the frame transitions, there occurs an overlap of systems to insure proper distribution of forces (Figures 6 and 7). Typical frame beams are ASTM A36 steel ranging from W24's to WTM36's (39 inches deep). There are transfer girders at the twenty seventh floor to pick up the frame offset and range up to the heavier WTM40 shapes.

A variety of column sizes are utilized from the database of sizes defined in the second phase of the preliminary design noted previously. Columns are typically ASTM A572 grade 50 steel. Non-frame columns are W14's with the exception of boxes with plates over 4" thick, which are ASTM A572 grade 42. Many of these non-frame columns carry transfer girders or support the transfer truss located at the seventh floor, which in turn carry frame columns. Rolled column members in the frame are WTM24's. In the lower floors and for tall stories the columns change to modified boxes with the addition of plates spanning from flange to flange.

FINAL DESIGN AND ANALYSIS PROCEDURE

For final design, a complete three dimensional model of the tower, including setbacks, transfers and disconnected diaphragms, was constructed using the CSI-ETABS+ computer program. Because of the complexity of the framing system virtually every column line in the building, including non-frame core columns were modeled so as to accurately represent the overall performance, including differential column shortening. Accurate dead and live loads were input. Live loads were reduced as permitted by the UBC. ANSI 75 mph wind loads were generated for low, medium and high damping and input as static loads. The fundamental period in the final design was 5.42 seconds. The governing load conditions were 55 mph wind with low damping for drift criteria and, based on the location of a specific level, gravity loads plus either the 75 mph wind loads with low damping or the code seismic loads for stress criteria. Final stress ratios in the columns are kept below those of the beams and the strong column-weak beam provision of the 1988 UBC is satisfied.

UNCOMMON DESIGN ISSUES

The primary analysis problems addressed as a result of the geometry of this framing system resulted in several design issues with significant cost and constructibility ramifications.

The major vertical load carrying element which impacted the lateral load resisting system is the presence of the transfer truss at the seventh floor. This truss is the result of the architects' desire to create a sixty eight foot clear atrium which, in turn, requires that one of the nine core columns be terminated. The truss is constructed entirely of W14, grade 50 shapes laid flat so as to utilize the increased allowable stresses (0.75Fy vs. 0.60Fy) and to mitigate the potential for laminar tearing as the majority of the complete penetration welds are now made end of flange (or web) to edge of flange. This truss essentially created a braced frame in the primary X-X direction between the seventh and eighth floors. A separate model was run including the truss to ascertain its effect, particularly in light of the fact that from the seventh floor down, diaphragm discontinuity played a major role in the frame performance. It was found that the truss helped in controlling drifts at this level, which is twenty two feet high, but due to its stiffness caused diaphragm problems by unloading the frame and reloading one level down.

Differential shortening between columns due to vertical load was significant, especially the box columns supporting the transfer truss. In a typical building, the core columns shorten more under vertical loads due to larger tributary areas relative to the perimeter frame columns. This shortening is compensated for during fabrication by adjusting the lengths of the columns. In this project, the core columns from the thirty second floor to the thirty sixth floor are also frame columns and at the twenty seventh floor, support transfer girders. Under full vertical load, analysis indicated differential shortening of up to 3" between the core and perimeter frame columns. This resulted in high stresses under vertical loads particularly in beams and columns sitting on transfers. All of the core columns were checked for stress by hand.

The analysis and design of the lateral load resisting system presented unique problems that, when addressed in conjunction with the transfer truss and differential shortening, resulted in multiple computer runs to properly ascertain the individual effect of each on the total system . The consequence of the two horizontal setbacks in the frames accentuated the differential column shortening and resulted in the building performing as if a spring were introduced, particularly at the twenty seventh floor setback. At the locations of the horizontal displacement of the frames, column and bays were overlapped so that the frame action at these specific locations would have a smoother transition.

On this specific project, there were two remaining issues which had direct impact on the overall cost of the structural steel. The first was the design issue of the weak axis effective length factor (K_y) for columns with beams framing at shallow angles. The second were the diaphragm discontinuities caused by the six story clear atrium.

The issue of the weak axis effective length factor was a problem in that the AISC procedure for calculating the effective length factor when beams frame in at shallow angles yields unrealistically high effective lengths. The task then became to resolve the out of plane moment vector so that the AISC provisions could be adapted to result in an effective length factor which was closer to reality. This analysis was done by hand for the typical 22.5 degree angle that the moment frame beams make with the columns at the buttress corners. This exercise saved a substantial amount of structural steel as during the preliminary computer analysis, the AISC procedures were followed exactly and an inordinate number of box columns were required which, in turn, shortened the fundamental period of the structure thereby increasing both the ANSI wind loads and the code specified seismic loads.

The remaining design issue of diaphragm discontinuities caused several analytical quirks due to the location in the structure. There is essentially no diaphragm to support the column weak axis at the second, fourth or sixth floors thereby resulting in unsupported lengths of twenty feet. This substantially lowered the allowable axial stress to a level where it became the predominant factor in the unity equation for combined stresses. Combining the lower allowable axial stresses with high axial loads due to the frame overturning moments created the situation where these lower level columns were all plated. Several options were investigated including inducing the low-rise annex frames to induce load out of the tower frame. However, it was decided that the performance of the tower frame would be easier to predict by simply increasing the column axial areas.

CONCLUSIONS

Architectural features that delineate the Minami Tower project required a finely tuned structural system to achieve the parameters as set forth by the architecture. This lateral load resisting system presented unusual analysis, design and detailing issues.

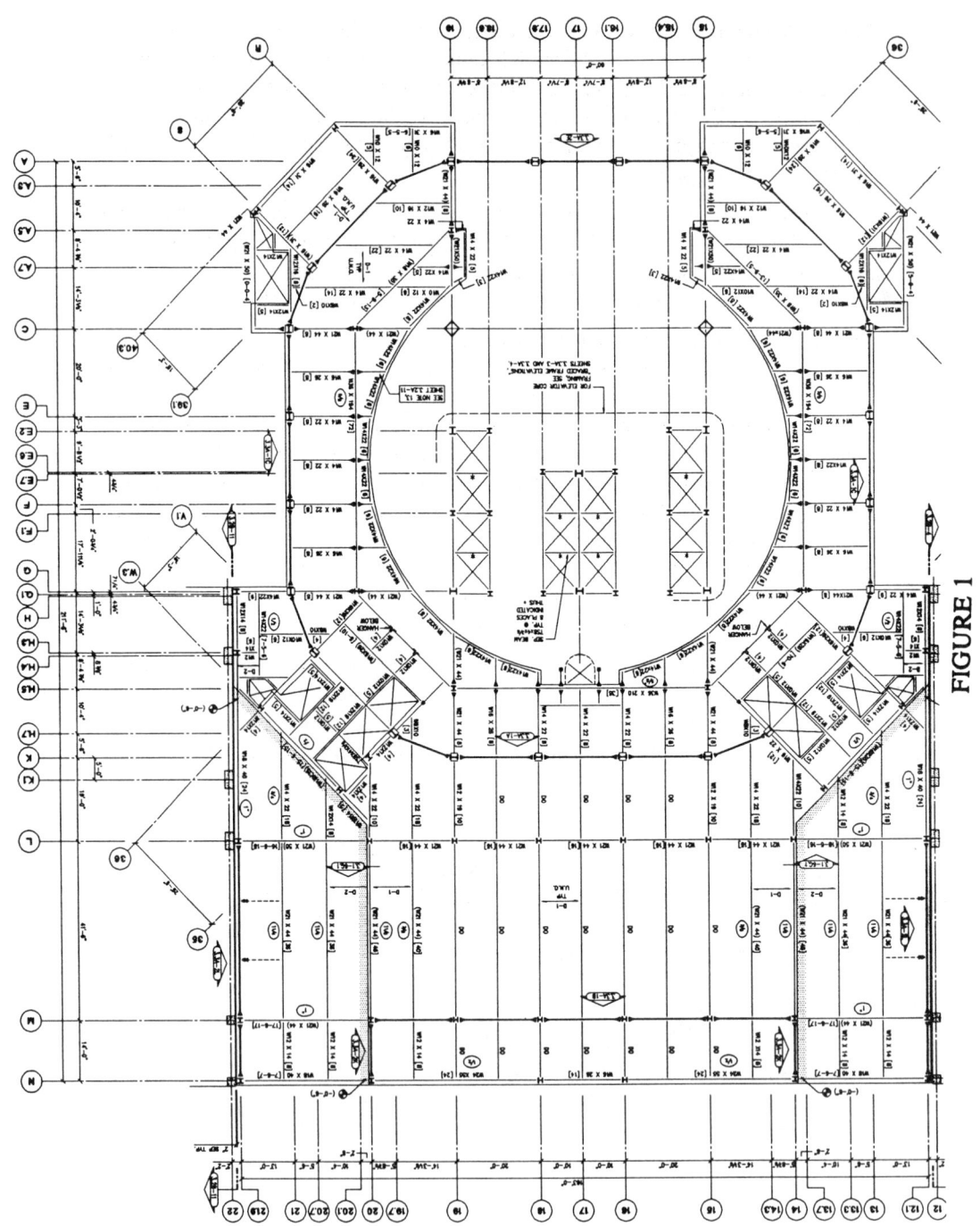

FIGURE 1

338

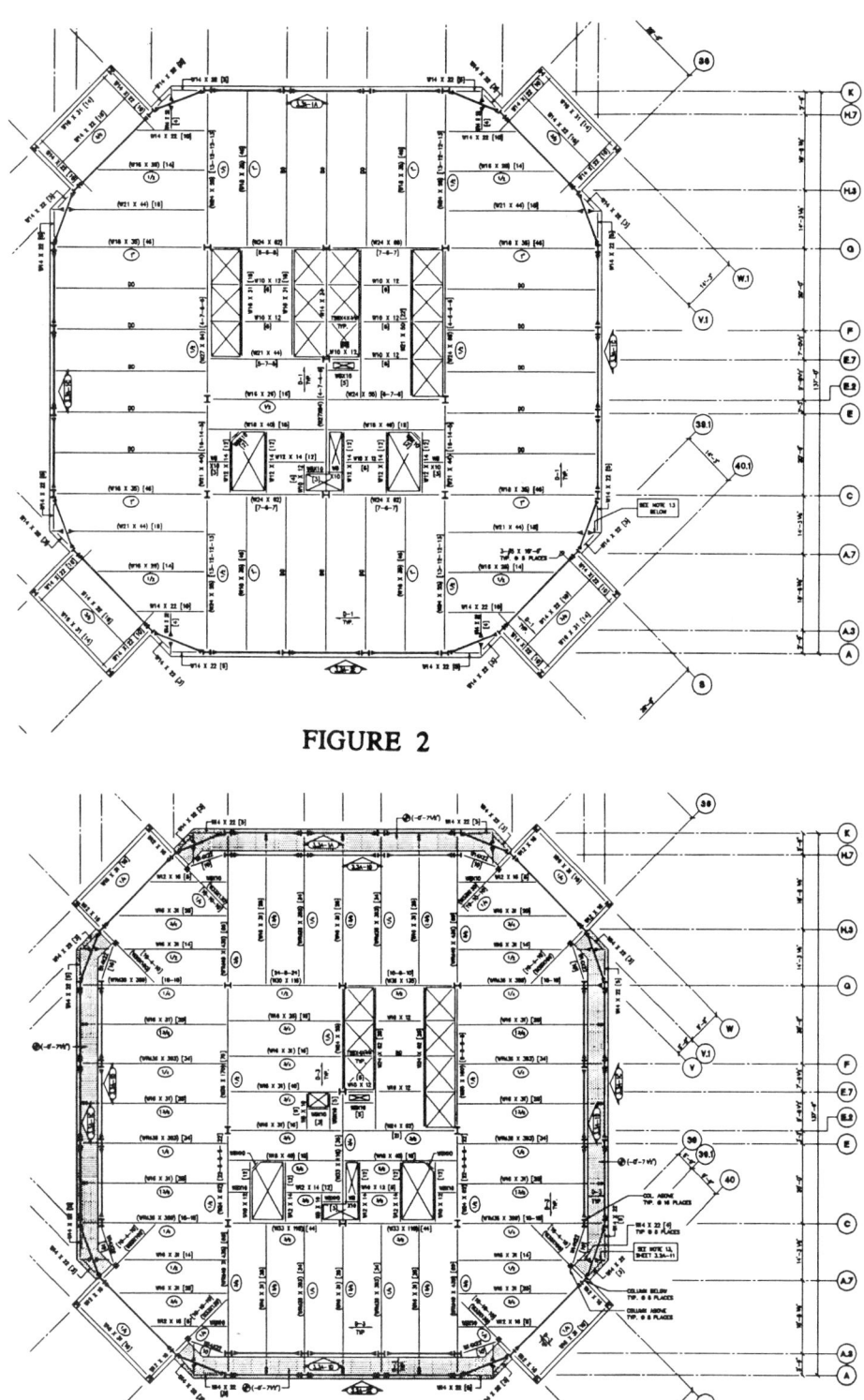

FIGURE 2

FIGURE 3

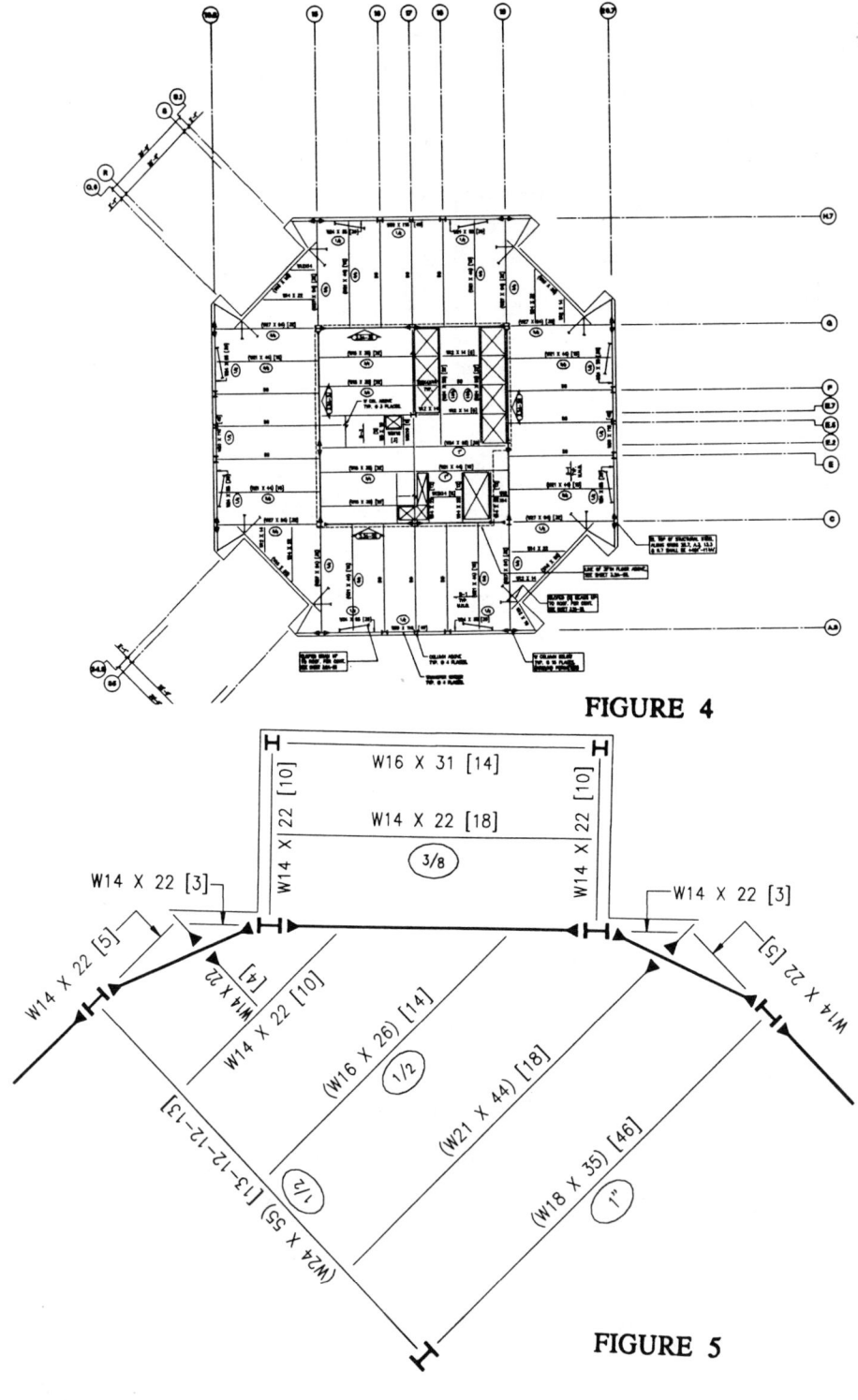

FIGURE 4

FIGURE 5

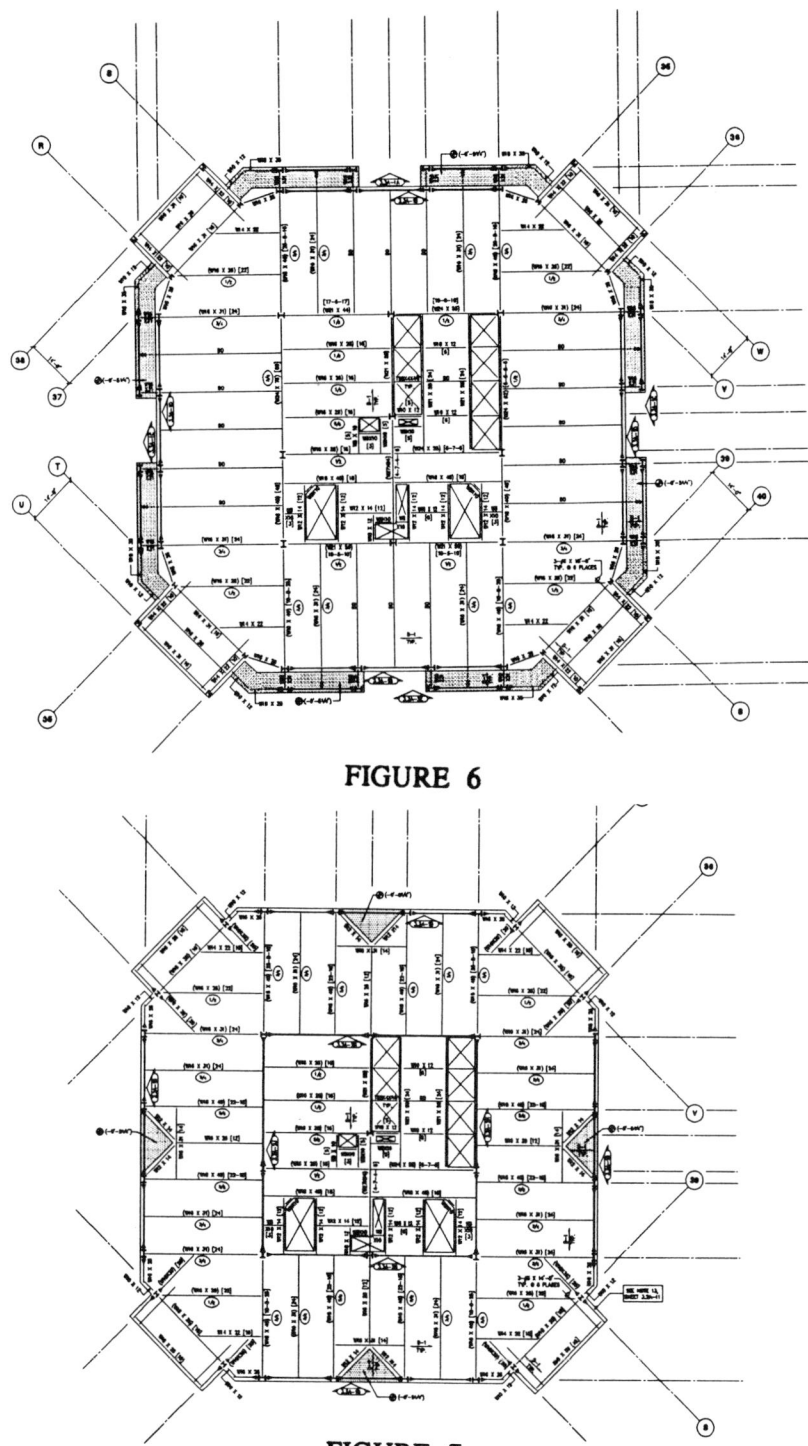

FIGURE 6

FIGURE 7

Proceedings of the Second Conference on Tall Buildings in Seismic Regions
55th Regional Conference
May 16 and 17, 1991, Los Angeles, California

Foundation Design of A High-rise Building on Reclaimed Land

Seiichi Oh'wada
Senior Structural Engineer
Structural Engineering Department
Nikken Sekkei Ltd

Naoki UCHIDA
General Manager
Structural Engineering Department
Nikken Sekkei Ltd

In Japan, which is a small island nation, there are many high-rise buildings constructed on reclaimed lands, especially in Osaka Bay and Tokyo Bay areas. Differential settlement of a ground and liquefaction of soil during an earthquake are grave issues in designing a high-rise buildings on a reclaimed land. We have accumulated experience of designing high-rise buildings which stand now on the reclaimed lands.

There were some cases in Japan in the past wherein buildings were damaged by soil liquefaction caused by earthquakes. The soil liquefaction in such cases drastically reduced the vertical bearing capacity of the soil, which tilted building structures and damaged their frameworks in the case of buildings having no piles, and broke piles in the case of buildings supported by piles.

This paper introduces the foundation design which was applied to a high-rise building to stand in the Makuhari Area of Chiba Prefecture which is a reclaimed land area in the Tokyo Bay area. (Fig. 1) In this reclaimed land area, soil liquefaction at the time of an earthquake poses more problems of engineering significance than settlement of the ground. In the case of the high-rise building in question, there is the possibility of liquefaction of the sand layers immediately below the building foundation if a severe earthquake were to occur. Therefore, the foundation soil was improved, and a foundation vibration analysis involving the building-pile-soil trinary system was performed, and the safety of the building and its foundation was confirmed. The following introduces the countermeasures taken against such liquefaction:

1. General Description of the Building Structure

The building in question is a hotel with 25 stories above ground and 1 basement. (Figs. 2 & 3) It is 97.2 meters high from the ground level. The upper stories in the above-ground portion of approximately 81 meters are made of structural steel, and the lower portion of approximately 16 meters and the basement level are of a steel reinforced concrete structure. The building base level was GL-8.7m, and many shear walls are provided in the basement to increase the rigidity and resistance in both horizontal and vertical directions. Foundation beams are constructed of reinforced concrete.

Seismic designs in Japan usually consider two types of seismic movement. One is seismic movement of moderate intensity likely to occur several times in a building's service life (hereinafter called "level 1 earthquake"), and the other is severe seismic movement of the highest intensity comparable to the Great Kanto Earthquake of 1923 (hereinafter called "level 2 earthquake").

Both "level 1 earthquake" and "level 2 earthquake" are taken into consideration in the seismic design of the upper frame of the said building. To be input into the seismic response analysis performed, the level 1 and level 2 earthquakes are standardized, from the following four actually recorded earthquakes, as the maximum ground surface velocity of "25 cm/sec" and "50 cm/sec," respectively. The seismic response analysis is conducted with the building foundation bottom level fixed, and by inputting those level 1 and level 2 earthquakes.

Resultingly, it was confirmed that the building framework was safe against earthquakes. The recorded earthquakes adopted in the analysis were as follows:

1) EL CENTRO NS   May 18, 1940

2) TAFT CALIF. EW July 21, 1952

3) HACHINOHE NS   May 16, 1968

4) TOKYO 101 NS   Feb. 14, 1956

2.   Study on Foundation Soil Liquefaction

As in the case of the upper framework of the building, the foundation soil liquefaction was also studied in relation to the level 1 and 2 earthquakes. Fig. 4 shows the flow of the study from consideration of foundation soil liquefaction to building foundation structural design.

2.1 Preliminary Investigation

This building is to be situated on an artificial land (reclaimed by using dredged soil) in the Makuhari Area of Chiba Prefecture during the period from 1969 to 1978. The possibility of foundation soil liquefaction of this site was investigated in light of the results of the soil tests performed on the ground of the neighboring area. The study of those tests suggested that there was a possibility of sand layer liquefaction immediately below the building foundation bottom in case a severe earthquake should take place.

2.2 Soil Investigation

The soil investigation, in addition to usually performed investigation subjects, included the following features to facilitate the soil liquefaction study:

- "Grain size analysis" was performed more often than usual for each different soil stratum.

- "Cyclic triaxial test" was repeated to know the resistance of the original soil formation against liquefaction.

- "Soil dynamic deformation characteristics test," "Resonant column test" and "Cyclic torsion test" were performed for each stratum to input the results thereof into the foundation soil dynamic non-linear response analysis. Fig. 5 shows the representative soil log, and Fig. 6 shows the representative results of the soil dynamic deformation characteristics test.

2.3 Liquefaction Judgement Method

Liquefaction judgement was based on the "Recommendation for Design of Building Foundation" of the Architectural Institute of Japan. This method judges liquefaction's possibility by comparing the intensity of shear stress in a ground which is generated at the time of an earthquake with that of shear stress in that soil layer which may resist liquefaction. Its outline is given in the attachment which follows.

The shear stress in a ground during an earthquake was sought by modeling the foundation soil and performing a dynamic response analysis. The dynamic behavior of foundation soil was obtained by 1-dimensional non-linear response analysis with reference to the results of the dynamic soil

345

deformation characteristics test performed in connection with the soil investigation. The recorded seismic waves shown in "1. General Description of the Building Structure" were used in the analysis. The level 1 earthquake and the level 2 earthquake were standardized at a maximum ground surface velocity of 25 cm/sec and 50 cm/sec respectively, and these values were input at GL-34m.

3. Original Foundation Soil Liquefaction Judgement and Countermeasures

The judgement of the original soil foundation liquefaction was made by method stated in 2.3 above using the result of the cyclic triaxial test of the soil investigation as the liquefaction resisting shear force of the foundation soil. Consequently, some liquefaction possibility was judged for layers AS1, AS2 and DS1 within the range of GL-8.7m to GL-19.2m at the time of a seismic movement larger than level 1 although there was no possibility of seismic movement of level 1 or less. Considering the situation comprehensively, the following anti-liquefaction countermeasures were conceived:

1) Soil layers AS1, AS2 and DS1, within the range of GL-8.7m to GL-19.2m, shall be improved to be free of liquefaction possibility as far as judged by the "2.3 Liquefaction Judgement Method" even under a seismic movement of level 2.

2) To support the vertical load of the building, the cast-in-place reinforced concrete piles shall reach to layer DS3 which has no liquefaction possibility and has a large bearing capacity.

4. Soil Improvement

The sand compaction method by sand piles is to be adopted for soil improvement considering the past record in the vicinity of this site as well as improvement efficiency and economy. It is intended to compact the liquefaction-fraught soil layers by these sand compaction piles to increase the resistance against liquefaction. (Fig. 7)

The sand compaction piles were designed by estimating the soil liquefaction resistance of the improved soil, using the "Recommendation for Design of Building Foundation" by the Architectural Institute of Japan so as to eliminate any possibility of liquefaction even under a seismic movement of level 2 earthquake. After all, it was decided that the N-values 6 to 17 of the layers fraught with liquefaction be increased to 14 to 27 by using sand compaction piles. Thus, the sand compaction piles were designed so that sand piles of 400mm∅ be compacted to 700mm∅ sand piles and such sand compaction piles be laid out at pitches of 1,600mm. (Figs. 8 & 9)

5. Design of Piles

In the pile cross section design, a seismic response analysis involving the building-pile-soil trinary system was performed. A cast-in-place reinforced concrete pile cross-section was determined to be sufficiently safe against the bending moment and shear force, which may be generated in the pile by a level 2 earthquake. The reinforcing steel bars on top of the pile head were firmly anchored to the foundation beam. The bending moment at the pile head section due to horizontal force was balanced by the bending moment in the foundation beam. (Fig. 10)

In many cases, piles are often designed without foundation soil improvement or considering the horizontal soil foundation reaction coefficient of the liquefied soil layer. However, in case of the building in question, the stress to be generated in the pile, without soil improvement, could be so extremely high as to require a larger pile cross-section area and steel pipes around the piles to attain high strength. Hence, it is an uneconomical design.

6. Postface

This report introduced an example of high-rise building foundation soil liquefaction countermeasures recently adopted in Japan although there are a number of factors concerning foundation soil liquefaction under seismic movement as well as behaviors of buildings still being studied and researched.

Fig. 1 Location of Makuhari District

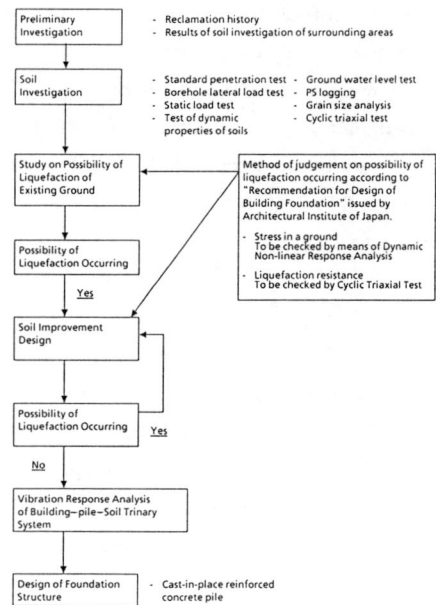

Fig. 4   Flow of Design Work from "Study on Possibility
of Liquefaction" to "Design of Foundation Structure"

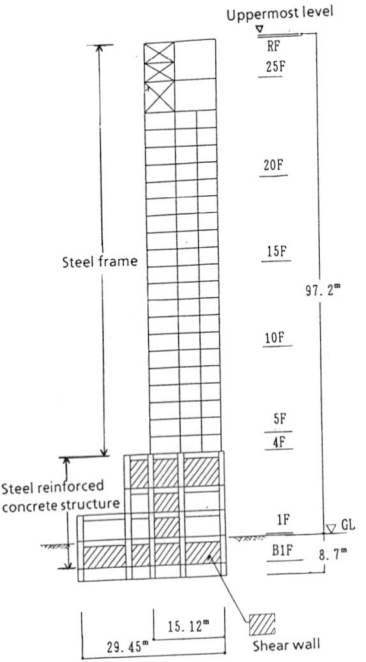

Fig. 2 Framing Elevation (at Section Ⓐ - Ⓐ')

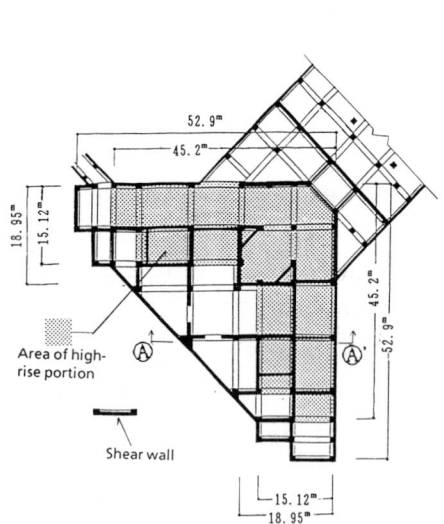

Fig. 3 Basement Framing Plan

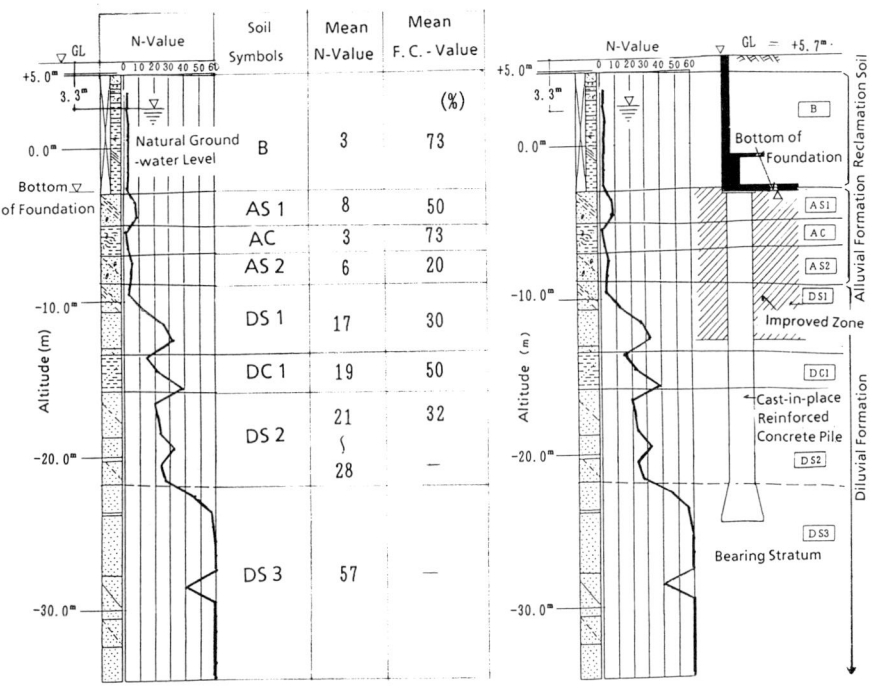

		Soil Symbols	Mean N-Value	Mean F.C.-Value
	N-Value			(%)
	Natural Ground-water Level	B	3	73
		AS 1	8	50
		AC	3	73
		AS 2	6	20
		DS 1	17	30
		DC 1	19	50
		DS 2	21 ∫ 28	32 —
		DS 3	57	—

Fig. 7 Illustration of Measures against Liquefaction

Notes

B   : Reclamation layers (Silty clay layer)
AS1 : Alluvial silty sand layer
AC   : Alluvial clay layer
AS2 : Alluvial sand layer
DS1 : Diluvial sand layer
DC1 : Diluvial clay layer
DS2 : Diluvial sand layer
DS3 : Diluvial sand layer

Mean N-Value
Mean F. C. - Value   } Mean value of 22 soil investigation borings

F. C. - Value   : Percentage of soil particles of which grain size is not more than 0.074mm

Fig. 5 Soil Boring Log

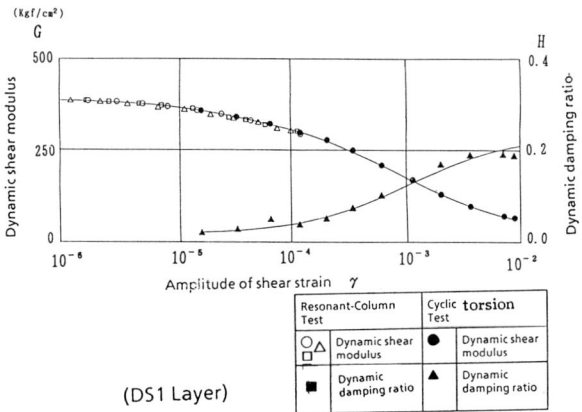

	Resonant-Column Test	Cyclic torsion Test
○△□	Dynamic shear modulus	● Dynamic shear modulus
■	Dynamic damping ratio	▲ Dynamic damping ratio

(DS1 Layer)

Fig. 6 Soil Dynamic Deformation Characteristics Test Results

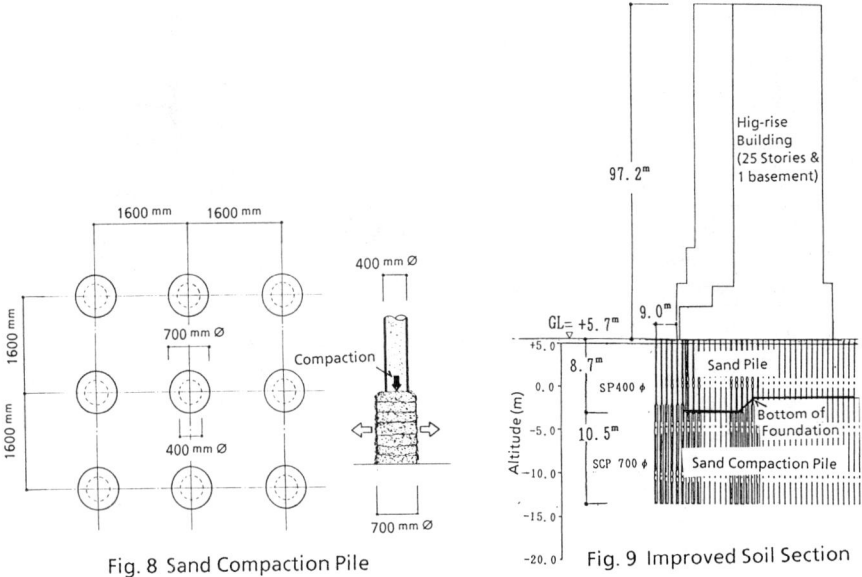

| 1600 mm | 1600 mm |
| 1600 mm | 1600 mm |

400 mm Ø

700 mm Ø

Compaction

400 mm Ø

700 mm Ø

Fig. 8 Sand Compaction Pile

Hig-rise
Building
(25 Stories &
1 basement)

97. 2ᵐ

9. 0ᵐ

GL= +5. 7ᵐ

+5. 0

8. 7ᵐ

0. 0

SP400 φ

Sand Pile

-5. 0

10. 5ᵐ

Bottom of
Foundation

Altitude (m)

SCP 700 φ

Sand Compaction Pile

-10. 0

-15. 0

-20. 0

Fig. 9 Improved Soil Section

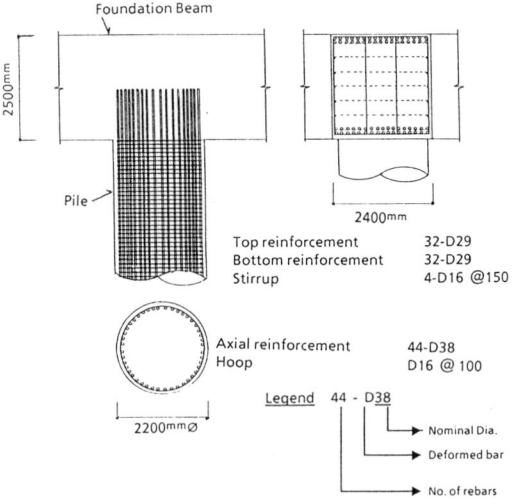

Foundation Beam

2500mm

Pile

2400mm

Top reinforcement   32-D29
Bottom reinforcement   32-D29
Stirrup   4-D16 @150

Axial reinforcement   44-D38
Hoop   D16 @ 100

2200mmØ

Legend   44 - D38

→ Nominal Dia.
→ Deformed bar
→ No. of rebars

Fig. 10 Typical Pile Section and Foundation Beam Section

(Attachment)

Extract of "Recommendation for Design of Building Foundation" by The Architectural Institute of Japan

Foundation soil liquefaction judgement method

Safety factor F_1 against liquefaction at a given depth is calculated by the following formula:

$$F_l = \frac{\frac{\tau_l}{\sigma_z'}}{\frac{\tau_d}{\sigma_z'}} = \frac{\tau_l}{\tau_d} \qquad (4.5.5)$$

No possibility of liquefaction will be adjudged for any soil layer whose F_1-value, obtained by the formula (4.5.5), is larger than 1. On the contrary, where F_1-value is smaller than 1, some liquefaction possibility will be adjudged. As it is smaller, the liquefaction occurance danger degree of that soil layer is higher.

Symbols τ_d : amplitude of equivalant constant repetitive shear stress occurring on horizontal surface (t/m²)

σ_z' : effective soil cover pressure at depth under investigation (vertical effective stress) (t/m²)

τ_l/σ_z' : liquefaction resistance ratio of saturated soil layer corresponding to the compensated N-value (N₁, as found from the shear strain amplitude 5% curve in Fig. 4.5.3.

Compensated N-value (Na) at each depth is calculated by using the following formulas and Fig. 4.5.2:

$$N_a = N_1 + \Delta N_f \qquad (4.5.2)$$
$$N_1 = C_N \cdot N \qquad (4.5.3)$$
$$C_N = \sqrt{10/\sigma_z'} \qquad (4.5.4)$$

Symbols N_a : compensated N-value

N_1 : converted N-value

ΔN_f : compensated N-value increment depending on the fine grain soil content in accordance with Fig. 4.5.2.

C_N : converted N-value coefficient (Unit of σ_z' = t/m²)

N : actually observed N-value

351

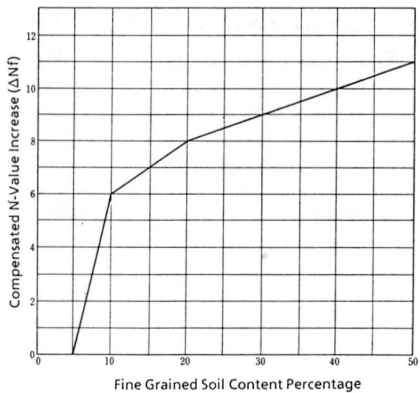

Fig. 4.5.2 Relationship between Fine Grained Soil* Content Percentage
and Compensated N-Value Increase "ΔNf"

* Not more than 0.074mm in grain size

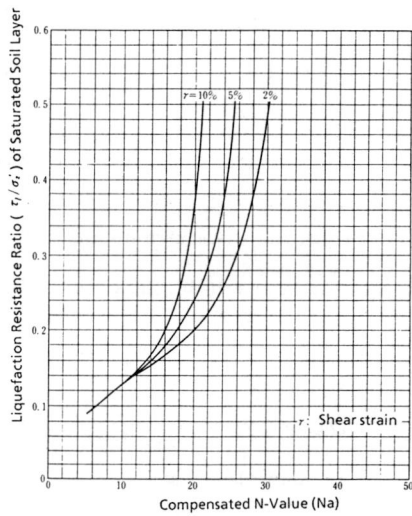

Fig. 4.5.3 Relationship of Compensated N-Value (Na) and Liequefaction
Resistance Ratio " τ_l/σ_z' " of Saturated Soil Layer

Proceedings of the Second Conference on Tall Buildings in Seismic Regions
55th Regional Conference
May 16 and 17, 1991, Los Angeles, California

SEISMIC EVALUATION

AND RETROFIT OF A

NINETEEN STORY STEEL BUILDING

ABSTRACT

The Hoge Building in Seattle, constructed in 1910, has withstood two significant seismic events during its lifetime. This paper will discuss an evaluation and seismic retrofit undertaken at the request of the building's owners to determine its probable response to future seismic events.

The Hoge Building is a slim structural steel, brick and terra-cotta clad tower (aspect ratio 3-1/2:1). Seismic evaluation noted numerous defects in its probable seismic response. However, of greatest concern were extreme drifts of up to 11" and member overstresses approaching ten times their acceptable level predicted under evaluation force levels in the soft first story. The building is not subject to mandatory retrofit and is in fact fully occupied. Consequently, the owner authorized the design and installation of a Drift Control Frame at the first floor. The object of the installation is to reduce drifts at the soft story to a level compatible with predicted drifts above without adding load to surrounding existing members which form the gravity and lateral support of the building. The Drift Control Frame was configured using an eccentric braced frame in three locations, designed to yield repeatedly at lower force levels than the surrounding existing members. The frame is weak compared with existing supports and was designed with drift criteria rather than strength dominating. The technique is extraordinarily cost effective and addresses the most severe detrimental performance characteristic of the Hoge Building while not aggravating other seismic weak points, thus enhancing seismic performance in moderate events.

Todd W. Perbix, P.E.
Vice President

John D. Hooper, P.E.
Technical Director

March 1990

Ratti Swenson Perbix Clark, 1411 Fourth Avenue, Suite 500, Seattle, Washington 98101.

SEISMIC EVALUATION AND RETROFIT OF
A NINETEEN STORY STEEL BUILDING

Todd Perbix, P.E.
John Hooper, P.E.

The evaluation and retrofit of the Hoge Building in Seattle is an opportunity to review the status of seismic evaluation and design technologies applied to existing tall buildings. It also offers a view of the policy questions inherent in seismic retrofit. Unreinforced masonry buildings (URM) are the clearest and most thoroughly investigated example of applied life and economic safety in the retrofit of older buildings. In larger structures constructed of non-ductile concrete or structural steel, these questions take on added dimension since the retrofit of complex, commercially viable, and often fully utilized, buildings is extraordinarily expensive and because these buildings may have the capacity for partially satisfactory performance. The performance level of retrofit as it is understood by regulators, engineers and the public is therefore the principal question clouding work in large frame structures.

The seismic design of new buildings is based on an explicit level of life and economic safety. Neither is absolute, but a high degree of life safety and a moderate level of economic security are the intention of contemporary code provisions applied to the systematic and highly redundant structures designed for new buildings.

Older buildings, however, were often designed without seismic systems. Their diverse structural and nonstructural systems often form configurations, discontinuities and structural weaknesses highly detrimental to satisfactory seismic performance. This diversity along with the inherent difficulties of designing within an old building's performance characteristics makes them highly resistant to prescriptive seismic retrofit codes. The question most often raised is, to what level of life and economic safety should hazardous older buildings be raised, or more to the point, how much money can be spent to improve earthquake performance if these economic demands actually discourage seismic repair and contribute to demolition or a deteriorating building stock.

In many respects these questions are unique for each structure. The Hoge Building in Seattle has experienced two significant earthquakes during its lifetime with relatively minor damage despite substantial configuration weaknesses. The question raised by this seismic evaluation was, how much and where can resources be applied to best improve the performance of this building while it is occupied, economically viable and not subject to

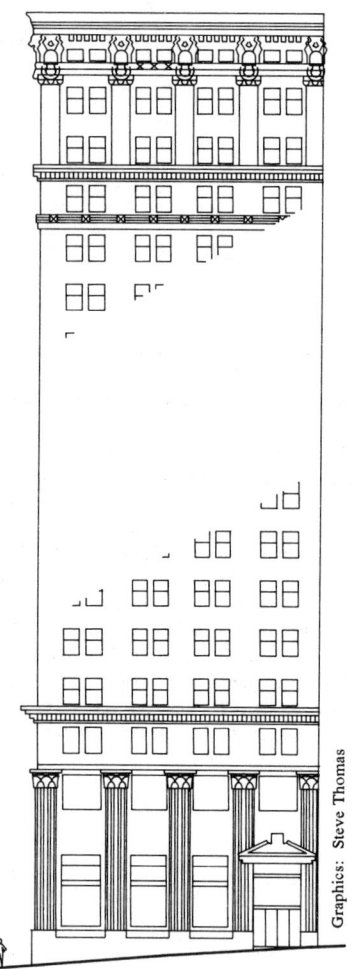

THE HOGE BUILDING
EAST ELEVATION
FIGURE 1

Graphics: Steve Thomas

mandatory seismic retrofit. The solution included a Drift Control Frame (DCF) geared towards improving in limited time its major weakness; a dramatically soft story with disrupted vertical continuity.

BUILDING DESCRIPTION

The Hoge Building is a slim, neoclassical tower with an aspect ratio of 3-1/2:1. When constructed in 1910, it was, at 19 stories, the tallest building in Seattle (Figure 1). Publications of the time noted the state of the art design provided by an exterior structural steel network of girders and columns (Figure 2) which in addition to supporting the structure's vertical loads were declared to improve both the wind and earthquake performance of the building. This system is comprised of latticed girders about 3' in depth and relatively small columns made up of 4 angles. The entire structure is clad, and the floors are cast, in concrete using crushed brick from a previous building on the site as large aggregate. The relatively weak concrete which resulted was believed to improve fireproofing performance.

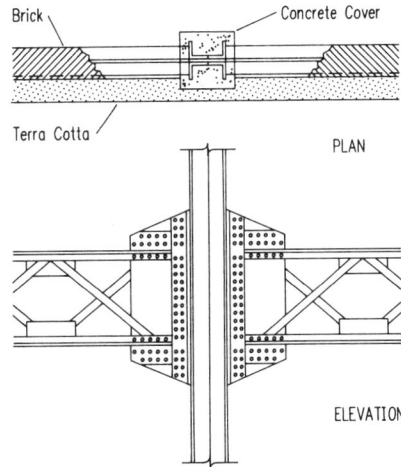

Typical Exterior Beam/Column Joint
FIGURE 2

The internal grid of steel columns which support the full height of the structure is largely interrupted by plate girder trusses at the second floor which span over a large bank
space planned for three quarters of the first floor. The relatively small floor to floor height of the average upper floor (10'4") compared with the substantial floor to floor height of the bank space (26') and the discontinuity in vertical load carrying elements were immediately noted during evaluation.

The building's relatively brittle but stiff exterior structural system, covered in concrete, is clad with a wonderfully ornate terra cotta and brick veneer typical of prominent buildings of the time. The design also provided large commercial windows. The Hoge currently has local landmark status and is subject to controls limiting improvements which affect its appearance.

EVALUATION

Following the 1989 Loma Prieta earthquake, the owners of the Hoge Building requested a seismic evaluation to determine the probable response of their building in a similar event. Their interest arose primarily from an interest in the safety of their tenants as well as in the continued economic viability of the Hoge following a moderate event. The building is a complex combination of structural steel frame and concrete and is certainly subject to many of the elemental hazards noted in buildings with terra cotta and brick cladding. The failure of these elements is however, related to poor frame performance. Therefore, analysis focused on considering these frames.

In order to model a moderate event and to link the analysis with established methodologies which recognize life safety, the ATC 14/22 evaluation methodologies were employed initially. The checklist format of ATC-22 was used to identify elemental and detail weaknesses as well as configuration and continuity problems within the steel frame. The analysis of the exterior frames useful for lateral load resistance was undertaken using ATC 22 force levels to establish a bench mark for recommendations.

Computer analysis of the Hoge Building was performed using a series of 2-D models on ETABS (Table A). The lateral loads which form the basis of this analysis were derived using ATC-22 "Seismic Evaluation of Existing Buildings, A Handbook (Preliminary)" contour maps. Acceleration A_a, and velocity related accelerations A_v, (figures 2-1 and 2-3 in the handbook) were established as .2. These figures have been subsequently revised by the U.S.G.S. for use with NEHRP provisions for the Development of Seismic Regulations for New Buildings but the revised figures are uncorrelated with these ATC provisions and result in significantly higher lateral forces. The acceptance criteria check made subsequent to analysis followed the approach outlined for non-compact sections described in Chapter 5 of the Handbook.

Considerable effort went into creating the member properties for the built-up columns and lattice girders designed for wind resistance in 1910. The concrete around these elements was ignored in estimating their stiffness and strength. This concrete was mixed using crushed brick as large aggregate and was intended

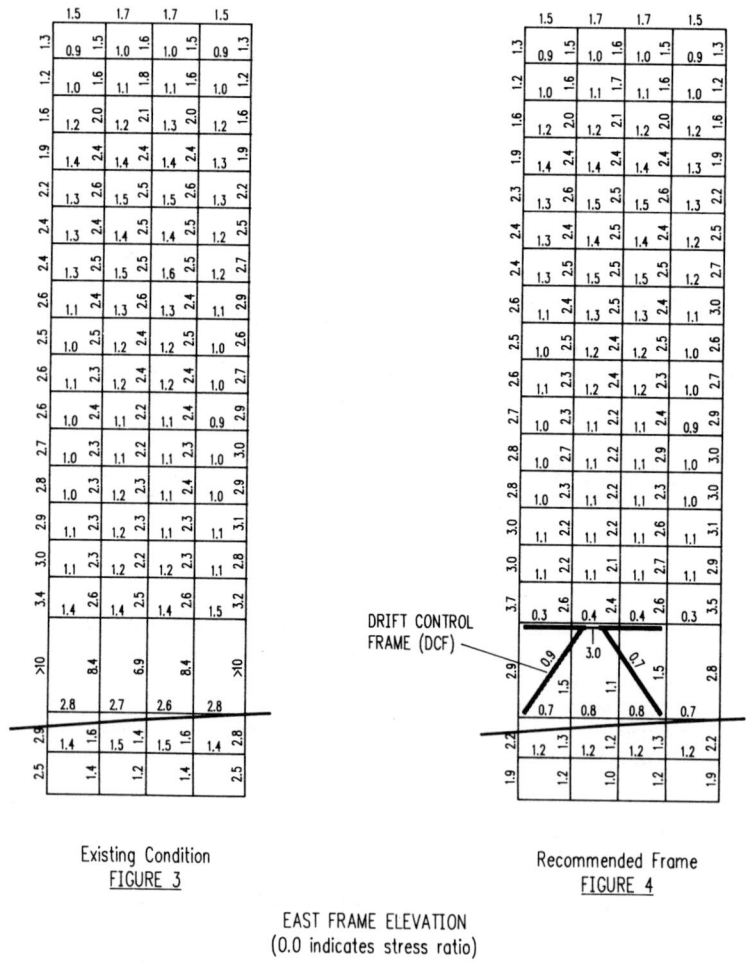

Existing Condition
FIGURE 3

Recommended Frame
FIGURE 4

DRIFT CONTROL
FRAME (DCF)

EAST FRAME ELEVATION
(0.0 indicates stress ratio)

356

TABLE A - SEISMIC DESIGN PARAMETERS

Acceleration Coefficient (A_a)	=	0.20
Velocity Related Coefficient (A_v)	=	0.20
Response Coefficient (R)	=	4.0
Building Period, seconds (T)	=	2.7*
Seismic Design Coefficient (C_s)	=	0.025
Building Dead Load, kips (W)	=	22536
Seismic Base Shear, kips (V)	=	563

*Maximum allowable in computation of base shear. Building period from the existing base steel frame was calculated as 8.2 sec. (P-delta effects considered). The Drift Control Frame (DCF) reduced the calculated period to 5.6 sec.

primarily as fireproofing. Confinement reinforcing for the concrete consists of light wire spaced at approximately 12" o.c. Hence, while the contribution of the concrete to building strength is substantial due to its great volume, composite effects are expected to degrade quickly. Equivalent web thicknesses were created to simulate the flexibility of the lattice angles connecting the flanges to the webs of the girders. This allowed the development of the standard member properties, A, A_v, and I used in the analysis and member check.

Load combinations for both existing structural performance as well as expected performance of the retrofit recommendation was made using STEELER, a post-processing program associated with ETABS. Results and recommendations were obtained in both building directions. In the interest of clarity, our discussion will be limited to transverse (east) frame.

The transverse frame registered 25.9" of drift under evaluation loading over the full height of the building (Figure 5). Of this, 10.1" occurred between the first and second levels (story drift ratio = 0.032). Beam stress ratios of up to 2.8 at the first floor and column stress ratios of over 10.0, between the first and second floors, were recorded. (The acceptance criteria is based on stress ratios less than or equal to 1.0.) Clearly, the soft story area represented the largest single hazard to life in this building.

RECOMMENDATIONS

Based on frame analysis, a number of possible retrofit solutions were proposed. Each had to be considered on its economic, technical and regulatory merits. When considering the existing steel frame available for the resistance of seismic loads, overstresses are evident throughout the building (Figure 3). Consequently, a complete seismic retrofitting option was proposed including installation of three full height structural steel braced frames, designed to reduce stresses in both existing and new members to acceptable levels. The level of overstress noted at the upper levels of the building, however, was not of great a concern as the soft story area. Clearly the stresses predicted in this area are significantly larger and the probable collapsed mechanism significantly more dangerous than those found in any other portion of the building. The structure also has a number of strengths not considered by analysis including the concrete cover, consequently, a minimal Drift Control Frame (DCF) system was proposed (Figure 4). The DCF reduces stresses in the soft story members and drift to amounts compatible with the rest of the building frame (from 10.9' to 1.4" between floor 1 and 2) (Figure 5) while working within the existing economic and occupancy possibilities.

The raw structural cost, that is, unfinished structure considered on an labor and materials basis only, varied from approximately $125,000 for the installation of the Drift Control Frames to $1,500,000 for installation of full height code conforming braces. Obviously the DCF offered the most dramatic increase in life safety per dollar while affecting only one tenant space already planned for remodel. The remaining height of the building would be unaffected by the work. Other advantages of the DCF included the opportunity to limit

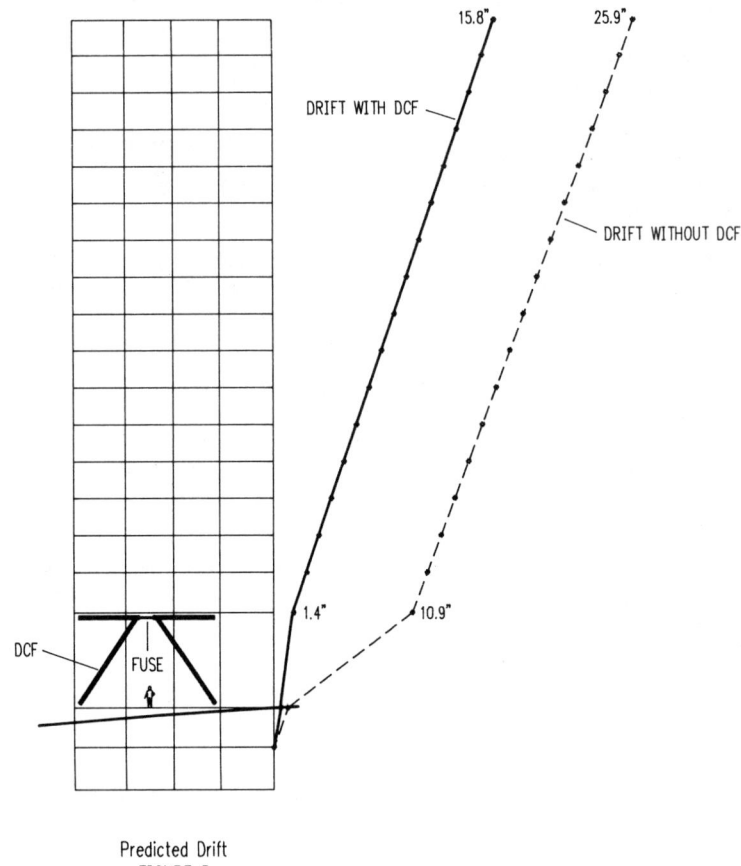

15.8"

25.9"

DRIFT WITH DCF

DRIFT WITHOUT DCF

1.4"   10.9"

DCF   FUSE

Predicted Drift
FIGURE 5

significantly the strength needs of critical frame elements in a manner which alter the basic response of the building to earthquake ground motion very little. The building period was naturally reduced along with building drift (Table A and Figure 5) but not enough to affect its basic response.

Due to the needs of the new tenant, a restrictive time of approximately two months was available to both design and construct the Drift Control Frames in the first floor banking space. Several commercially available damping devices were considered but in the interests of time an eccentric brace fuse (Figure 6) was detailed to act as the weak link forestalling increased stress levels in existing members. From a design standpoint this fuse is dramatically weaker than either the surrounding new brace elements or existing beams and columns. The intent being to allow the DCF to improve the performance of the building frame without increasing stresses in any existing members.

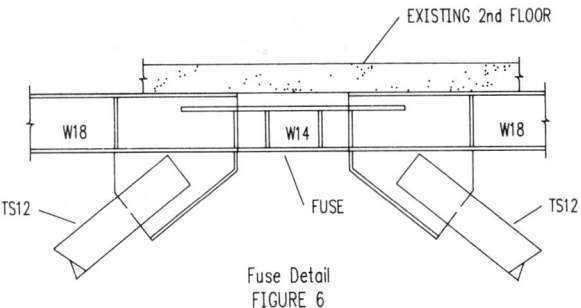

Fuse Detail
FIGURE 6

CONCLUSION

Installation of the Drift Control Frame was completed in February of 1991 at a cost of $115,000. The Hoge Building's owner completed this work in an effort to improve the performance and safety of their building. No mandatory regulatory upgrade requirements were in force, but the owners were faced with an older building securing a marginal return on investment which was nonetheless fully tenanted and in good condition. The Hoge Building has survived two significant seismic events in the past and while certainly no predictors of its behavior in the next, they are testimony to the performance of the building as a combined system of structural and nonstructural elements.

The Drift Control Frames were installed in a way which did not disturb the exterior appearance of the building and indeed will reduce the possibility of racking failure of the cladding. The frame dramatically reduces drifts in the first and second story area allowing the exterior building frame to act consistently throughout its height, thus reducing danger of collapse. For a relatively small price, therefore, a dramatic improvement in safety can be achieved by attacking hazardous building configurations directly. For a dramatically higher cost, marginal improvements could be realized but not without vacating the building.

BIBLIOGRAPHY

ATC-14 Evaluating the Seismic Resistance of Existing Buildings. (1987). Applied Technology Council, Redwood City, CA.

ATC-22 A Handbook for Seismic Evaluation of Existing Buildings. (Preliminary). (1989). Applied Technology Council, Redwood City, CA.

Bergman, D.M. and Hanson, R.D. (1990). *Viscoelastic Versus Steel Plate Mechanical Damping Devices: An Experimental Comparison*, Proceedings of Fourth U.S. National Conference on Earthquake Engineering, Palm Springs, CA, Volume 3, pp. 469-477.

Building Code Requirements for Reinforced Concrete. (ACI 318-89). First printing, November 1989.

Englekirk, R.E. and Sabol, T.A. (1990). *Strengthening Buildings to a Life Safety Criterion*, Proceedings of Fourth U.S. National Life Safety Conference on Earthquake Enigneering, Palm Springs, CA, Volume 3, pp. 315-321.

Habibullah, A. (1989). *ETABS -- Three Dimensional Analysis of Building Systems.* Computers and Structures, Inc., Berkeley, CA.

Habibullah, A. (1989). *STEELER -- AISC Stress Check of Steel Frames, A Post Processor for ETABS.* Computers and Structures, Inc., Berkeley, CA

Manual of Steel Construction. (1980). Eighth Edition

NEHRP Recommended Provisions for the Development of Seismic Regulations for New Buildings, Part 1, Provisions. (1988). Federal Emergency Management Agency, Washington, DC.

NEHRP Recommended Provisions for the Development of Seismic Regulations for New Buildings, Part 2, Commentary. (1988). Federal Emergency Management Agency, Washington, DC.

Perbix, T.W. and Burke, P. (1989). *Toward a Philosophy of Seismic Retrofit; the Seattle Experience.* Earthquake Spectra, Volume 5, No. 3, 1989.

Scholl, R.E. (1990). *Improve the Earthquake Performance of Structures with Added Damping and Stiffness Elements,* Proceedings of Fourth U.S. National Conference on Earthquake Engineering, Palm Springs, CA, Volume 3, pp. 489-498.

Su, Y-F. and Hanson, R.D. (1990). *Comparison of Effective Supplemental Damping Equivalent Viscous and Hysteretic,* Proceedings of Fourth Annual U.S. National Conference on Earthquake Engineering, Palm Springs, CA, Volume 3, pp. 507-516.

Uniform Building Code. (1988 edition). International Conference of Building Officials, Whittier, CA.

Proceedings of the Second Conference on Tall Buildings in Seismic Regions
55th Regional Conference
May 16 and 17, 1991, Los Angeles, California

REDUCE EARTHQUAKE DAMAGE TO TALL BUILDINGS USING ADDED DAMPING AND STIFFNESS ELEMENTS

Roger E. Scholl *

ABSTRACT

Earthquakes impose severe lateral forces on structures well above U.S. code levels that can result in large and damaging interstory drift deformations. Presently structural designers can reduce drift deformations by increasing either lateral strength and/or damping of structures. These two options are reviewed by performing detailed earthquake response analyses in connection with seismically retrofitting a 24-story steel moment-frame building. Added Damping and Stiffness (ADAS) elements, which have been recently introduced as a new innovation for reducing earthquake damage, are used in the analyses to provide supplemental damping to the structure. Results of the study show that using ADAS elements reduced direct damage to the building and contents to one-third of that expected for the asbuilt moment-frame building. Results also show that for a reliable concentric braced frame retrofit, direct damage is reduced somewhat, but the column and foundation capacities must be increased, thus making this option uneconomical.

INTRODUCTION

Tall buildings involve a very broad spectrum of dynamic response behavior, particularly because of the wide range of story heights included in this class of structures. Tall buildings include structures that are 20 stories in height or greater, with fundamental periods ranging from about two to ten seconds.

Early observations of the earthquake performance of tall buildings in the 1950s revealed that little damage was inflicted on this class of buildings. As structural analysis and design technology progressed, and as instrumental recordings of ground motion and building response were made, rational explanations were put forth to explain the lack of damage to tall buildings. Simply put, earthquake demand spectral acceleration diminishes with increasing period and, in many cases, earthquake demand spectral displacements do not exceed the drift capacity of tall buildings. The collapse of the 24-story steel frame Pino Surez building in the 1985 Mexico earthquake shed new light on this matter. That is, under certain circumstances, demand earthquake spectral accelerations and displacements can exceed the strength and/or drift capacities of tall buildings.

Typically, if earthquake demands exceed structure capacities, designers have increased structure capacities. This involves increasing the sizes of the foundations, columns and beams (or shear walls, or bracing) in a new or retrofit building design. A new and innovative option available to designers is to reduce earthquake demand by increasing the damping in structures. Added Damping and Stiffness (ADAS) elements can be implemented in the design of new or existing structures to increase damping and thus reduce earthquake demands (both forces and drifts).

This paper summarizes the results of a detailed earthquake response analysis of a 24-story steel frame building. Earthquake response analyses were performed for three structural configurations of the building: 1) The as-is moment frame; 2) the building retrofitted with ADAS elements; and 3) the building retrofitted with concentric bracing.

---

* CounterQuake Corporation, Redwood City, California

ADDED DAMPING AND STIFFNESS ELEMENTS

Added Damping and Stiffness (ADAS) structural elements are recently developed devices that can be practically installed in structures to: 1) substantially increase overall damping and 2) increase overall stiffness. Figure 1 shows a prototype ADAS element used in conjunction with shake table testing of the device. ADAS elements are engineered assemblies of x-shaped steel plates that are deformed in double-curvature flexure. The plates are made of common ductile mild steel and are sized to respond elastically for wind forces, and to deform inelastically, dissipating energy during earthquakes. ADAS elements have no mechanically moving parts and require no maintenance. These Bechtel Corporation patented devices have been developed for application to structures in joint association with CounterQuake Corporation.

ADAS elements can be easily incorporated in new designs, or in retrofitting existing buildings. Figure 2 shows an example installation in a flexible moment frame. The only requirement is that the devices must experience moderate amounts of relative displacement to deform the plates. ADAS elements can be installed anywhere within the architectural framework of a building, at both interior or exterior walls.

Figure 1: Photograph of ADAS Element Used in Earthquake Simulation Testing

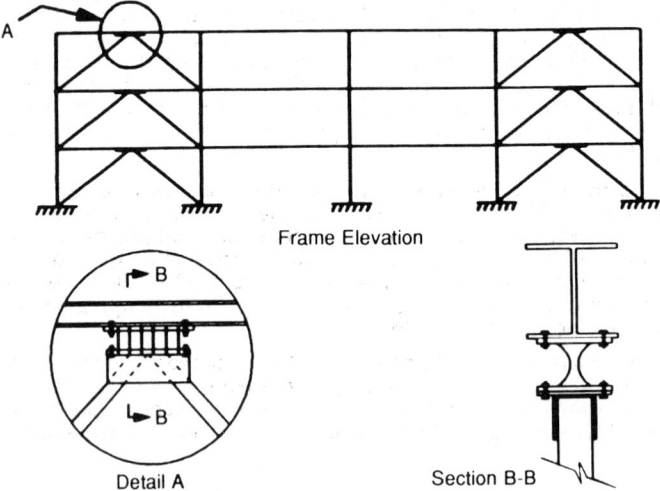

Frame Elevation

Detail A

Section B-B

Figure 2: Example Installation of ADAS Elements in a Three-Story Frame Building

The engineering innovations involved in ADAS elements are the unique "X" shape of the steel plates, and the manner in which the plates are required to deform. The advantages of the x-shaped plate over a rectangular-shaped plate are shown in Figure 3. A rectangular steel plate in double curvature will yield at its ends only. Conversely, the x-shaped plate in double curvature yields uniformly over the entire height of the plate. The large volume of material that is deformed plastically for the x-shaped plate maximizes the energy dissipation of the plate.

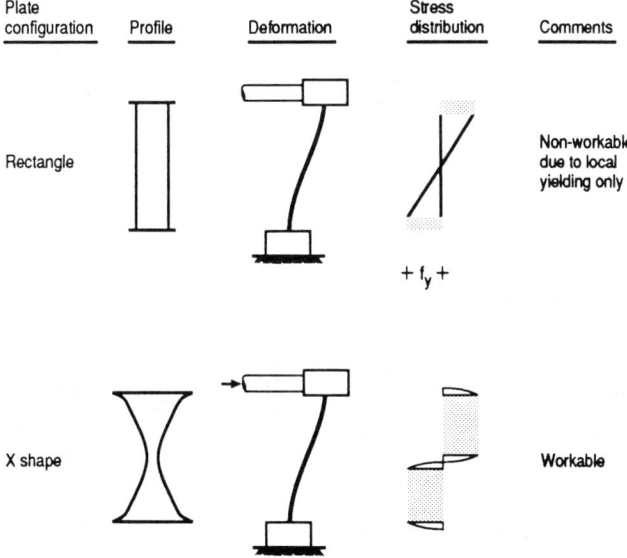

Figure 3: Concepts of Steel Plate Energy Absorbers

EXPERIMENTAL VERIFICATION

The applicability of the basic concept of using x-shaped plates for improving the earthquake response of structures was established in connection with the patented Bechtel Energy Absorber (BEA) used for nuclear power plant piping supports [1]. Extensive cyclic and shake table testing of the BEA has demonstrated its effectiveness and reliability. The device has been approved by the U.S. Nuclear Regulatory Commission for plant-specific applications, and BEA's have been installed in nuclear power plants as pipe supports and to support a reactor building dome service crane.

Although ADAS elements and the BEA share the same basic concept, namely x-shaped steel plates, they differ in their physical configuration and in the manner in which they are utilized. Accordingly, cyclic and shake table testing was performed to verify the reliability of the ADAS device configuration, and to confirm analysis procedures and design principles for building applications.

The energy dissipation capability was verified by tests performed at the University of Michigan [2]. Similar hysteresis loop plots were obtained (Fig. 4) from cyclic testing of ADAS elements conducted at the University of California at Berkeley [3]. Figure 4 shows representative cyclic characteristics of ADAS elements for three displacements: first, 5 cycles at a displacement of 2-1/2 times the yield displacement; second, 50 cycles at 10 times the yield displacement; and finally, 8 cycles at 14 times the yield displacement. Considering the additional displacement cycles experienced by this specimen, there were a total of 124 cycles at 8 times the yield displacement or greater which demonstrates considerable toughness.

Figure 5, a steel frame 17 feet high (one bay wide and three stories high), was tested on the earthquake simulator at the University of California's Earthquake Engineering Research Center (EERC) in 1988. The steel frame was subjected to a series of earthquake and sinusoidal motions. The most severe was the 1985 Llolleo, Chile, earthquake which was scaled to produce a peak table acceleration of 0.56g. By testing the frame with and without ADAS devices, their ability to improve the frame's performance was clearly demonstrated, and analytical procedures were verified. The improved performance resulted from both the added damping produced by yielding the x-shape plates and the added stiffness and strength produced by the ADAS devices and associated bracing. Detailed results of this testing are described by Whittaker, et. al., [3].

BUILDING STUDIED

The building studied is a 24-story moment steel frame constructed in the mid-1960s. Figure 6 shows the elevation framing in the transverse direction, which is the focus of this study. Figure 7 shows the plan layout of the building frames, having a gross area of about 29,100 sq. ft. per floor. Note that for conciseness purposes, Figures 6 and 7 show the proposed locations of ADAS elements in the building.

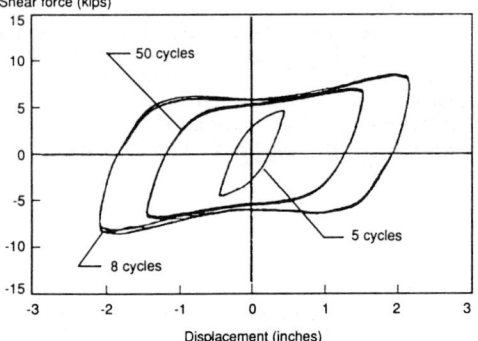

Figure 4: Example Cyclic Test Results from Tests Conducted at the University of California [3] $\Delta y \sim 0.15$ inches

Figure 5: ADAS Retrofitted Three-Story Frame on the Earthquake Simulator

The steel frame weight is about 26 psf. The dead load and seismic live load for a typical floor used for the analyses are 80 psf and 20 psf, respectively. The fundamental period of the building in the transverse direction is about 3.3 seconds.

Consistent with typical construction of the mid-1960s, the building has only a modest amount of fixed partitioning. The building has a single basement level. Because the building is located on a soft clay layer that is about 45 feet thick, it is pile supported.

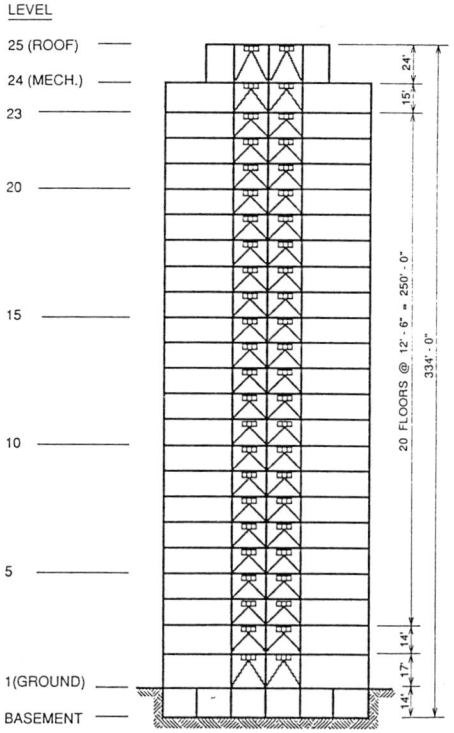

Figure 6: 24-Story Building Transverse Elevation Showing ADAS Locations

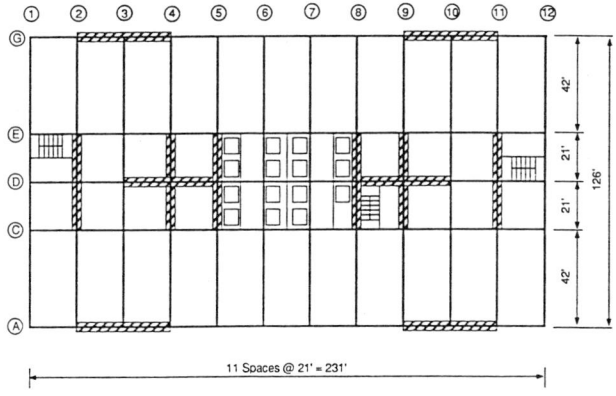

Figure 7: 24-Story Building Plan Showing ADAS Locations

365

EARTHQUAKE RESPONSE ANALYSES AND RESULTS

Earthquake response analyses were performed for three building configurations: 1) the as-is moment frame, 2) the building retrofitted with ADAS elements, and 3) the building retrofitted with bracing that does not yield or buckle. The analyses were performed using the ground motion record shown in Figure 8. This is a synthetic ground motion record that was created to approximate the amplitude and frequency content of motion for the soft soil at the building site from a major earthquake. Figure 9 shows response spectra of the ground motion record for 5 and 20 percent damping.

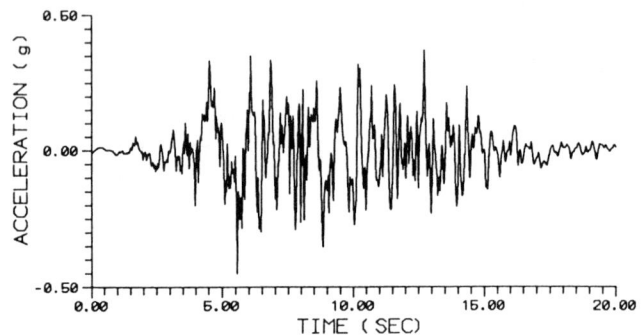

Figure 8: Soft Soil Record

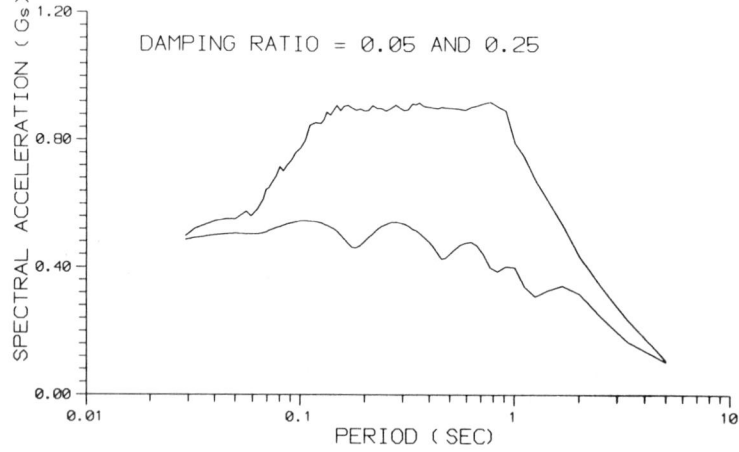

FIgure 9: Response Spectra for Soft Soil Record

366

The earthquake response analyses of the building were performed using the Drain-2D (D2D) computer program [4]. The D2D program is a general purpose nonlinear dynamic analysis program which is very useful for earthquake response analyses. 5% damping, for the linear response, was used in performing the analyses.

Results of the earthquake response analyses for the three building configurations are summarized in Table 1. The various forces and deformations shown in the table clearly demonstrate the improved earthquake performance of the building that can be realized by including ADAS elements in the retrofit. The primary benefit of the ADAS devices is seen by comparing the deformations and forces for the frame with ADAS and frame with concentric bracing conditions. The base shear coefficient for the braced frame is 0.19, while that for the frame with ADAS elements is only 0.13. Correspondingly, the roof displacement for the braced frame is 26.5 inches while that for the frame with ADAS elements is only 19.6 inches. The elastic fundamental periods for both of these building conditions is the same and is about 2.6 seconds. The reason for the differences in forces and deformations is revealed in Figure 8 which shows that for higher damping values, spectral accelerations and displacements are reduced at the 2.6 second period.

Another important earthquake response parameter to consider is the column axial forces. Table 1 shows the maximum column forces in the basement at column line C or E. No tension forces occur for either the bare frame or the frame with ADAS element conditions. Note that P_{TOT}/P_Y reaches 0.91 for the frame with concentric bracing. P_{TOT} is the combined axial load including gravity loads and earthquake forces from transverse, and estimated longitudinal and vertical motions. P_Y is the simple yield axial force of the column. The P_{TOT}/P_Y of 0.91 is very high, and is generally unacceptable. In addition, the maximum compression force of nearly 6,000 kips for the concentric braced frame condition exceeds the capacity of the pile foundation, and foundation strengthening would be required if this retrofit strategy were to be implemented.

Table 1

DYNAMIC ANALYSIS EVALUATION RESULTS *

		FORCES		
	BASE SHEAR		COLUMN FORCES	
BUILDING CONDITION	COEFFICIENT	TENSION (K)	COMPRES-SION	P_{TOT}/P_Y
BARE FRAME	0.10	—	2,073	0.38
FRAME w/ADAS	0.13	—	3,293	0.57
FRAME w/CONCENTRIC BRACING	0.19	2,337	5,752	0.91

		DEFORMATIONS		
	ROOF DISPLACE-MENT		STORY DRIFTS	
		MAXIMUM		AVERAGE
BUILDING CONDITION	(inches)	(in.)	(%)	(%)
BARE FRAME	31.2	2.10	1.39	0.98
FRAME w/ADAS	19.6	1.20	0.79	0.61
FRAME w/CONCENTRIC BRACING	26.5	1.30	0.86	0.78

* Accidental torsion not included.

Direct damage and loss of functionality after a major earthquake are of primary concern for a building like the one under consideration here. Both are typically estimated based on interstory drifts and floor accelerations. Ferrito [5], for example, shows that damage to moment frame buildings begins at interstory drifts of about 0.4 percent. A general methodology for estimating post-earthquake loss of function, based on building damage, is given in ATC-13. Table 1 shows that the average interstory drift for the moment frame and frame with ADAS building conditions are 0.98% and 0.61%, respectively.

Table 2 summarizes the estimated earthquake losses for the moment frame and frame with ADAS building conditions. Note that the building and contents have an estimated value of $105 million, and 2,500 persons work in the building.

The cost of retrofitting the building with ADAS elements is estimated to be about $4.1 million. This cost includes the ADAS elements, associated hardware, installation, and required architectural finish work.

Table 2
EARTHQUAKE LOSS SUMMARY

	Moment Frame	Frame w/ADAS
Direct Damage	$17.1 million	$ 5.6 million
Business Interruption	$32.7 [1]	$ 8.4 [2]
Total Loss	$49.8 million	$14.0 million

Notes

[1] 71 days required to return to 100% building functionality

[2] 14 days required to return to 100% building functionality.

SUMMARY

Extensive cyclic test data of x-shaped steel plate ADAS systems have shown that the concept is constructable and structurally reliable. The results of earthquake simulation tests of ADAS elements in a three-story frame have demonstrated the effectiveness of the devices in improving response performance, and have verified analytical procedures and principles used for designing ADAS elements.

Results of the parametric analyses conducted here show that ADAS elements provide an effective means for reducing deflections and forces in buildings. When ADAS elements are included in a structural system, much of the energy dissipation is provided by elements designed for that specific purpose.

The stiffness and strength of ADAS elements incorporated in a structure can be varied to provide optimum conditions between controlling building deflections and forces developed in the structural system. By controlling the deformation of a structure, earthquake damage is reduced and the building can be returned to service more quickly.

REFERENCES

[1] Khalafallah, M. and H. M. Lee, "Technical Basis for the Use of Energy Absorbers as Supports of Nuclear Power Plant Piping System", Technical Report prepared by Bechtel Western Power Corporation, January, 1985.

[2] Bergman, D. M. and S. C. Goel, "Evaluation of Cyclic Testing of Steel-Plate Devices for Added Damping and Stiffness", Report UMCE 87- 10, The University of Michigan, Ann Arbor, November 1987.

[3] Whittaker, A., V. Bertero, J. Alonso, and C. Thompson, "Earthquake Simulator Testing of Steel Plate Added Damping and Stiffness Elements", Report No. UCB/EERC - 89/02, Earthquake Engineering Research Center, University of California at Berkeley, January 1989.

[4] Kanaan, A. and G. Powell, "General Purpose Computer Program for Inelastic Dynamic Response of Plane Structures", EERC 73-6, Earthquake

[5] Ferrito, J. M., "Economics of Seismic Design for New Buildings", American Society of Civil Engineers, Journal of Structural Engineering, Vol. 110, No. 12, December 1984.

Proceedings of the Second Conference on Tall Buildings in Seismic Regions
55th Regional Conference
May 16 and 17, 1991, Los Angeles, California

INTEGRATED SYSTEMS FOR ACTIVE CONTROL OF INTELLIGENT BUILDINGS

by H. A. Smith[1], Y. Takeuchi[2], and H. C. Shah[3]

Department of Civil Engineering
Stanford University
Stanford, California

ABSTRACT

The concept of actively controlled intelligent buildings has been the focus of recent research in both the U.S. and abroad [Chong, et. al., 1990]. This research typically has considered vibration control of structures undergoing external excitations where the goal is to control system behavior within the limits of structural safety and occupant comfort. However, the ideas on which intelligent buildings are based go beyond the traditional applications in earthquake and wind engineering. This paper discusses the concept of integrated systems for active control of intelligent buildings.

Integrated control systems can be developed which combine the control of structural behavior with control of the structure's environmental state and energy needs. In addition to monitoring the behavior of the primary structure (i.e the building itself) and any secondary systems (i.e. equipment, computer systems, etc.), an intelligent structure should be capable of monitoring its environment (air quality, noise levels, etc.) and energy consumption. The objectives for such an integrated control system are to maintain structural safety and occupant comfort levels while maximizing the energy efficiency of the system.

The integrated control systems of individual structures can be networked to the control systems of adjacent structures such that the behavior, environment, and energy consumption of entire regions can be monitored. This "control network" could recognize given stimuli (i.e. wind patterns, earthquake frequencies, pollution levels, impending power surges, etc.), send the information to proceeding networks, and notify intelligent buildings of impending excitations or changes in environment. Advance notification helps eliminate the effects of time-delays on both the control systems and emergency management procedures and, hence, greatly increases building safety.

1   INTRODUCTION

Systems which use combinations of sensors and actuators to actively monitor and control their behavior or which are composed of intelligent materials possessing adaptive capabilities are typically referred to as "intelligent systems." An idealized intelligent system has the ability to adapt to internal and external stimuli such that an optimal system state is maintained.

---

[1] *Assistant Professor,* [2] *Graduate Student,* [3] *Professor and Chairman*

This concept of actively controlled intelligent buildings has been the focus of recent research in both the U.S. and abroad [Chong, et. al., 1990; McClelland, 1988]. The majority of this research has focused on control of system response due to external excitations with the primary objective of maintaining structural and occupant safety. Other research in the development of intelligent buildings has been concerned with control of a system's internal environment (i.e. temperature, acoustics, air quality, etc.) such that occupant comfort and an optimal working environment is maintained. However, the concept of "intelligent buildings" extends beyond these traditional applications and integrates both internal and external control systems. This paper discusses the concept of integrated systems for active control of intelligent buildings.

2   CONCEPT OF INTELLIGENT SYSTEMS

The ideas of applying control theory as an approach to the safety problems associated with structural engineering has been around since the beginning of this century. Yao [1972] poineered the recent efforts in control of civil engineering systems, and since that time, various conceptual and practical ideas have been suggested in the literature for vibration control due to wind and earthquake loading. Research in this area has focused on two specific applications: (1) vibration control of buildings due to wind or earthquake excitations and   (2) control of a building's internal environment. This section discusses these two applications and the state-of-the art developments in this field of research.

2.1   Vibration Control Systems

Most of the past and present research efforts in intelligent systems has focused on the development of methods for controlling structural vibration response due to external excitations where the goal is to control system behavior within the limits of structural safety. Several control methods have been developed including the active tuned mass dampers, active tendon control systems, active base isolations systems, and other active and hybrid control devices. Despite the recent advances made in control theory, no structures are in existence which rely on control methods as the only system for prohibiting undesired structural behavior. Questions are yet to be answered concerning reliability of control methods, cost of installation and maintenance, effects of time delay on control algorithms, energy requirements for control systems as well numerous other complex issues. Recent literature reviews discussing the state-of-the-art in vibration control systems are presented by Reinhorn and Manolis [1989], Soong [1988], and Yang and Soong [1988].

The first full-scale building which utilized active control methods is the Kyobashi Seiwa Building, an office building in Tokyo which uses an active mass driver to control system vibration [Kobori, 1990]. The active mass driver consists of an auxiliary mass whose movement is controlled by a set of sensors which continuously monitor ground motion and structural response. The movements of the auxiliary mass applies counter-acting control forces to minimize a given performance index which is a function of structural response. Based on analytical and experimental results, the active mass driver works very effectively in reducing undesirable system behavior. However, the building is designed such that structural integrity would not be lost during an earthquake in which the control system fails.

2.2   Internal Environment Control Systems

Building environment control systems have been developed recently due to an increased demand for an optimal working environment. The objectives of this internal environment control system are to maximize employee productivity, minimize cost and inconveniences associated with maintenance, and reduce energy demands for environmental considerations. To meet these objectives, these control systems monitor building air quality, temperature, humidity, air circulation, and lighting in an effort to optimize energy consumption while maintaining the optimal working environment for the building occupants. Shoureshi et. al. [1989, 1990] discusses an building control system which integrates the control of all internal systems with advances in artificial intelligence and fuzzy control theory.

A system which utilizes control methods to maintain control of its internal environment is the main office building for Obayashi Technical Research Institute, a super energy conservation building [Obayashi, 1990]. This control system uses numerous sensors to monitor and control the HVAC system, lighting, and power sources (electricity, solar collectors, and thermal storage tanks) such that energy efficiency is optimized without causing ill effects on the desired internal environment. As a result of this elaborate control system, this building consumes approximately 25% of the energy required by a conventional building of equal size. In 1984, this building was given a design award by ASHRAE for its innovations in energy conservation.

3   INTEGRATED CONTROL OF INTELLIGENT BUILDINGS

Past studies on active control have focused on the primary objective of monitoring and controlling the system's internal or external behavior such that structural safety and occupant comfort are maintained at an optimal level. It is the first and foremost objective of any building control system to maintain a safe environment for building occupants by ensuring structural integrity during crisis situations. However, a fully integrated control system elaborates beyond this primary objective to include additional goals which help define the system as "intelligent." The purpose of integrating the control systems is to reduce the redundancy of the control issues associated with monitoring building behavior, both the external response and the internal environment, so that a widely optimal solution can be found.

3.1   Objectives of Integrated Systems for Active Control

This study has formulated the following objectives for integrated control of intelligent buildings:

(1)   To monitor and control the system's behavior, environment, and energy consumption

(2)   To identify system changes or problems and to adapt to these modifications

(3)   To maintain system security and emergency management procedures

(4)   To maintain structural safety and occupant comfort

The first and last of these objectives were discussed previously in this paper and have been the focus of much recent research. The second objective involves incorporation of techniques for system identification and diagnosis, but includes the ability of the system to "self-correct" for any changes deemed undesirable. Maintaining system security and emergency management procedures is crucial during crisis events. Therefore, the third objective involves a control system designed to detect breeches in system security and to activate emergency management procedures in the event of a disaster.

3.2   The Fully Integrated Control System

The objectives previously discussed can be satisfied using a control system composed of five parts: 1) vibration control system, 2) security control system, 3) building environment and energy control system, 4) information collection database, and 5) network for sharing information. A diagram presenting the relationship between these control systems is shown in Figure 1 and described in the following paragraphs.

3.2.1   Vibration control system

The conventional vibration control system has the ability to monitor both external excitations and structural response to those excitations. However, the algorithm producing the control forces is usually designed based on specific structural parameters and is not modified as the structural parameters change over time. During the lifetime of the structure or during major events, there may be a decrease in structural stiffness or an increase in damping which can affect the performance of the vibration control system. Studies have shown that the effects of parametric uncertainties on the performance of the control system are dependent on the control algorithm employed, and in some cases can be extreme [Yang and Akbarpour, 1990]. Therefore, system identification and diagnosis techniques should be utilized to monitor these parameters and adapt the control system to any changes that take place.

System identification techniques typically have identified the effects of a given stimulus on system parameters after the application of the stimulus was completed. Recently, real time techniques have been developed which allow system identification and diagnosis to take place during the stimulus enabling the system to recognize potential degradation and damage as it occurs [Lin, et. al., 1990]. Such an identification system can be integrated into the control system allowing the control algorithms to adapt the actuator-induced forces to the changes in the system parameters. Filtering techniques also could be employed to aid the convergence of these system identification problems [Yun and Shinozuka, 1980; Shinozuka, et. al. , 1982; and Hoshiya and Saito, 1984]

3.2.2   Security control system

In this study, the meaning of building's security includes every aspect ensuring the safety and security of building occupants (except for safety requirements monitored by the vibration control system). Therefore, in addition to the conventional security systems such as the security monitoring system (against intruders) and the fire monitoring and alarm system, the integrated system includes use of a structural diagnostic system, an emergency management system, and a networked telecommunications systems.

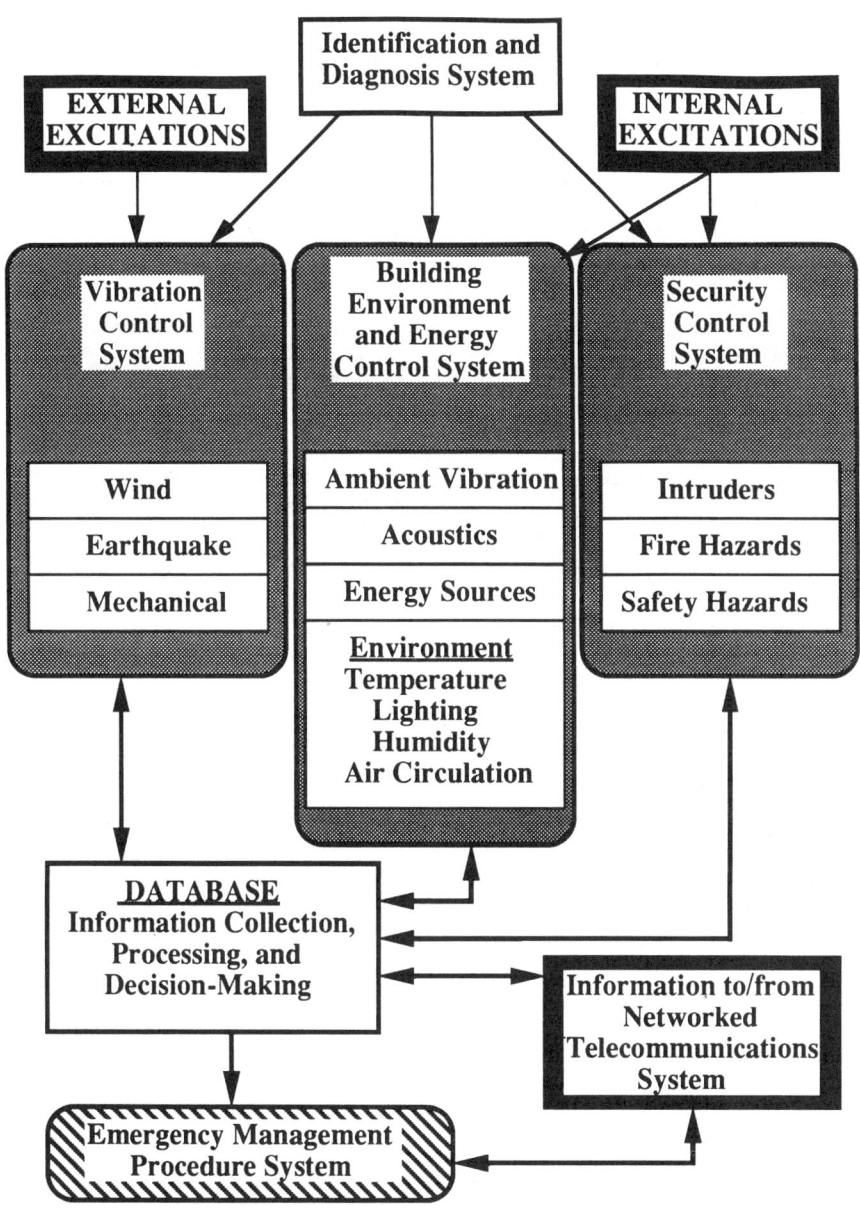

FIGURE 1: Integrated Control System

375

In addition to continuously monitoring and diagnosing modifications to the system stiffness and damping as discussed in **3.2.1**, a structural diagnostic system could monitor the internal system environment and its energy consumption. Such a system could detect leaks in water or power supply lines, potential fire hazards, and faulty mechanical equipment (air conditioners, industrial equipment, etc.)

The emergency management system concept covers all factors which affect security of the building including detection of fire hazards, improperly functioning emergency exits, faulty wiring, or leaks in natural gas mains and any other hazardous substance housed in the building. This system would activate a pre-planned emergency management procedure in the event of a crisis and would help ensure that building occupants could safely exit or remain in the building (as appropriate for the given situation). As an example, this system would detect potential fires, automatically seal-off the endangered portion of the structure, shut-off natural gas sources leading to the building, activate all emergency exits, and inactivate all potentially dangerous electrical systems (industrial equipment, elevators, etc.).

The networked telecommunications system (discussed in detail in **3.2.5**) would be used to ensure that important information is not lost or uncommunicated in the event of a crisis. Several critical buildings in a given region (i.e. hospitals, utility stations, police and fire departments, etc. as well as less important facilities) can be networked via a telecommunications system such that crucial information is shared during an emergency. This system (called a disaster information system) has been proposed as a research initiative by the Japanese Ministry of Construction [1990] with the objective "to develop an information system which can gather, transmit, and process disaster information consistently for the purpose of preventing the spread of damage and to aid in the efficient restoration in the damaged areas."

3.2.3 Building environment and energy control system

This building environment and energy control system has two major objectives: maximization of occupant comfort and minimization of energy consumption. This system monitors and controls HVAC systems, lighting systems, and energy conservation devices. In addition, this system can include: (1) ambient vibration control for occupant comfort which recognizes and controls (via the vibration control system of **3.2.1**) any structural motion that is uncomfortable to occupants but does not threaten structural safety, (2) noise control which monitors noises from internal or external sources and is capable of controlling acoustics if noise levels become uncomfortable to occupants, (3) energy source control which monitors supply and demand of energy and identifies an optimal usage strategy for all energy sources (solar, electrical and natural gas), and (4) building environment control which monitors temperature, humidity, and air circulation to maintain a given level of occupant comfort.

3.2.4 Information collection database

The integrated control systems discussed in the preceding section need a sophisticated database to collect information and maximize performance. This database should collect information from the vibration, security, and building environment control systems, process the information, and make decisions concerning the activation of the various control algorithms. Figure 2 illustrates the decision making process for the integrated control system.

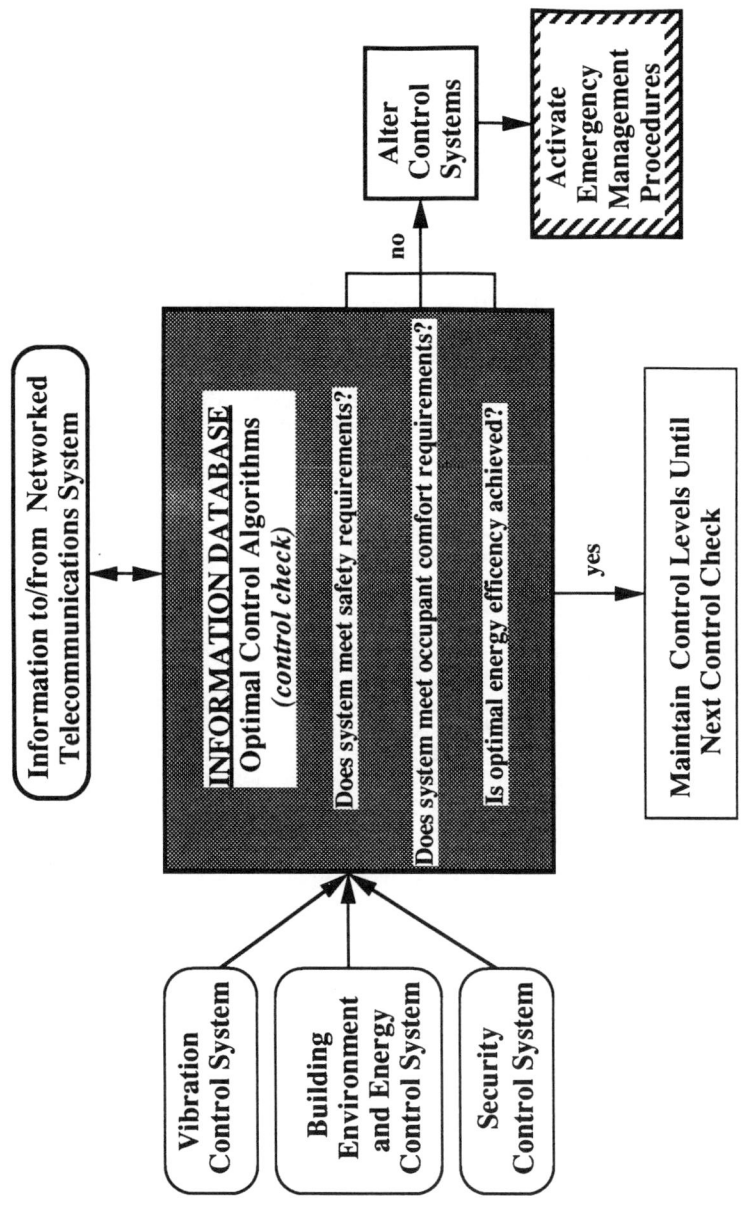

FIGURE 2: Decision-Making Process in Database

377

The vast amount of information involved with an "intelligent" building consists of both quantitative and qualitative data; therefore, the database should be capable of collecting, processing, and integrating both types of data. Time histories of responses can be collected to aid in diagnosing structural degradation over the lifetime of the structure. Fuzzy set theory can be utilized in the decision-making process [Dong, et. al., 1990; Furata, et. al., 1985], and filtering techniques can be utilized to optimize the processing of information.

3.2.5 Networked telecommunications system

As mentioned earlier, the critical buildings in a given region can be networked together for the purpose of sharing information. The buildings in the network system could recognize various stimuli in the region and notify each individual building of excitations or changes of environment. In addition to transmitting valuable information after an event (as discussed in **3.2.2**) the telecommunications system could inform systems of impending events such as potential power surges, loss of water pressure, potentially troublesome wind patterns, and even earthquakes. Since the transmission of information takes place at the speed of light, messages concerning incoming earthquakes would arrive a fraction of a second prior to the majority of the earthquake waves. Though, this prior knowledge may seem trivial, it could potentially eliminate the ill effects of time delays on vibration control systems and emergency management systems. These time-delays have been shown to be of significant concern in designing control algorithms [Yang, et. al., 1990].

In developing the optimal control strategy for the building environment, the network system also would be capable of recognizing the regional environment and energy conditions and predicting any future changes or fluctuations. Information on temperature, humidity, air pollution, and wind patterns would be gathered and processed with the predictions of any potential fluctuations (based on data from the weather bureau). Using this information, each intelligent building develops its own optimal operation mode. This network also processes relationships for the supply and demand of energy and water for each building in the region so that optimal use is made of resources.

4   SUMMARY AND CONCLUSIONS

This paper presents and discusses the concepts for a fully integrated, intelligent building capable of actively controlling its response to both internal and external stimuli. The purpose of integrating the control systems is to reduce the redundancy of the control issues associated with monitoring and controlling the external and internal building behavior so that a widely optimal solution can be found. Integrated control systems can be developed which combine the control of structural behavior with the control of the structure's internal environment and energy needs. In addition to monitoring the behavior of the primary structure (i.e. the building itself) and any secondary systems (i.e. mechanical equipment, computer systems, etc.), an intelligent structure should be capable of monitoring its environment (air quality, temperature, acoustics, etc.) and energy consumption. The proposed integrated control system fulfills the primary objective of maintaining structural safety and occupant comfort, but has the additional capability of identifying system changes and adapting its control algorithms to account for these modifications. Large databases collect and process this vast amount of both qualitative and quantitative data.

The integrated control systems of individual structures can be networked to the control systems of adjacent structures such that the behavior, environment, and energy needs of an entire region could be monitored. This "control network" recognizes given stimuli (i.e. wind patterns, energy and water needs, earthquake frequencies, etc.), sends the information to proceeding networks, and notifies other buildings connected to the network of impending excitations, changes in environment, or reallocation of resources. Advance notification of these events helps eliminate the effects of time-delays on both the control systems and the emergency management procedures. Thus, building safety is increased and optimal modes of operation are maintained. The additional functions of the active control system discussed here extend beyond the traditional functions of the control systems studied previously, and help define the structure as "intelligent."

REFERENCES

Chong, K. P., Liu, S. C., and Li, J. C. (editors), Intelligent Structures (proceedings from the International Workshop on Intelligent Structures, July 23-26, 1990, Taipei,Taiwan), Elsevier Applied Science, 1990.

Furata, H., Shiraishi, N., Fu, K.-S., and Yao, J. P. T., "Evaluation of Structural Serviceability Based on Fuzzy Reasoning," Fuzzy Mathematics in Earthquake Researches, pp. 218-234, (proceedings of International Symposium on Fuzzy Mathematics in Earthquake Researches, Beijing, China), edited by Deyi, F. and Xihui, L., 1985.

Dong, W., Chiang, W-L., Shah, H.C. and Wong, F.S., "Assessment of Safety of Existing Buildings Using Fuzzy Set Theory," Proceedings of International Conference on Structural Safety and Reliability, pp. 903-910, 1989.

Hoshiya, M. and Saito, E., "Structural Identification by Extended Kalman Filter, " Journal of Engineering Mechanics, Vol. 112, pp. 1757 - 1770, 1984.

Japanese Ministry of Construction, Designing Toward the Future: Construction Technology and Development in Japan, 1990.

Kobori, T., "Technology Development and Forecast of Dynamical Intelligent Building (D.I.B.)," Intelligent Structures, pp. 42-59, (proceedings from the International Workshop on Intelligent Structures, July 23-26, 1990, Taipei,Taiwan), edited by Chong, Liu and Li, Elsevier Applied Science, 1990.

Lin, C. C., Soong, T. T., and Natke, H. G., "Real-Time System Identification of Degrading Structures," Journal of Engineering Mechanics, Vol. 116, pp. 2258-2274, 1990.

McClelland, S. (editor), Intelligent Buildings: An IFS Executive Briefing, IFS Publications, UK, 1988.

Obayashi Corporation Technical Research Institute Environmental Laboratory Letter, Operation Record of Super Energy Conservation Building at Obayashi Corporation Technical Research Institute, Tokyo, Japan, 1990.

Reinhorn, A. M. and Manolis, G. D., "Recent Advances in Structural Control," Shock and Vibration Digest, Vol. 21, pp. 3-8, 1989.

Shoureshi, R. and Rahmani, K. "Intelligent Control of Building Systems," Intelligent Control Systems, DSC-Vol.16, ASME, 1989

Shoureshi, R., Rahmani, K., and VanDoren, V., "Intelligent Building Control Systems," Intelligent Structures (proceedings from the International Workshop on Intelligent Structures, July 23-26, 1990, Taipei,Taiwan), edited by Chong, Liu and Li, Elsevier Applied Science, 1990.

Shinozuka, M., Yun, C.-B., and Imai, M., "Identification of Linear Structural Dynamics Systems," Journal of Engineering Mechanics, Vol. 108, pp. 1371-1390, 1982.

Soong, T. T., "Active Structural Control in Civil Engineering," Engineering Structures, Vol. 10, pp. 74-84, 1988.

Yang, J. N. and Soong, T. T., "Recent Advances in Active Control of Civil Engineering Structures, " Journal of Probabilistic Engineering Mechanics, Vol. 3, pp. 179-188, 1988.

Yang., J.N., Akbarpour, A. and Askar, G., "Effect of Time Delay on Control of Seismic-Excited Buildings," Journal of Structural Engineering, Vol.116, pp. 2801-2814, 1990.

Yang., J.N. and Akbarpour, A., "Effect of System Uncertainty on Control of Seismic-Excited Buildings," Journal of Engineering Mechanics, Vol.116, pp. 462 - 478, 1990.

Yao, J. T. P., "Concept of Structural Control," Journal of Structural Division, ASCE, Vol. 98, pp. 1567-1574, 1972.

Yun, C.-B., and Shinozuka, M., "Identification of Nonlinear Structural Dynamics Systems, Journal of Structural Engineering, Vol. 8, pp. 187 - 203, 1980.

Proceedings of the Second Conference on Tall Buildings in Seismic Regions
55th Regional Conference
May 16 and 17, 1991, Los Angeles, California

DESIGN AND CONSTRUCTION OF THE NATIONAL BANK CENTRE, AUCKLAND

Kevin C.F. Spring   **G.K. Sidwell**

DESIGN AND CONSTRUCTION OF THE NATIONAL BANK CENTRE, AUCKLAND, NEW ZEALAND

1. Introduction

The new National Bank centre, recently completed, is a focal point for commerce and retailing in the central business district of Auckland. The site is situated on the corner of Queen and Victoria Streets (Fig.1) and as such comprises a full city block.

The new development comprises twin towers located diagonally opposite one another (see Fig.2). The tower in the south west corner of the site described as Tower 2 comprises 27 stories and the other in the north east corner - Tower 1 - 20 stories. The on-site branch of the National Bank will be located in Tower 1 at Ground Floor level and will also occupy a further floor of this building. Shopping malls cover the remaining ground floor areas and link to each of the towers. Below ground floor are two levels of basement used for carparking and also for the location of the Bank Vaults. Located in each tower are two levels of plant, one immediately above the podium level and the other at the top of each tower. Both towers in plan, form the quadrant of a circle and were architecturally designed as such to have the maximum impact on what is the business centre of the city.

2.   Structural Form

Each of the towers have the same plan area of 696 square metres per floor. To maximise rentable area and facilitate building function the services areas have been located centrally alongside the curved sections of the towers.   Before arriving at the final structural solution numerous alternatives were investigated and cost estimates for each were evaluated.   The final form of a perimeter concrete frame and structural steel flooring together with structural steel internal columns provided the best answer both for building function and structural analysis. The only penetrations required in the secondary beams are through the diagonal beams (see Fig 3).

By adopting a steel deck flooring, composite steel beams and internal structural steel columns, and lightweight partitions about the lift and stair shafts the building mass was reduced significantly which in return greatly reduced the effect of earthquakes on the building. While it is common practice both in New Zealand and overseas to combine structural steel as the gravity support medium with concrete shear walls and to use this form for the perimeter, the reverse was adopted for the project.   This resulted principally from limitations on available core structure for either a braced frame or concrete shear walls. Additionally architectural limitations made it impossible to place walls or bracing around the perimeter.   The final decision was then whether to adopt a perimeter steel or concrete frame. The latter was chosen because a steel frame would have required considerable on-site welding which in recent years had made steel a cost prohibitive building medium.

As noted earlier the partitions around the lift and stair shafts are lightweight in form and their rigidity is such that they are not capable of altering the structural behaviour of the building as a whole when subject to seismic excitation. The stairs for the towers are precast concrete and have been articulated at the midheight landings, so that no strutting effect can take place under lateral load.

With the decision made that the perimeter frame would be reinforced concrete, and following discussion with the contractor it was decided that as much as possible of this concrete frame would be precast so that the advantage gained with steel for speedy erection would not be lost. The result was that the majority of the beam elements were precast, with the columns, together with the beam column joints being the only insitu work (see Fig 4).

Again with speed of construction in mind, secondary beams were placed at approximately 2.5 metre centre to centre so that propping for the steel decking was eliminated. However to achieve economy of material the supporting steel beams were designed to act compositely with the steel/concrete deck and required propping during construction until the concrete slab had reached the specified strength.

The fire rating for the structural steel has been obtained by spraying the columns and beams with mineral fibre and by placing additional steel in the floor slabs. For some of the internal columns concrete encasing was required to give additional capacity over that provided by readily available UC column sections.

The design yield stresses for beams and columns in structural steel were 250 MPa and 350 MPa respectively. The reinforcement in the concrete beams and columns had specified minimum yield stresses of 275 MPa and 380 MPa respectively.

3.   Foundations and Temporary Works

A detailed geotechnical study and investigation was carried out at the site--

.   To establish soil conditions and feasible foundation types
.   To verify ground water levels and
.   To determine realistic lateral loads for the basement wall design

The information was obtained was through drilling a total of 5 boreholes in a vertical alignment extending to approximately 20 metres. Samples were then recovered and forwarded for laboratory testing. The results of the analysis indicated that the footings for the building should be drilled piles belled as required to provide the resulting bearing or in the extreme seismic overload to resist nett tensions. The bearing material for the piles is the "Waitemata Formation" a sedimentary deposit which was encountered at depths of 9 to 18 metres below ground level. The ultimate bearing capacity of this material was established as 6000 kPa.

During the early design stage of the superstructure it was decided that the horizontal shears resulting from seismic attack should be transferred to the basement walls at pavement level and then in turn with soldier piles. Because of the loose fitting nature of the precast concrete soldiers and planking between, to obtain the required stiffness it was necessary to cast an insitu concrete wall between the inner flanges of the soldiers. The connection between the soldiers and insitu work is achieved by bending starters from the soldier into the wall (see Fig. 5). To complete the transfer of horizontal load an insitu capping beam is cast on top of the perimeter basement wall.

As stated all horizontal load is transferred to the perimeter walls and as a result the drilled caissons supporting the towers were only designed for axial loads. The tie beams linking the tops of the caissons have only been designed for the NZ code minimum of 10% of the column axial load.

ANALYSIS AND DESIGN

Lateral Load Resisting System

Although a twin towered development, the lateral load analysis was made on a model of each tower in isolation. Above ground level the podium, is subdivided by seismic gaps and hence was considered part of each tower as appropriate. Below ground level, the extensive boundary walls were assigned to each tower by considering the nature of the podium construction.

Each of the towers was analysed using the 1986 version of ETABS this program providing an efficient input-output format for the analysis of building frames. Preliminary analyses indicated that the principal axis of vibration are skewed at 45o to the right angled corner of the structure (see Fig 6).

With this in mind modal analyses were undertaken with the ground excitation considered in each of these principal directions. From these modal analyses a set of static loads that produced the same storey shears as the modal analysis were derived and considered to act on various accidental eccentricities from the centre of the mass (see Fig 6).

Considering these principal directions of loading, while simplifying many aspects, does result in most beams and columns participating in the framing action for each principal response direction. Accordingly for any member the total design action (Ad) was determined by the nonlinear combination:

Ad is the greater of A_{12} or A_{21} where

$$A_{12}=\sqrt{A_1^2+0.6A_2^2}$$

$$A_{21}=\sqrt{A_2^2+0.6A_1^2}$$

A_{12}, A_{21} being the combined action (moment, shear, axial loads) and A_1, A_2 being the actions in the principal directions 1 and 2 respectively. This combination was currently proposed in the draft revision to the New Zealand Loading code for buildings of 1986.

Inherent in design loadings assumed for most New Zealand Buildings is the consideration that seismically induced loadings are in the extreme too large to be resisted by strength alone. This leads to a philosophy of requiring ductility within the structure in order that it may "ride out" the severe event.

In order that ductility may be relied upon with confidence the New Zealand design philosophy requires a capacity design to be undertaken. For the National Bank centre this translates into designing the flexural reinforcement in the beams for the most critical moment combination after due allowance for moment redistribution.

The moment redistribution used for this project was to adjust the bending moment diagrams such that the flexural reinforcement at each of the beam hinge locations were identical for both top and bottom reinforcement. This is shown diagrammatically in Fig 7. On the curved face this process was carried out as a piecewise process with the total face being considered as groups of 2 bays, 1 bay and a further 2 bays.

After the design of the beam's longitudinal reinforcement, their shear reinforcement and all column reinforcement for flexure, shear and confining actions are governed by considerations of the actual overstrength moments of resistance within the beams and potential collapse mechanisms of the structure.

For other than low rise buildings it is considered unacceptable to have collapse mechanisms that occur by column hinging within a single storey, commonly termed a column sway mechanism. The moment at the beam design location requires to be computed as its overstrength value.

This overstrength arises from--

i)   the ratio of the yield stress which only 10% of all reinforcement will exceed and the design value which will be exceeded by 90% of all reinforcement.

ii)   the removal of the material under capacity factor when calculating the strength of the section

iii)   any reinforcement excess to that originally computed as necessary.

The overall overstrength factors $\emptyset_o$ for the joint is then defined as the ratio of the overstrength beam moment input to the total analysis seismic beam moment input. Non linear analyses have indicated that the column bending moment diagrams can vary quite markedly from those determined from an analysis that is dominated by the fundamental mode. Accordingly the normal New Zealand practice is to allow for a dynamic magnification factor which increases with the fundamental period of the building. For the National Bank centre its value was considered to be 1.8. The column design moment is determined by factoring the analysis moment by the term $\emptyset_o$.

Column axial load effects are derived by considering the combination of the gravity load effects and the summation of the sway induced component of beam shears for all beams framing into the column above the level under consideration. Since the moment used in the column design includes higher mode effects a reduction factor is applied to the summation as the same higher mode effects will preclude the total formation of the fundamental mode mechanism.

The New Zealand concrete code recognises that beam column joint action is enhanced if flexural yielding of beam reinforcement is prevented from occurring at the column face. For all columns except those at the corners of the tower this was implemented and flexural reinforcement from each beam anchored in the span of its neighbour on the opposite face of the column. The associated detail is shown in Fig. 4 and provides a ready means of joining the precast components and displacing the beam hinge the requisite distance.

In the beams that join the column pairs at each of the corners rather than yielding within a flexural hinge in the beam a jointing element of a large steel plate yielding in shear was adopted. This was done as the flexural reinforcement would be dictated by the adjoining beams and the very short span would then produce very high shears. In turn these could result in higher capacity axial loads on the columns into which they frame.

Gravity Support System

As stated in the introduction, a great deal of emphasis has been placed on speed of construction and the need to reduce mass and hence minimise the seismic impact on the buildings.

With these two thoughts in mind the decision was made to make the gravity only members structural steel. The beams forming the floor slab and some of the columns at the lower levels were designed to act compositely. The secondary beams for the floors have been placed typically at 2,500 c/c and within this spacing the 0.75 mm Hi-bond metal deck has sufficient capacity to support the construction loads without temporary shoring. The reinforcement for the concrete slabs comprises 665 mesh to control shrinkage and the additional reinforcement shown (Fig. 8),*has been added to give an hour and a half fire rating. By adding the fire reinforcement no further material, needed to be placed on the underside of the metal deck. The fire design for this reinforcement has been carried out in accordance with a European Standard.

At the time of the design, New Zealand did not have a composite design code and the Australian Standard 2327 was used for the beams and approved literature for column design.

Where it was necessary to increase the capacity of the columns by concrete encasing the initial design was based on the cased strut formulae of AS1250 (2). However as an ultimate strength check on the design method proposed by Basu and Somerville (3) was used.

The squash load of the member is typically

$$Nu = As\ fys + Av\ fyr + Ac\ fc$$

The squash load then is modified for slenderness by a factor K1. Curves relating K1 to Lr. are included in ref (3) with the basic equation for r being

$$r_2 = (fysIs + 0.24\ fc\ Ic)/Nu$$

The yield stress for the structural steel columns was nominally 350 MPa. The concrete cylinder stress at 28 days was 30 MPa.

---

*Fig. 8 was missing from Authors manuscript submittal. Contact the manuscript Author for copy of Fig. 8.

To give sufficient tolerance, when fixing to differing material such as the perimeter precast concrete beams and the internal structural steel, the cleats for the beam connection were slotted horizontally (see Fig.9). The slotting of these holes also minimised the end rotation effects for the comparatively long spans.

With the initial design philosophy being to keep mass to a minimum, it was necessary to ensure that floor vibrations were kept at acceptable levels. The criteria use to check the human response scale was the Murray (6) criterion where

$$D \geq 35 \text{ Ao f} + 2.5$$

The conservative estimated damping D for the floor was calculated as

Bare Floor	1.5%
Ceiling	1.0%
Ductwork	2.0%
	4.5%

With beams calculated as having a frequency of 5.6 hz and a spacing of 2.5 metres the right hand side of the equation came to 4.05 hz. In the case where the secondary beams are supported on the splayed girder the results gave a value of 4.2 hz.

Proposed Method of Construction

The proposed method of construction and indeed the final design system resulted from discussions involving the contractor Mainzeal NZ Ltd and the architectural and engineering consultants.

The contractor having had a great deal of experience in precast concrete and having proprietary forms to support beams two levels above finished floor level was keen that as much as possible of the beams forming the perimeter frame be precast.

The initial approach was to look at precasting both beams and columns, however weights of members and crane limitations mitigated against using such an approach.

From the capacity of the crane, member sizes were fixed so that about 70% of the beams would be precast and the remaining section of the beam and the columns would be cast insitu with the floor.

This construction method means that the beams are held in position and the column and beam column joint transverse steel inserted through an open face on the column form. The column longitudinal steel is then placed through the beam reinforcement and through into the column and joint ties. These are then spaced out and tied in their correct positions. In order that the longitudinal reinforcement from each of the beams can intermesh the beam stirrups are open and closed with separate links after erection of the beams.

To maximise what is believed to be a very quick method of construction it is essential that all subcontractors should be fully coordinated in the construction programme, otherwise the initial advantages will be lost.

General Features

The physical arrangement of the building are two identical towers protruding from a podium with set backs for pedestrians footways, arcades and gardens. Below street level are two levels for carparking in which the bank vaults are housed.

The cladding for each of the towers is fluorocarbon coated aluminium glazing system with double tinted glass. The glass will appear slightly variegated with shades of green, grey and blue showing.

The floor finish for the bank and office entries is to be granite with the tower generally being heavy grade commercial carpet. On the plaza floors at ground level porphyry blocks will be laid.

Transportation for the buildings is provided by 5 high speed lifts in each tower. Serving the plaza are a series of escalators. All the building areas are fully air conditioned with plant rooms in two locations in the tower height. Together with material coatings to the building elements fire resistance is provided with a full sprinkler system throughout the complex.

Principal Consultants to the Project

Project Manager	-	Realty Development Corporation Ltd
Architect	-	Glossop Chan Partnership in association with Kann Finch and Partners (Pty) Limited, Australia
Engineers	-	Civil and Structural Brickell Moss Limited Mechanical and Electrical, B & R Consultants Ltd
Quantity Surveyors	-	Knapman Clark and Company

Acknowledgements

The authors would like to express their gratitude to the Realty Development Corporation Ltd for their permission given to publish the paper. Additionally they would like to thank Mr. John Rankine and their colleague, Mr. Norman Lea for their valuable comments and encouragement.

References

(1)   European Convention for Constructional Steelwork - Publication No. 32

1983, "Calculation of the fire resistance of composite concrete slabs with profiled steel sheet exposed to the standard fire".

(2)   Standards Association of Australia - Australian Standard 1250-1981 "SAA steel structures code".

(3)   Johnson, R.P. "Composite structures of steel and concrete" 1975

(4)   Standards Associationn of Australia - Australian Standard 2327 Part 1, 1980 "SAA   Composite construction code Part 1   - Simply supported beams".

(5)   Redwood: "Tables for plastic design of beams with rectangular holes". American Institute of Steel Construction, Engineering Journal Volume 9, No. 1, January 1972.

(6)   T.M. Murray "Building floor vibrations" American Institute of Steel Construction, 3rd Conference on steel developments, 1985.

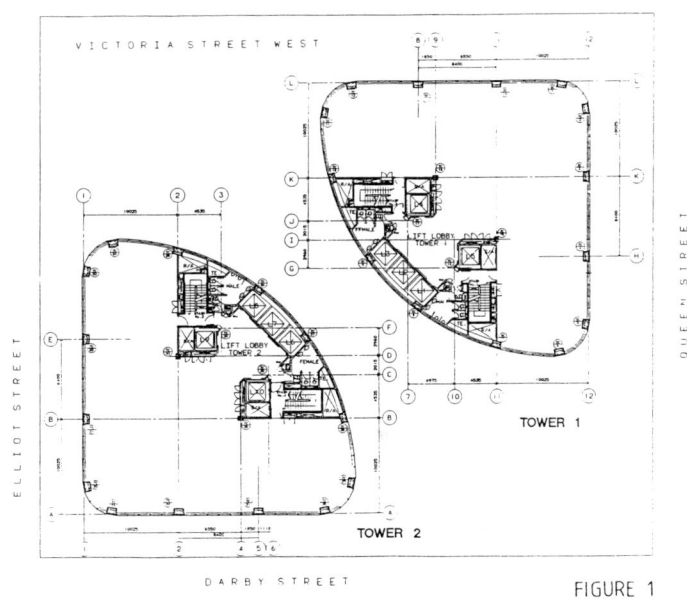

FIGURE 1

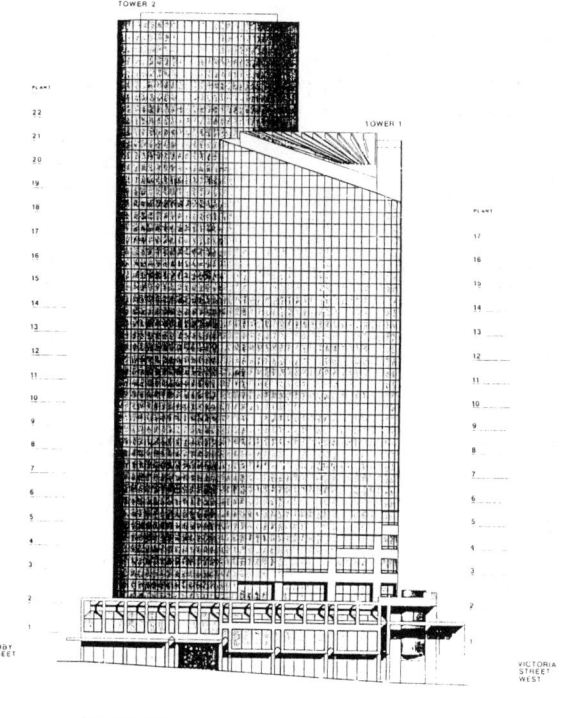

TOWER 2

TOWER 1

EAST ELEVATION

FIGURE 2

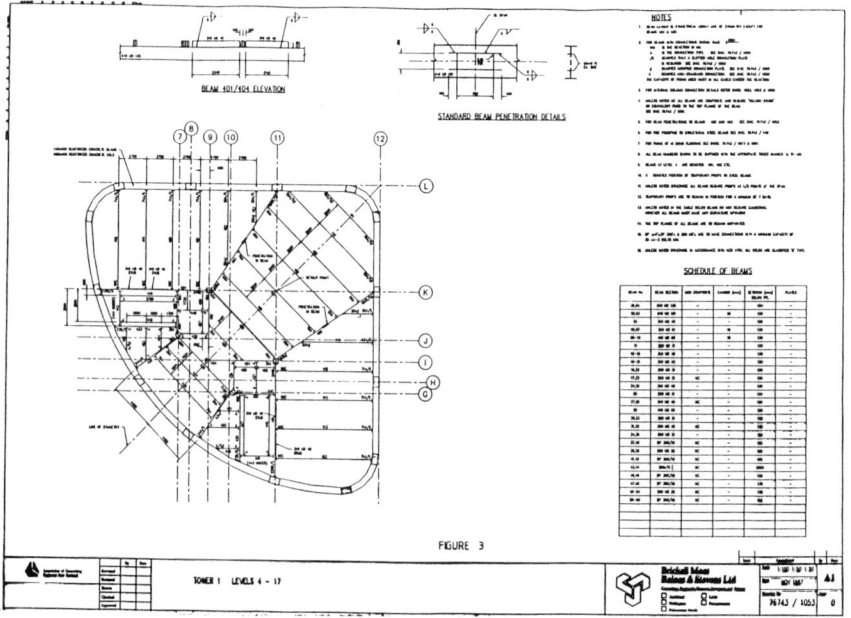

FIGURE 3

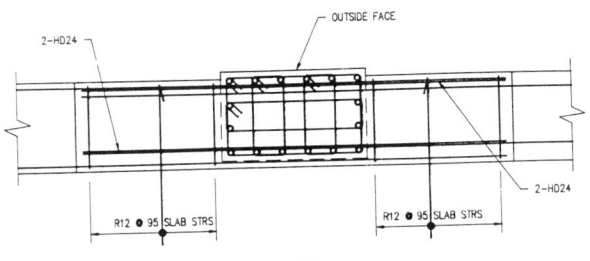

PLAN

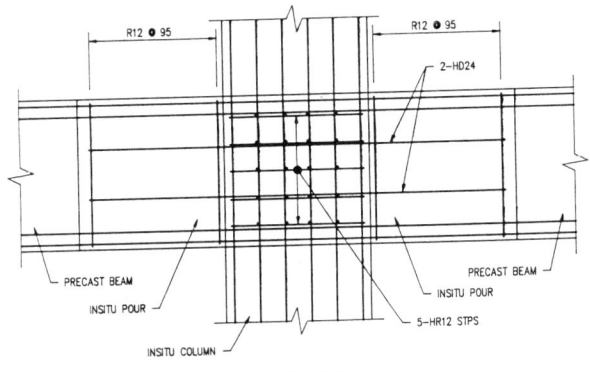

ELEVATION

TYPICAL BEAM-COLUMN JOINT DETAIL
FIGURE 4

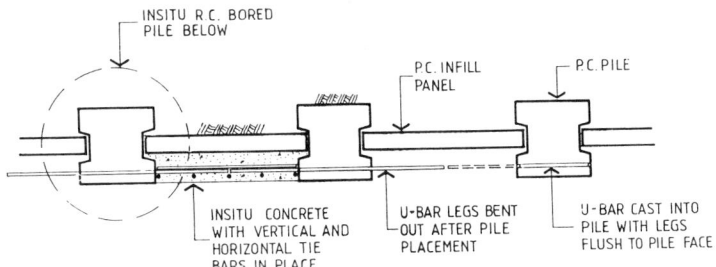

PERIMETER BASEMENT WALL PLAN DETAIL
FIGURE 5

391

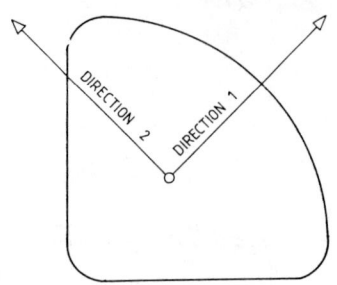

DIRECTION OF ANALYSIS

FIGURE 6

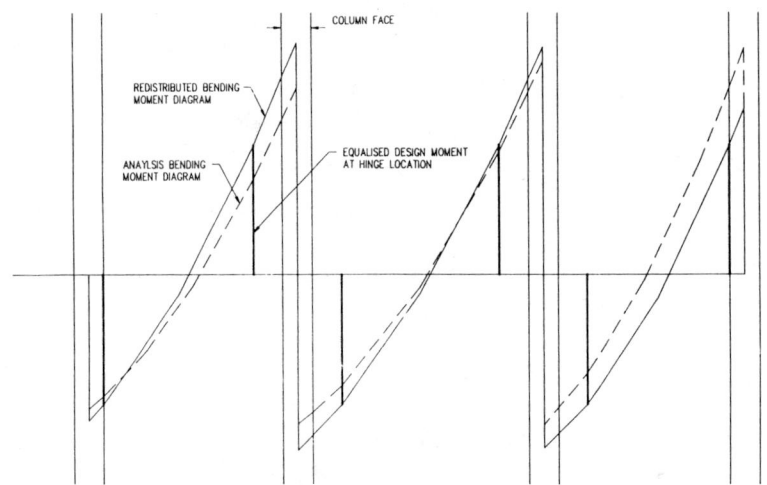

COLUMN FACE

REDISTRIBUTED BENDING
MOMENT DIAGRAM

ANAYLSIS BENDING
MOMENT DIAGRAM

EQUALISED DESIGN MOMENT
AT HINGE LOCATION

REDISTRIBUTION OF BENDING MOMENTS

FIGURE 7

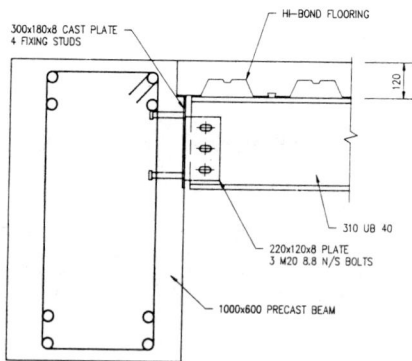

300x180x8 CAST PLATE
4 FIXING STUDS

HI-BOND FLOORING

120

310 UB 40

220x120x8 PLATE
3 M20 8.8 N/S BOLTS

1000x600 PRECAST BEAM

TYPICAL STEEL TO PRECAST BEAM CONNECTION DETAIL

FIGURE 9

392

ENHANCING ELEVATOR PASSENGER SAFETY AND
MITIGATING ELEVATOR DAMAGE
DURING EARTHQUAKES

D.A. Swerrie, PE *

With more than 500,000 passenger elevators in the United
States moving people at all times of the day and night,
every day of the year; with cities, transportation systems,
hospitals and buildings in general so dependent on the
functioning of those elevators; with everyone afraid of
being trapped in an elevator during an earthquake, is it
any wonder a community would enforce measures to enhance
elevator safety during earthquakes? California did so in
1975.

During May, 1990, the Elevator Safety Unit of Cal-OSHA
called a meeting to review a summation of elevator damage
resulting from the October 17, 1989, Loma Prieta
Earthquake.

This paper discusses some of the measures taken to mitigate
earthquake caused elevator problems in California, the
results of the post earthquake review and some new
recommendations.

When considering the responses to the Division's
questionnaire, the recommendations made and the discussions
summarized, the reader should keep in mind that in 1975
California did mandate that all existing elevators had to
be retrofit to comply with a somewhat lesser version of the
code that was then being made applicable to all new
elevators. The reader should also keep in mind that the
Division's questionnaire did not differentiate between
elevators installed prior to 1975 and elevators installed
since.

According to Division records, there were a total of 19,634
elevators in the area affected by the earthquake. That
number did not include elevators in federal buildings or
structures, nor did it include elevators in private
residences. About 95% of the elevators in the area were
being maintained by elevator companies.

● The response indicated that 7,666 electric (cabled) and
7,248 hydraulic elevators were under a service contract by
the respondents. It is concluded that the information
received represents what happened to about 75% of the
elevators in the affected area.

● It was indicated that 3,956 elevators were removed from
service by an earthquake protective device. Of the cabled
elevators 51.6% were removed from service by activation of
an earthquake protective devices. They are: the Collision

---

*D.A. Swerrie, PE, Swerrie, Inc., 244 San Felipe Way, Novato, CA 94945

Switch; the Derailment Switch; the Seismic Switch.
Hydraulic elevators are not required to be equipped with
earthquake protective devices, although some may be, where
all elevators in a facility are connected to one seismic
device.

● Respondents indicated that 1,286 electric (cabled)
elevators had earthquake damage. This represents 16.7% of
the cabled elevators. The type of damage, the extent,
whether in the hoistway, in the machine room, or whether it
was safety related, was not indicated.

● None of the respondents knew of any injuries to an
elevator passenger at the time of the earthquake. It was
concluded that safety had been achieved during that
earthquake. However, a case was discussed in which a man,
trapped for an extended time was distressed when extricated
and suffered a fatal heart attack a few days later.

● The respondents effected rescues from 84 elevators. They
were aware of other rescues from 49 elevators in which
passengers were trapped. The number of individuals
involved is unknown. There have been accounts of persons
being trapped who extricated themselves, by whatever means;
numbers unknown.

● Nineteen respondents were aware of 296 instances where
"unauthorized persons reset earthquake devices." Eight
respondents said additional damage was caused. Nine said
no additional damage was caused. "Unauthorized" should be
understood to mean other than elevator company personnel.
"Resetting the earthquake device" means placing the
elevator back in service. Building owners and managers may
train and "authorize" persons at their own discretion. It
was noted that not all units placed back in service
suffered additional damage.

● Respondents reported a total of 249 counterweights came
out of guide rails. If this figure is compared to a
previous response it is obvious that it was not just
displaced or free-swinging counterweights that caused
damage. More specific damage information is necessary if
code revisions are to be contemplated.

● Seven respondents noted that derailments tended to occur
when the counterweight was high in the structure. One
observed failures in which the counterweights were at both
the top and bottom of the building.

● Counterweights came out of guide rails of nominal weights
and material as indicated:

```
Wood  . . . . . . . . . 3
8-1b  . . . . . . . . . . 20
11, 12-1b  . . . . . . . 1
15-1b  . . . . . . . . 5
```

As would be anticipated -- 8-1b guides. Yet, many in 8-1b
guides were retained. Possibly the retrofitting with
intermediate brackets was not all it should have been. The
information is that all new (since 1975) installations
utilize 15-1b counterweight guide rails. Were the five that
came out installed since 1975? More specific information
is necessary to further evaluate what, if anything, should
be done with respect to guide rails. New? Existing?

● A question related to the comparison of elevator damage
in buildings, such as Federal or private residences, where
they were not required to comply with State of California
regulations. Four respondents said the damage was more
extensive. Eight respondents said the damage was about the
same. Again, more specific information is necessary to
make a comparison.

Although federal installations are not required to comply
with state regulations, in many instances they do. In some
instances they exceed what the state requires. Some
federal installations in the affected area experienced no
elevator damage or loss of service due to the earthquake,
itself. Apparently all new federal elevator installations
in California after 1975 were specified for earthquake
damage mitigation.

● Fifteen respondents indicated cases of earthquake
protective devices not operating as intended. No total
number of failures was reported.

● Eighteen respondents found the derailment device most
effective. Six found the seismic switch most effective. In
California the requirement is that either a seismic switch
or a derailment switch be provided. A better approach is
to specify that both be provided.

A derailment device, arranged to continually monitor the
position of the counterweight and, therefore, act as a
collision switch, may be used in lieu of a collision switch
on elevators arranged to operate under emergency conditions
following activation of either a seismic or derailment
switch. In virtually all cases where a collision switch is
desired, such is the arrangement used. Derailment
(displacement) switches indicate that the counterweight has
come out of the guide rails. Derailment switches are not
intended to indicate impending earthquake movement. The

quake is usually taking place, or has taken place, when that switch activates.

Properly installed, adjusted and maintained seismic switches do have the ability to give some warning. When they function as intended, it is possible, depending on the time interval which is related to the distance from the epicenter, to remove the elevator from service and park it with the doors open before the horizontal motion that does the damage, arrives.

Seismic switches have the ability to detect the vertical waves and/or the horizontal waves eminating from the epicenter of an earthquake. California orders are permissive. If a seismic switch is used, it may be of the horizontal type, the vertical type or may detect both types. The writer is of the opinion that safety would be better served if both the vertical and the horizontal types were required and provided. Both switches need not necessarily be in one unit. The location of the switch(es) in a building is as important as the switch itself, its calibration and its maintenance. Probably, to be most effective, each building should be individually analyzed and the seismic switch(es) location specified by someone with appropriate expertise.

It is universally agreed that causing an elevator to park at a floor with the doors open prior to the quake is the best solution for personal safety and earthquake caused trauma.

● Fourteen respondents found the required fastenings to be adequate. Nine said they were not. Again more specific information is required to analyze the response. Fastenings include those that hold machines, generators, controllers and selectors in place, as well as guide rails and hoistway equipment.

● Twenty-two out of the twenty-three respondents found the rope guards (retainers) to be adequate.

● It was indicated that 313 hydraulic elevators suffered damage from: falling debris; oil lines cracked; elevators (5) totally demolished (along with the buildings themselves); structural damage in hoistway; damage due to ground settling; twisted rail brackets; damaged roller guides; doors damaged; cylinder damaged; cab damaged; rail alignment out; pipe and muffler separated; leaking at gaskets and fittings; jack failure; pipe breakage; cabs came loose; door operator problems; jacks scraping; jack (1) came up out of ground.

No questions about escalators were included in the questionnaire. During the meetings several problems were mentioned. One escalator came off its lower support and dropped. Luckily it was at ground level. A number of escalators shut down. Many were subsequently found to be so out of alignment that some safety switches would not work. Misalignment was serious enough in many cases so that wear will be affected unless corrective measures are taken.

● Some respondents provided general comments regarding the elevators in general:

1. Car (1) came out of guides;
2. Counterweight guide rail brackets bent;
3. Guide rail (8-pd) tie brackets not adequate;
4. Traction elevators (2) demolished (buildings also);
5. Cars (4) with bent CWT guide brackets;
6. Cars (3) where fire-proofing in hoistway fell;
7. Cars (14) where leveling was affected;
8. Battery-operated switches failed;
9. Seismic switches were not operative;
10. Activating switch (1) on an earthquake protective device (EPD) defective;
11. Actuating switch (1) on an EPD out of adjustment;
12. Some hoistway doors actually bounced open during the shaking. (Car away from floor?) Door closers did reclose and lock the doors when the shaking ceased;
13. One rope hitch plate on the car moved and twisted;
14. Sulfurred bolts pulled loose;
15. On 214 structurally certified elevators, six had CWTs leave their guides. There was no damage except for a few bent rail brackets. NOTE: This item evidently relates to those elevators which were not required to have anything done because of an engineer's certification. This item was adopted in order that owners of older buildings which had little or no earthquake resistance in the design would not have to go to the expense of upgrading the elevator beyond the capability of the building.

Although not requested, several general recommendations were submitted with the questionnaires:

1. Two rings should be required on "ring on string" derailment switch devices;
2. Rail fastening should be improved;
3. The California code should be upgraded;
4. Reset buttons on controllers should be removed;
5. Provide more light in elevator machine rooms that will function under emergency power conditions;
6. On derailment switch devices put one ring at the top of

the CWT's and another at the bottom (two rings);
7. Stronger supports on rails of hydraulic elevators;
8. Better CWT guide rail supports needed;
9. Governor rope retainer and guards are needed.

What, if anything, should be done? It would be easy to
conclude that since no one was injured little action is
necessary. With every day that passes it becomes easier to
follow that course. Revisions are hard work.

Various recommendations were made. Some of them are as
follows.

1. Develop a more comprehensive questionnaire. In a
number of instances more specific information would have
been beneficial.

2. The Division should request the legislature to enact a
law requiring that questionnaires be completed as
accurately as possible and submitted to the Division. The
law should provide the right to aggregate the result totals
and eliminate individual reports in order to insure
anonymity.

3. The questionnaire should be in the hands of the
elevator companies prior to the incident to be analyzed.
The questionnaire should be used whenever there is a
reportable occurrence. Statistics can be developed from
which predictions are possible and codes justified.

4. Fires often accompany earthquakes. The Authority
should consider reports and questionnaires regarding the
activation of "Fireman's Service" in a similar manner to
earthquakes.

5. The committee should work to develop guidelines for
mechanics and others to follow prior to placing an elevator
back in service following an earthquake; possibly different
guidelines for different types of elevators. The
guidelines would be for that first restoration of service
and during the emergency period of the earthquake.

6. A sign, or signs, with instructions relating to the
restoration of elevator service should be developed; and
posted in the machine room or near the earthquake reset
switch.

7. The Authority should initiate an educational program.
Everyone must be made aware of the hazards associated with
restoring an elevator to service following an earthquake.
Owners, managers, mechanics, inspectors must know the

hazards prior to the earthquake.

8. A major source of concern expressed by many was the number of elevators placed back in service by someone other than elevator company personnel. Each elevator controller is required to have an identified, momentary button or switch, the resetting of the which places an elevator back in service. The problem arises when the individual resetting the device, does so without first making certain the elevator can be operated safely.

The idea of removing the reset from the controller has been proposed. Also mentioned, is the possibility of putting the reset on the car top, or even a second reset on the car top that must be reset prior to resetting the controller switch.

It is always possible to use jumpers in the controller to return elevators to service. There is no way to prevent an owner from using jumpers or manipulating the elevator controls if he so desires. It is his elevator! However, individuals seldom take foolish chances, especially with their own equipment, when they understand the hazards. Everyone who can gain access to an elevator machine room must be educated in the procedures to be followed if the earthquake reset has been tripped.

9. The committee should develop a second inspection procedure for more in-depth investigation of the condition of the elevator equipment. It is important that, as conditions return to normalcy, a thorough, equipment survey be made. Conditions indicating a need for repair, or more study, must receive appropriate action.

10. The committee should develop an ongoing inspection procedure for inspectors. It is not enough to take preventative measures once. Maintenance must assure that all is kept in readiness.

11. The code should indicate the need for periodic inspections; that an inspection is required following each activation of an earthquake protection device and/or earthquake; that it is necessary to have the entire installation inspected by competent personnel following resumption of normal service after an earthquake. The code should mandate such inspections.

12. Problems of nuisance activation of seismic switches have been noted. The condition causing such must be corrected! If they are not, the seismic switches soon will be inactivated. Not least was the problem of devices with

component failure, dead battery or out of calibration.
There is an order covering calibration. Evidently it is not
being enforced. Who is responsible? Which devices do
require calibration? Who is keeping the records? It is
important that seismic switches (that require it) be
periodically inspected and documentation kept. Where
batteries are required, who checks to ascertain they are in
working order. What will occur if the batteries lose their
charge?

"Fail Safe": There is an order requiring that earthquake
protection devices must be of the "fail safe" type. A
problem -- mechanics and inspectors do not necessarily
agree what the term "fail safe" means. Either define it,
or provide a description of what is meant by "fail safe" in
each instance the term is used. The latter might be the
best approach. In that manner the idea of "fail safe" is
left intact and its meaning is specified for each of the
devices as used. As new devices come on the market (as
they are) the manufacturer or inventor is alerted to the
fact that a "fail safe" specification must be established.

13. Where damage was experienced more information is needed
in many cases. Damage occurring during start-up procedures
is one thing. Damage because of failure of an earthquake
protection device to activate is another. One must
consider, if known, what forces were generated in a
particular installation. Damage caused when the forces
were greater than the equipment was designed to withstand
should be delegated to committees established for that
purpose. They will determine if the code design criteria
should be changed.

Where damage was experienced because of older type
fastenings, a committee must evaluate the damage and make
appropriate recommendations. Eight-pound counterweight
guide rails were a source of damage. Yet, there were many
installations where eight-pound guides performed well.
Wood guide rails are on very old elevators. Should they now
be declared obsolete? To what extent? More information and
committee work is indicated.

14. Some hospitals and other facilities where elevator
operation is critical following a catastrophe have older
elevators and elevators with 8-lb counterweight guides.
They might well decide to have their installation analyzed
by a structural or mechanical engineer. Evidence indicates
that the 8-lb guides can do the job when adequately
fastened, reinforced and bracketed.

15. To be confined in an elevator during an earthquake is

frightening. If the car is in motion during the earthquake or at the time the seismic switch activates, arrangements should not be made to immediately stop the elevator. Such has been done in California on older elevators with unsophisticated operation that did not lend itself to the establishment of a second call, response and shut down. That order should be reconsidered. It's bad enough when entrapment occurs due to conditions beyond anyone's control.

16. Elevators in some locations should be kept in operation during earthquakes and other emergencies if at all possible. Some committee work should consider a criteria to allow elevators in selected locations to continue to operate, at a reduced speed, following the activation of a seismic switch. Use of a derailment switch and a collision device could be mandatory. Possibly prior approval by the jurisdictional authority could be required. The potential of entrapment is something that must be clearly understood and considered.

17. Handicapped individuals are entitled to the same safety consideration as is the general public. The committee should extend seismic considerations to the private residence type elevators that the authorities allow to be installed in commercial buildings for the limited use of the handicapped.

18. Machine room illumination should be addressed. Power outages following an earthquake are common. When mechanics go to machine rooms to get elevators running following an earthquake, they need adequate illumination.

19. Wiring Diagrams: Authorities should address the problem of missing wiring diagrams, illegible wiring diagrams and inaccurate wiring diagrams in the machine room or on the premises. Following the earthquake, other than the regular service mechanic may be charged with placing the elevator in service. It could possibly be emergency-related so time is of the essence. Everyone in the industry can relate to the hazards connected to working on an elevator controller without accurate, up to date prints.

20. Contingency Planning: Building owners and managers, elevator companies, jurisdictional authorities, all should have instructions in place, and personnel should understand how to respond during an earthquake. There are not enough competent persons to handle all the elevator emergencies that may occur. Communications fail. Power fails.

Individuals must respond on their own. The services of no
competent individual should be lost or wasted. It takes
pre-planning.

It might be noted that enforcement, education, planning and
accumulation of data are given highest priority at this
time. That and ongoing inspections should go a long way
towards improving future results.

The serious reader desiring more detailed knowledge of what
is recommended to enhance elevator safety during
earthquakes, is referred to the "Safety Code for Elevators
and Escalators," ASME/ANSI A17.1-1987, Appendix "F," which
is some 10 pages dedicated to those specific requirements.

With adequate regulations in place, with the elevators in
compliance, being prepared is dependent upon maintenance
and inspection: maintenance of the complete system;
periodic calibration of devices in need of such; ongoing,
competent inspections of the entire system; planning and
readiness to respond; reports on the results of earthquakes
as they occur. Willingness and ability to modify or revise
the regulations as reports of occurrences indicate such
modifications are appropriate and prudent. When the
foregoing has been done, the community may realistically
claim that elevator safety in the event of an earthquake
has been enhanced, the potential for earthquake damage to
the elevator system has been mitigated and that a sincere
effort has been made to protect people and property from
the effects of earthquakes.

D.A. Swerrie, PE
Elevator Safety Consultant

*Published in the January 1991 issue of ELEVATOR WORLD, the Trade
Journal of the International Vertical Transportation Industry.*

Proceedings of the Second Conference on Tall Buildings in Seismic Regions
55th Regional Conference
May 16 and 17, 1991, Los Angeles, California

ULTIMATE HIGH-EFFICIENCY SYSTEMS FOR TALL BUILDINGS

AN OVERVIEW

B.S. TARANATH Ph.D., S.E.*
JOHN A. MARTIN & ASSOCIATES, INC.
LOS ANGELES, CALIFORNIA 90057

The advent of tube system with its myriad variations appears to have changed for ever the glass box architecture of high-rise buildings. Both the public taste and the demands of the real estate market have subdued the moral arguments and belief that clean stark lines of international style would in some way be good for the community. What was once considered as gospel by prominent architects is now seen as an outdated dogmatism. Virtually every large skyscraper built since the beginning of 1980's is something other than a box of generation before. This metamorphosis first occurred in seismically benign regions. This is understandable because the telltale sign of post-modern architecture with its rather abrupt change in stiffness has not been a cause of great concern in the wind design of tall buildings. Wind design of tall buildings has been more daring because structural failures resulting from service level dead, live, and wind loads are quite rare and are usually the result of gross error, deteriorated structural condition or gross over load.

Structural design for earthquakes is not governed by the same principles as for wind design. Although the design is often performed for service level earthquake loads, substantial inelastic response with excellent ductility and energy-absorptive capacity is expected of structures designed in high seismic zones.

However, the dramatic change in architecture has begun to establish gradually in regions of high seismicity requiring innovative and daring response from structural engineers. Computers have given the structural engineers of today the tools to come-up with economical systems within the confines of architectural requirements.

* Dr. Taranath is the author of "Structural Analysis and Design of Tall Buildings", published by McGraw-Hill Company.

It is of great interest to examine different bracing
systems for tall buildings in the context of present day
architecture especially as related to seismic design . A
seismic system not only should possess the strength and
stiffness characteristics of its counterparts, but should also
exhibit ductility or the ability to take excursions into
cyclical inelastic zones without collapse or serious damage.

RIGID FRAMES

Because of the continuity of members at the joints, the rigid
frame responds to lateral loads primarily through flexure of
beams and columns. The lateral deflection components of a
rigid frame may be thought of as being similar to the
deflection components of a cantilever beam. One component can
be likened to the bending deflection and the other to the
shear deflection of the beam. The component analogous to the
beam shear deflection dominates the deflection picture in a
normally proportioned frame and may amount to as much as 80
percent of the total deflection, while the remaining 20
percent comes from the bending component.

For a normally proportioned rigid frame, as a first
approximation, the total lateral deflection can be thought of
as a combination of three factors:

1. Deflection due to axial deformation of columns (15 to 20
    percent).

2. Frame racking due to beam rotations (50 to 60 percent).

3. Frame racking due to column rotations (15 to 20 percent).

   Rigid frames are perhaps the most popular in both wind
and seismic designs. Provided that local instability is
inhibited by adding lateral bracing to compression flanges,
its reserve strength is well suited for energy absorption
capacity. The basic difference between the wind and seismic
designs is in the detailing of the beam-column joints. In
seismic design, almost always, doubler and continuity plates
are required to beef-up the panel zone. However, continuity
plates are sometimes used in wind designs even when not
required by calculations to reduce panel zone deformation and
hence, the overall building drift.

   For medium-rise buildings in the 20-30 story range, rigid
frame offers an economical structural solution. Implicative
in its preferred use in seismic regions is the fact that it
possesses a high level of redundancy and ductility in addition
to offering maximum flexibility in architectural planning.

BRACED FRAMES

A braced frame attempts to improve upon the efficiency of pure rigid frame action by virtually eliminating the column and girder bending factors. This is achieved by adding truss members such as diagonals between the floor system. The diagonals carry the lateral forces directly in predominantly axial action, providing for nearly pure cantilever behavior. All members are subjected to axial loads only, thereby creating an efficient structural system.

Any rational configuration of bracing can be used for bracing systems. Bracing types range from a concentric brace between two columns to knee bracing and eccentric bracing with complicated geometry.

Since there is no undue penalty in its design for wind, the efficiency of braced system which is directly related to the separation between the windward and leeward columns may be increased by simply increasing the width of the bracing. Examples are shown in Figures 1 and 2 in which the bracing is spread to the full width of the building. Note that the bracing will not interfere with the exterior architecture, but the presence of interior sloped columns within the lease space has to be acknowledged architecturally as a trade-off for structural efficiency.

The system, however, lacks energy absorbing capacity. Its use in seismic design is almost always in conjunction with a ductile system such as a rigid frame.

STAGGERED TRUSS SYSTEM

Because the staggered truss resists major gravity and lateral loads in direct stresses, the system is quite stiff. In general, no material needs to be added for drift control, and high-strength steels are conveniently used throughout the entire frame. The system has been used for buildings in the 35 to 40 story range.

The behavior of a staggered truss system is similar to that of a system with offsets in vertical bracing system. The essence of structural action is the progressive transmission of lateral loads across the floor system to trusses on adjacent column lines. Between the floors, lateral forces are resisted by the truss diagonals, and at each floor these forces are transferred to the truss below by the floor system acting as a diaphragm (Fig. 3). The columns between the floors receive no bending moments, resulting in a very efficient and stiff structure.

ECCENTRIC BRACING SYSTEMS

Eccentric bracing is a unique structural system that attempts to combine the strength and stiffness of a braced frame with the inelastic behavior and energy dissipation characteristics of a moment frame. The system is called eccentric bracing because deliberate eccentricities are employed between the beam-to-column and beam-to-brace connections in an effort to force shear yielding of the eccentric beam element. This offset or eccentricity, promotes formation of an energy-absorbing hinge in the portion of the beam between the two connections. Thus, the system maintains stability even under large inelastic deformations. The required stiffness during wind or minor earthquakes is maintained because no plastic hinges are formed under these loads and all behavior is elastic.

INTERACTING SYSTEM OF BRACED AND RIGID FRAMES

Unbraced or rigid frames deform in a predominantly shear mode with the relative story deflections depending on the magnitude of shear applied at each story level. Although near the bottom, the story deflections are somewhat larger and near the top, somewhat smaller as compared to the braced frames, the floor-to-floor deflections can be considered more nearly uniform. The deformations of a braced frame, on the other hand, increase rapidly near the top, mainly due to the cumulative effect of chord drift. This combination of different deflection patterns between braced and rigid frames is very helpful in forging stiff structures (Fig. 4).

OUTRIGGER AND BELT TRUSS SYSTEMS

A relatively new concept which has evolved within the past two decades is the technique of using a cap truss on a braced core combined with exterior columns. In this system, columns are tied to the cap truss through a system of outrigger and belt trusses. Therefore, in addition to the traditional function of supporting gravity loads, the columns restrain the lateral movement of the building. When the building is subjected to lateral forces, tie-down action of the cap truss restrains the bending of the core by introducing a point of inflection in its deflection curve. This reversal in curvature reduces the lateral movement at the top. The belt truss functions as a horizontal fascia stiffener and engages the exterior columns which are not directly connected to the outrigger trusses. A general improvement of 25 to 30 percent in stiffness can be realized in contrast to the same system without such trusses because instead of individual columns acting as tie-downs, all the facade columns participate in resisting the lateral loads.

The magnitude of beneficial effect of tying down the exterior
columns to the core is a function of two distinct
characteristics, the stiffness of the equivalent spring and
the magnitude of the rotation of cantilever at the spring
location due to external loads. The stiffness of the
equivalent spring, for example, is at a minimum when located
at the top and a maximum when at the bottom. The strain
energy that can be stored in the spring is a function of
stiffness and the rotation of the cantilever at its location.
The rotation of the free cantilever for a uniform wind load
varies parabolically from a maximum value at the top to zero
at the bottom. Therefore, from the point of view of spring
stiffness alone, it is desirable to locate the outrigger at
the bottom, whereas from a consideration of rotation, the
converse is true. It is obvious that the optimum location is
somewhere in between.

With certain simplifying assumptions, the analytical model for
determining the optimum location reduces to that of a tied
cantilever as shown in Figure 5. The optimum location can be
shown to be approximately at the mid-height of the building.
For a twin-truss system, it can be shown that several truss
locations give solutions very nearly equal to the optimum
location as shown in Figure 6. It should be relatively easy
to choose a combination that satisfies simultaneously the
structural, mechanical, and architectural requirements.

Concentrically braced outrigger and belt trusses lack the
ductile performance beneficial in the seismic design of
structures. An additional concern regarding the concentric
system is the abrupt change in stiffness along the height of
the building brought about by the large rigidity of the truss.
A sudden kink or reversal in curvature typically occurs in the
deflection curve signifying an undesirable dynamic response.

Seismic considerations dictate a modified form of the
outrigger concept that mimics the flexural behavior of a
ductile moment frame. A vierendeel system for the trusses
presents a logical solution. An added bonus is the absence of
diagonal elements which can be more easily accommodated in an
open floor area. An eccentric braced truss is another version
which has the desirable ductile behavior and an excellent
stiffness characteristics at service loads.

ULTIMATE HIGH-EFFICIENCY STRUCTURES

Super-tall buildings are generally defined in architectural
terms as skyscrapers with a silhouette whose proportion in
height to width is of the order of 6:1. To an engineer, a
super tall building is one in which the engineering demands
imposed by lateral loads far exceed those of gravity,
requiring a very significant premium in their design. The

ideal structural form to resist the effect of bending, shear
and vibration is a system possessing vertical continuity
located at the farthest extremity from the horizontal center.
A perfect form is a chimney but as a windowless structure it
is inadequate as an architectural model. A more practical
model is a skeletal structure cantilevered from the earth.

Modern high-rise technology has largely replaced the heavy
cladding and interior masonry systems with relatively
lightweight counterparts. The holding-down power of these
systems is no longer present in modern construction. Let us
examine how we can employ the relatively lightweight materials
to help in providing resistance to the lateral loads. The
tube system, with its characteristic deployment of the columns
at the building perimeter, certainly provides the much
required separation between the windward and leeward faces of
the building for resisting the overturing moments. However,
since the exterior columns, especially in a framed tube, are
placed relatively close to each other, their tributary areas
for collection of gravity loads are rather small. Therefore,
the beneficial effect of gravity load in counteracting the
tensile forces of the columns is somewhat limited, first
because of the relatively light materials used in current
construction practice, and second by the limited tributary
area for the exterior columns. The main premise behind the
ultimate high-efficiency structure is to transfer as much
gravity load as practicable to the columns resisting the
overturning moments.

An approach is to eliminate as many interior columns as
possible, perhaps all of the interior columns, and collect the
total weight of the building on the exterior columns. This
way the holding-down power of gravity loads is put to use in
the most efficient manner. It must be recognized that there
is a certain amount of trade-off in the floor framing system
because it is economically prohibitive to clear-span the floor
members using traditional approaches. Accommodation has to be
made to achieve the transfer of entire building load into the
exterior columns without paying a significant premium.

It is, of course, necessary to tie the windward and the
leeward columns of the tube with a structural system capable
of resisting the shear forces caused by the lateral loads.
This can be achieved with a system of deep spandrel beams when
the perimeter columns are closely spaced, or with a system of
diagonal bracing when the columns are spaced apart as in a
trussed tube system. By progressively shifting the exterior
columns to the corners of a rectangular trussed tube, the
efficiency of the system can be greatly improved. An ultimate
structure for a rectangular building, then, will have just
four corner columns inter-connected by massive diagonals (Fig.
7).

The efficiency of a building to resist lateral loads can be increased further by using interior bracing with a structural system in which the total gravity load of the building is made to bear on a limited number of exterior columns. To increase the uplift capacity, interior columns are eliminated within the building envelope. A system of interior bracing that performs the dual function of channeling the loads from the interior columns to the exterior while at the same time functions as a shear element between the windward and leeward columns is likely to be the optimum system.

Such a concept is shown in Figure 8 in which the total gravity and lateral loads on the building are resisted by only four composite columns located in such a way as to give almost limitless architectural freedom on the two short faces of the building. The lateral resistance in the short direction of the building is provided by two diagonal inverted K-braces spanning between the exterior composite columns. The lateral resistance in the long direction is provided primarily by the full height vierendeel trusses.

To collect the gravity loads on the columns, a story high truss is used at every twelfth floor. Shown in the figure are the K-braces running the full depth of the building between the composite columns. The primary function of the interior vierendeel frame is to collect and distribute the floor gravity loads to the columns via K-braces. However, because of its geometry, it also will take part in resisting the lateral load.

The current flamboyance in the architecture of tall buildings in seismic regions has required the structural engineer to revisit the structural systems which have been used successfully to resist wind forces. However, structural systems for seismic design must exhibit ductile performance in addition to providing a balance of stiffness and ductility. Economy in the use of material quantities without undue compromise in the dynamic response under strong dynamic excitation and safety provided by redundancy in the lateral resisting systems are the key factors in the structural system selection process.

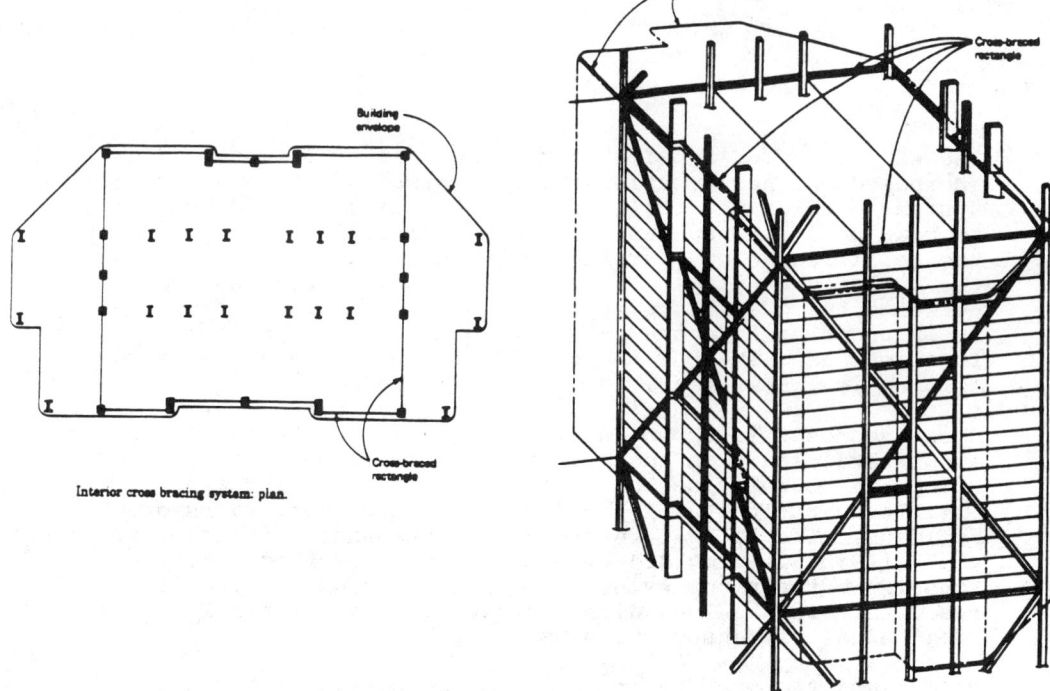

Interior cross bracing system: plan.

Isometric view of interior cross bracing system.

Figure 1

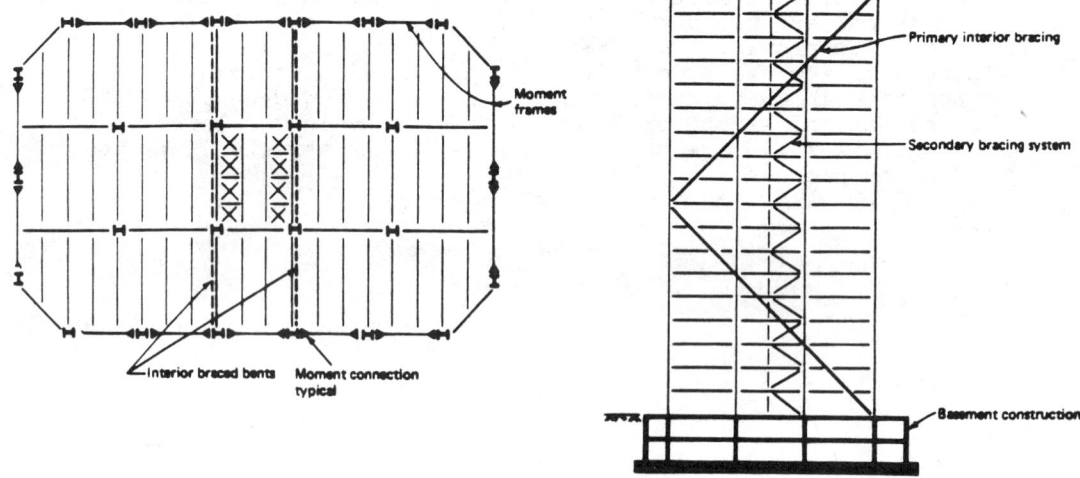

Figure 2

410

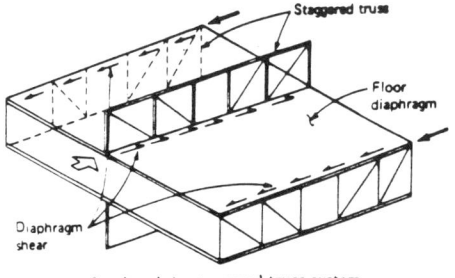

Load path in staggered truss system.

Figure 3

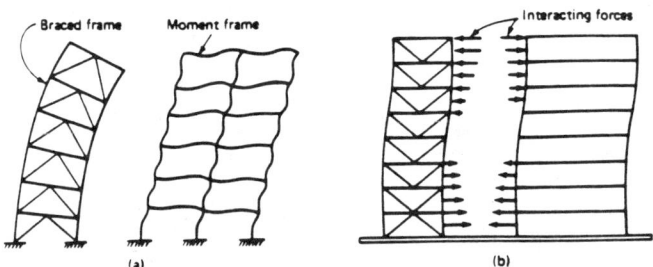

Interaction between braced and unbraced frames. (a) Characteristic deformation shapes; (b) variation of shear forces resulting from interaction.

Figure 4

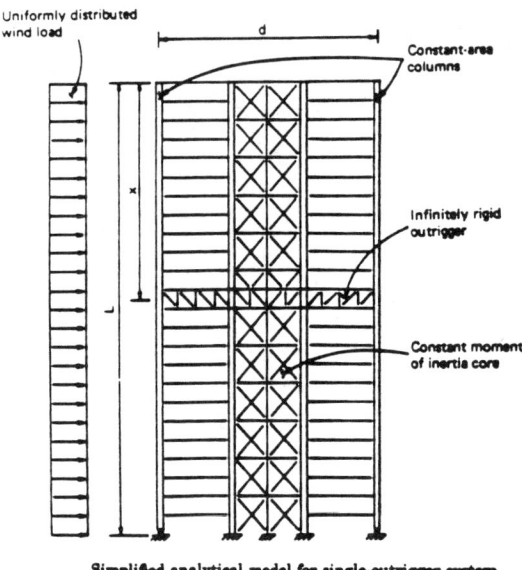

Simplified analytical model for single-outrigger system.

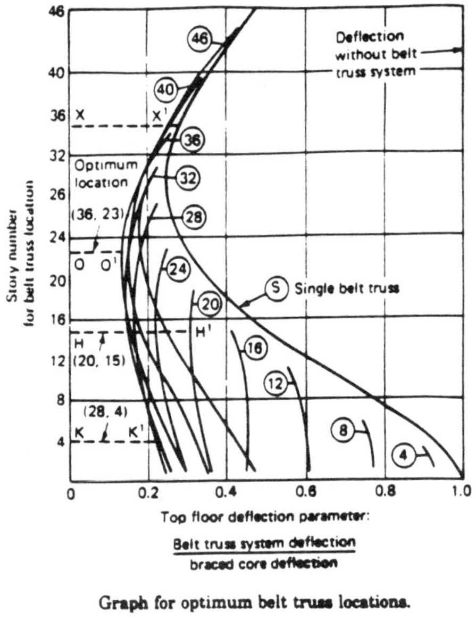

Figure 5

Graph for optimum belt truss locations.

Figure 6

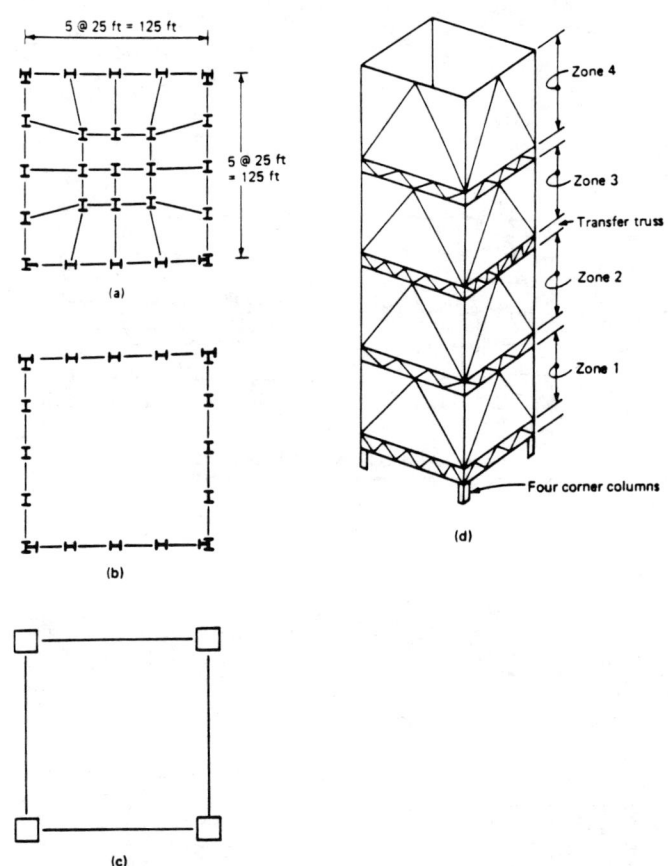

Figure 7

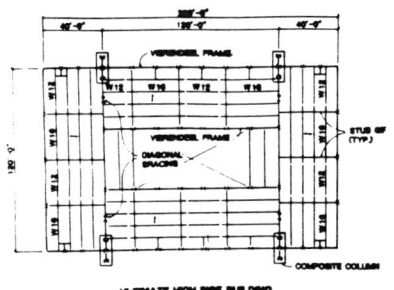

ULTIMATE HIGH RISE BUILDING
STRUCTURAL CONCEPT : PLAN

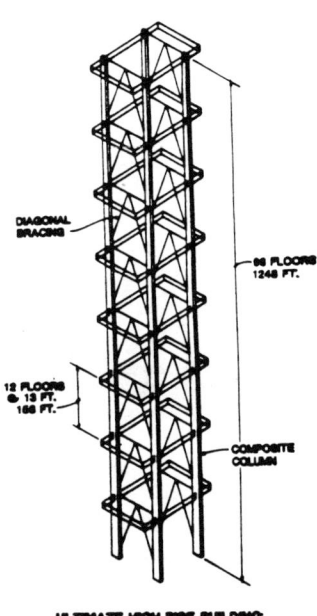

ULTIMATE HIGH RISE BUILDING
STRUCTURAL CONCEPT

MEGA MODULE
INTERIOR VIEW

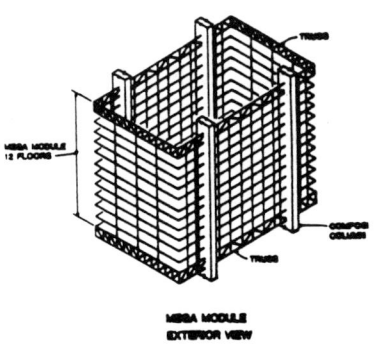

MEGA MODULE
EXTERIOR VIEW

Figure 8

413

Proceedings of the Second Conference on Tall Buildings in Seismic Regions
55th Regional Conference
May 16 and 17, 1991, Los Angeles, California

HIGH STRENGTH STEEL SHAPES –
A NEW DEVELOPMENT FOR ECONOMICAL
HIGH–RISE CONSTRUCTION

by Frank van Rest, M.Sc.
TradeARBED, Inc. New York

INTRODUCTION

The use of HSLA (High Strength Low Alloy) structural steel (F_y > 50,000 psi) in high-rise construction has been limited. The reasons for this are varied. Stiffness considerations and code restrictions have been important factors, but a limited availability, a poorer weldability, a lower toughness and a high price have been equally important in limiting the acceptance of HSLA steels.

With the development of its Quenching and Self-Tempering (QST) process, the Luxembourg steel manufacturer ARBED S.A. has succeeded in producing a steel for structural shapes that overcomes most of the limitations common to HSLA steel shapes.

QST PROCESS AND PROPERTIES

Traditionally, HSLA structural shapes are produced by adding alloying elements to the steel and rolling at controlled temperatures. However, the addition of alloying elements must be limited, as it has a negative effect on the weldability and toughness of the steel. Controlled rolling is restricted by the mechanical power of the rolling mills. Subsequently the production is limited to the lighter shapes.

These limitations are overcome by the QST process. In this process the desired yield strength is obtained by an in-line heat treatment after the last rolling pass. An intense watercooling is applied to the whole surface of the shape so that the skin is quenched. Cooling is interrupted before the core is affected by quenching and the outer layers are tempered by the flow of heat from the core to the surface during the temperature homogenization phase. Fig. 1 shows the QST process schematicly and in Fig. 2 the quenching of a Jumbo section is shown.

Because less (expensive) alloying content has to be added to the steel to obtain the high yield strength, the QST process is a cost effective process.

415

Due to its metallurgical principle, the QST process results in steel properties that were considered incompatible:

* a high yield strength (up to 70,000 psi for the entire section range);
* an outstanding low temperature toughness;
* an excellent weldability due to a low carbon equivalent.

Fig. 3 shows the Transition Temperature for different yield strengths for both the classical TM (Thermo Mechanical = controlled rolling) process and the QST process. Fig. 4 shows the required carbon equivalent for different steel grades for both processes.

The excellent weldability of the QST shapes is illustrated by a series of welding tests. Using different welding processes and heat inputs, QST shapes up to 5 1/2 inches thickness in Grade 50 and Grade 65 have been butt-spliced using full-penetration single bevel groove welds.

The tensile tests and Charpy V-Notch impact tests performed on the welded material show that for all heat inputs up to 200 kJ/inch, the tensile properties of the joints are not influenced and the toughness of the Heat Affected Zone (HAZ) remains on a high level. The welding tests have been carried out without preheating the material. Fig. 5 shows the improvement in weldability of QST shapes for a 2 inch thick section in Grade 65.

Fig. 6 shows the results of toughness tests performed on the core area (web-flange intersection) of a 5 1/2 inch thick QST shape in Grade 65 in both the as rolled and as welded condition. Again the welding has been performed without preheating. Note that the toughness requirement of 20 ft-lbs. at 70 °F, specified by AISC to guarantee the reliability of splices using full penetration welds in heavy shapes subject to primary tensile stresses, is easily met in the as rolled and the as welded condition.

The welding test results obtained at ARBED's research facility in Luxembourg are currently being assessed by AWI (American Welding Institute), an independent scientific research institute. Grade 65 Jumbo sections have been spliced using full penetration groove welds. Different welding processes (SMAW, FCAW and SAW) have been used, without applying preheat. The first results confirm the previously obtained data and a full report will be published shortly.

ECONOMY

An economical use of HSLA steel shapes is limited to the members that can be designed to higher allowable stresses. As higher stresses result in larger deformations (the E-modulus of HSLA steel is virtually equal to that of its lower strength equivalents), the shapes whose sizes are determined by deformation criteria are not suited to be designed in HSLA steel. In high-rise buildings these members will be the found in the floor framing and most likely in the lateral resisting steel frame.

Elements most economically designed in HSLA steel are tension and compression members, for which the deformation consists of axial lengthening and shortening only. In Fig. 7 the economy of ASTM A 572 Grade 50 and HISTAR Grade 65 is compared for a heavy tension member. An important part of the cost savings in this example originates from a decrease in weld volume when splicing, caused by a smaller cross sectional area of the tension member. Fig. 8 shows a similar comparison for a heavy compression member Again note the considerable fabrication cost savings.

The economy of HSLA steel in moment resisting frames has not been thoroughly analyzed. Naturally, drift is an important factor. However, ASTM A 572 Grade 50 is currently being used in moment resisting frames, indicating that there might be use for HSLA steel.

An interesting concept in moment resisting frame design is the omitting of web doublers and continuity plates or stiffeners in the columns by using a higher strength steel. The column is kept the same size, so the lateral resistance will remain equal; the high yield strength is used only in the connection design.

The code specifically excludes steel with $F_y > 50,000$ psi in Special Moment Resisting Frames. Certainly this is due to concerns regarding the ductility of higher strength steels. All mechanical tests performed on QST shapes indicate that the material has a structural behavior that is similar to that of lower strength steels. The tests include tensile tests, toughness tests, full scale buckling tests and full scale bend tests.

CONCLUSION

The development of the QST process has led to the availability of a steel that overcomes most of the disadvantages of the classical HSLA steels. Excellent weldability, outstanding low temperature toughness and a competitive price are guaranteed. As deformation is not a decisive criterium, QST shapes can be used today for compression and tension members. Weight savings will vary from 15 % to 25 %. Cost savings can be higher because of fabrication cost savings.

The economy of QST shapes in moment resisting frames needs to be analyzed further, but there are certainly indications that

HSLA steel shapes lead to savings for that type of application as well. The code restriction of $F_y \leq 50,000$ psi for special moment resisting frames needs to be addressed based on available data and possible future research.

TANY, 03/01/91

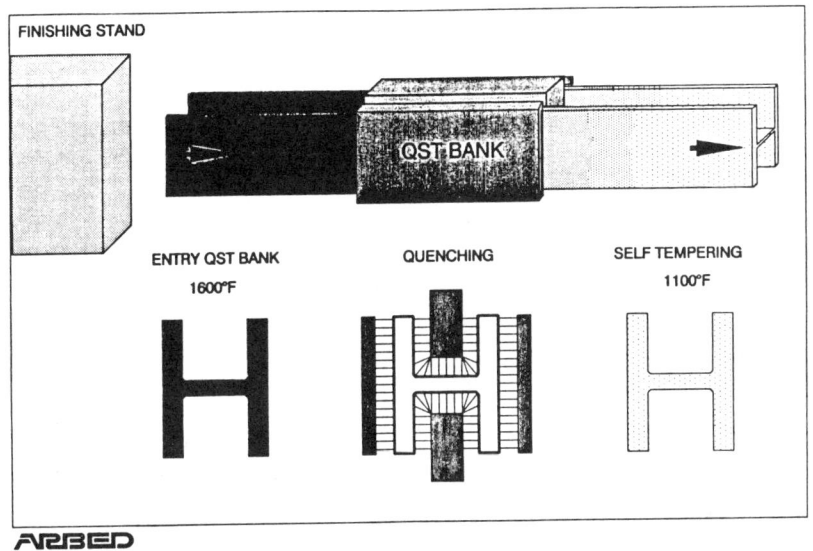

FINISHING STAND

QST BANK

ENTRY QST BANK
1600°F

QUENCHING

SELF TEMPERING
1100°F

ARBED

Figure 1: Quenching and Self Tempering (QST)
Treatment of Beams in the Rolling Heat

Figure 2: QST Treatment of a Jumbo Section

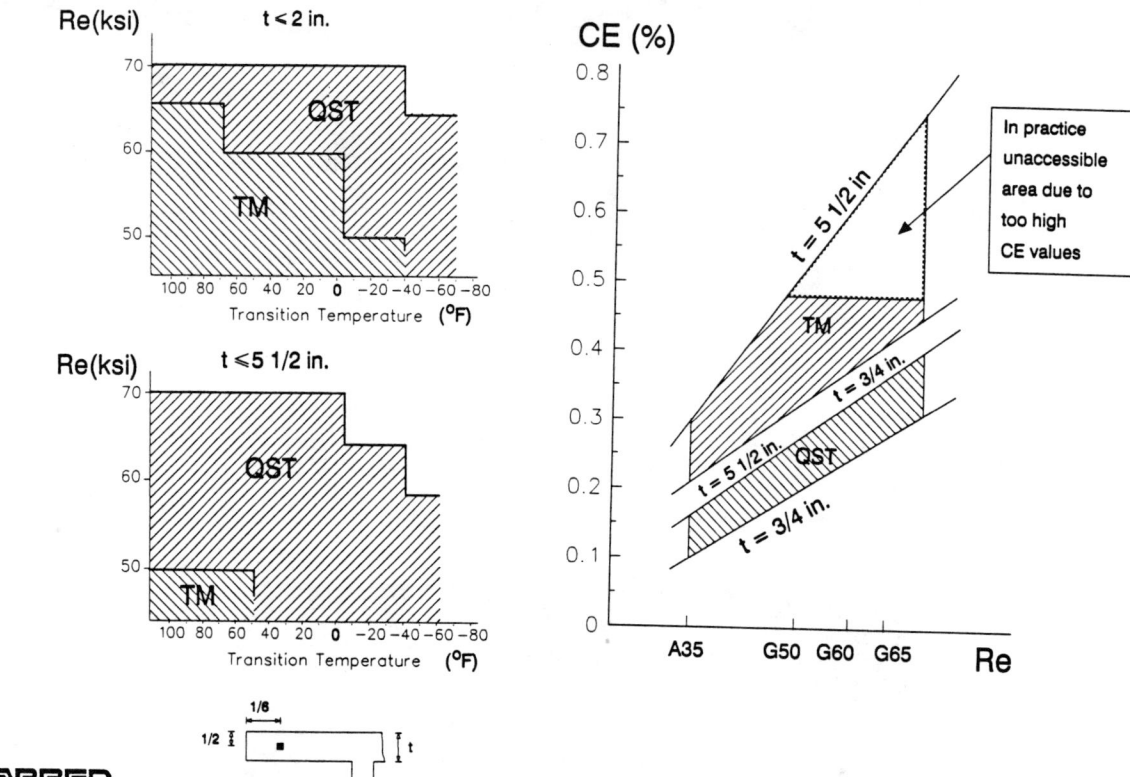

Figure 3: Yield Strength (Re) and Toughness of Steel obtainable by QST Treatment and TM Rolling

Figure 4: Carbon Equivalent (CE) and obtainable Yield Strength (Re) for different Product Thicknesses (t)

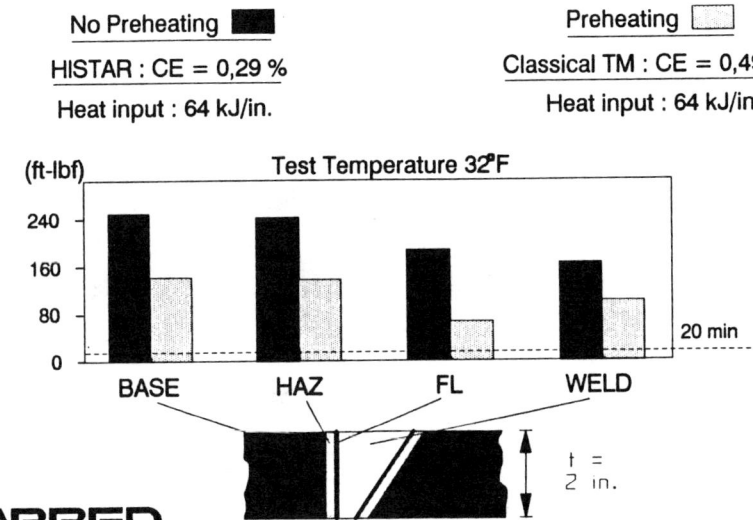

Figure 5: Charpy V-Notch Impact Testing of Welded
2 Inch Thick W-Sections in HISTAR Grade 65

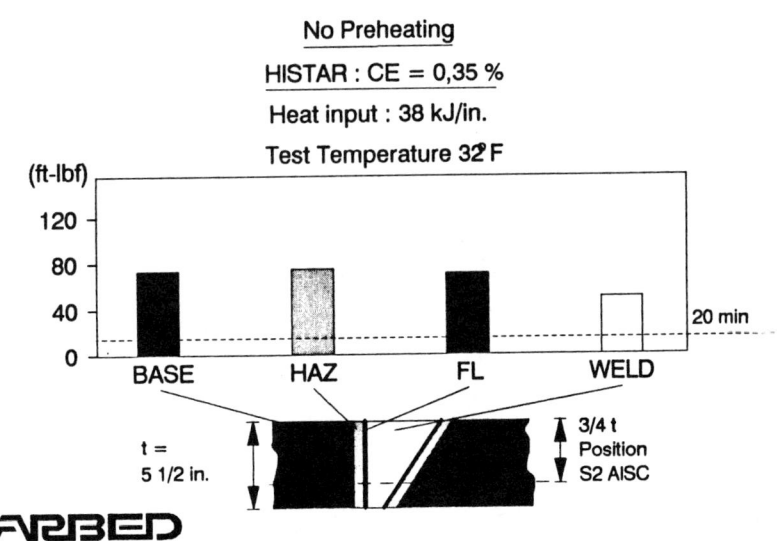

Figure 6: Charpy V-Notch Impact Testing of Welded
Jumbo Sections in HISTAR Grade 65

A ARBED

Grade	A 36	A 572 Grade 50	HISTAR Grade 65
	5.5" / 3.9"		
Shape	WTM 14 x 873	W 14 x 665	W 14 x 500
Design Strength	8312 kips	8820 kips	8600 kips
Weight	131 %	100 %	75 %
Cost	119 %	100 %	87 %
Weld Volume	160 %	100 %	59 %

(spliced using full penetration welds)

Figure 7: Material Weight, Material Cost and Weld Volume
Savings for Tension Members Depending on Steel Grade

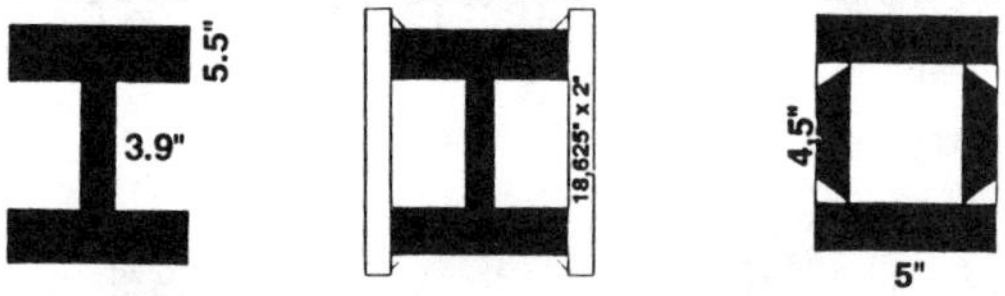

Grade	HISTAR Grade 65	A 572 Grade 50	A 572 Grade 50
	5.5" / 3.9"	18.625" x 2"	4.5" / 5"
Shape	WTM 14 x 873	W 14 x 730 + plates	Box Column
Design Strength	12670 kips	12571 kips	12486 kips
Weight lbs/ft	873	1053	1055
Weight	83 %	100 %	100 %
Fabrication Cost	62 %	100 %	118 %

(buckling length: 14 ft.)

Figure 8: Material Weight and Fabrication Cost
Savings for Heavy Column Sections

Proceedings of the Second Conference on Tall Buildings in Seismic Regions
55th Regional Conference
May 16 and 17, 1991, Los Angeles, California

THE MULTI-PURPOSE TALL STRUCTURE

By Thomas R. Vreeland, FAIA
Albert C. Martin & Associates
811 West Seventh Street
Los Angeles, California 90017

I

In the history of tall structures the single purpose tower has largely been the norm: the Pharos lighthouse, the Italian campanile, the Eiffel Tower, the Empire State Building, the modern high rise office. However, the less well-know multi-purpose tower deserves equal recognition. Given detectable new social and economic tendencies, it may quite possibly be more important than its purer and more homogeneous sister. The single use tower has the clear advantage of structural simplicity. With its unity of purpose and simple, repetitive spans, its structure can be simple, elegant and uniform, an appearance which has perfectly suited the mid-20th century aesthetic norms of modernism. This domination, however, has recently shown signs of weakening and this paper addresses itself to the probable outcome of the break up of the present hegemony by looking at predecessors of the multi-purpose tall structure.

II

Multi-purpose tall structures were quite common through the 19th century and early 20th. They came about naturally, due to the steep rise of land values in the city, by the grafting of high rise components containing offices, apartments or hotel rooms on to existing or desirable low rise land uses such as concert halls, ball rooms, banking halls or memorial halls. One thinks of Adler and Sullivan's Auditorium Building in Chicago as an example. In a city which was making history with the new high rise commercial structures it was erecting at the turn of the century, there was built this extraordinary building which combined in a single tall structure an opera house, a hotel, an office building and apartments. The Waldorf Hotel in New York somewhat later combined over 2,000 guestrooms, a dozen ball rooms of assorted sizes, and a platform for Mr. Astor's private railroad car. The Penn Athletic Club of 1928 contains within a single structure a ballroom, a gymnasium, a swimming pool, an outdoor tennis court and eight floors of bedrooms. The five-hundred foot high U.S. Custom House in Boston of 1915 incorporated into its base intact an earlier customs house built in the 1840s. There has been more than one example of an office tower built over a church or place of worship, a form which is strongly suggested by Cass Gilbert's Woolworth building in New York with its incredible ecclesiastical lobby. Such a building was actually proposed by Bertram Goodhue in his sketch for the Convocational Tower of 1921. Goodhue's Nebraska State Capitol stands as arguably the most eloquent expression of this hybrid building type.

III

The origin of this hybrid type can be found in the attempts of gothic church builders to place a tower directly over the crossing, the ultimate achievement of which can be seen in the great English cathedrals of the Middle Ages, notably Lincoln and Salisbury. Since it entails the extraordinary structural feat of supporting the weight of the tower over the void of the church nave below it, it symbolizes all subsequent attempts to build above banking halls, theaters or hotel ballrooms. It is precisely the structural complexity of such an endeavor, the need to resort to deep trusses or sloped abutments, combined with new zoning laws which prohibited several uses in a single building, which discouraged their development after the first quarter of this century. Le Corbusier's visionary city, with its pure towers of finance set quite apart and distinct from residential structures, arrives at this point to replace the earlier dream and dominate the rest of the century.

The easing of these zoning restrictions, combined with the economic desirability today of combining commercial, residential and institutional uses into the same structure, herald a return to the development of multi-purpose tall structures. The increasingly complex appearance of high rise office buildings today compared to the aesthetically purer buildings of the last four decades probably indicates the architect is readying himself for the next generation of mixed-use skyscrapers.

Proceedings of the Second Conference on Tall Buildings in Seismic Regions
55th Regional Conference
May 16 and 17, 1991, Los Angeles, California

Drift Control Method for Structural Design of Tall Buildings

Akira Wada

Professor, The Research Laboratory of Engineering Materials
Tokyo Institute of Technology
Nagatsuta Midori-ku Yokohama 227 Japan

1. Introduction

Drift problem as the horizontal displacement of tall buildings is one of the most serious issues in tall building design, relating to the dynamic characteristics of the building during earthquakes and strong winds. Drift shall be caused by the accumulated deformation of each member, such as a column, beam, brace and shear wall. Therefore, when we want to control the quantity of displacement by changing its design, we can not figure out which member of the structure should be changed, only from one result of computer calculation.

This paper aims at proposing a new method on drift control in structural design. This method can be regarded as a kind of design sensitively analysis which is newly used for structural analysis programs. Finally, I apply the method to a tall building structure.

2. The change of displacement caused by changing section of members

2.1 Unit Load Method for Calculating the Displacement of Structures

When we perform structural analysis with computers, we generally use Matrix method which defines the displacement of each node as the unknown. In this case, the displacements of a structure are printed out automatically and we can check them easily.

On the other hand, the displacement of a structure shall be caused by the accumulation of each member's deformation as mentioned before, The computer output can not indicate which member's deformation has a large share in its displacement.

I would like to mention Unit Load Method for calculating displacements derived from the principles of Virtual work.

According to References [2], the process of this method are as follows.

"The procedure for calculation a displacement by means of the unit-load method using Eq.1 may be summarized as follow : (1) determine the stress resultants N_L, M_L, and V_L in the structure caused by the actual loads : (2) place a unit load on the structure corresponding to the displacement D that is to be found : (3) determine the stress resultants N_U, M_U, and V_U caused by the unit load : (4) form the terms shown in Eq.1 and integrate each term for the entire structure : and (5) sum the results to obtain the displacement D."

$$D = \int \frac{N_U N_L}{EA} dx + \int \frac{M_U M_L}{EI} dx + \int \frac{\alpha V_U V_L}{GA} dx \qquad \text{Eq. 1}$$

Additionally, Prof. S.P.Timoshenko described the validity efficiency of this method as follows.

"The unit-load method can be used not only for beams, trusses, and other simple kinds of structures, but also for very complicated structures having many members. Furthermore, the unit-load method is suitable for finding all types of displacements, including the deflection of a point in the structure, the rotation of the axis of a member, the relative displacement between two points, and others. Theoretically, it may be used for either statically determinate or indeterminate structures, although for practical purposes the method is limited to determinate structures because its use requires that the stress resultants be known throughout the structure."

For these reasons, this method has been used only for hand calculation exercises at universities. In practical design, computers usually calculate structural deformation. There are very few engineers who take the trouble to calculate it by themselves using this method.

Today, Stress resultants of statically indeterminate structures can be obtained by computers. Unit load method has the advantage that it can analyze the source of displacement of certain point of a structure.

Therefore, this method can be practically used not only for statically determinate structures, but also for any type of structures.

2.2 Application to Statically Determinate Structures

As the simplest model, series springs are shown in Fig. 1.

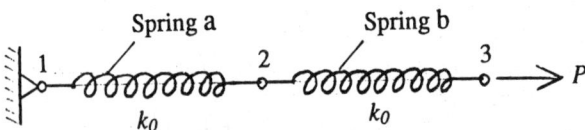

Fig. 1 Simple Statically Determinate Structure

Spring a, b are assumed to have the same spring constant k_0 ($= EA_0 / l_0$) : It is apparent that the axial forces of spring a, b can be represented by P : $N_{La} = N_{Lb} = P$. When a unit load is applied to point 3, the axial forces are expressed as, $N_{Ua} = N_{Ub} = 1$.

426

From Eq.1 of Unit load method, the displacement of point 3, defined as D_3, can be written as

$$D_3 = \int_{l_0} \frac{1}{EA_0} \frac{P}{EA_0} dx + \int_{l_0} \frac{1}{EA_0} \frac{P}{EA_0} dx$$

$$= \frac{Pl_0}{EA_0} + \frac{Pl_0}{EA_0} = \frac{P}{k_0} + \frac{P}{k_0} \qquad \text{Eq. 2}$$

By setting $d_a = P/k_0$, $d_b = P/k_0$, this equation becomes

$$D_3 = d_a + d_b \qquad \text{Eq. 3}$$

We can think of d_a, d_b as constituents of D_3.

Then, I consider a modified case in which spring constants of spring a, b are multiplied by α_a, α_b respectively.

The displacement of point 3, defined as D_3', can be similarly obtained from Eq.2.

$$D_3' = \frac{P}{\alpha_a k_0} + \frac{P}{\alpha_b k_0} = \frac{d_a}{\alpha_a} + \frac{d_b}{\alpha_b} \qquad \text{Eq. 4}$$

This model is a statically determinate structure. Even if spring constants are changed, the stress resultants of them shall stay unchanged. Therefore the result of this calculation is a right solution.

2.3 Application to Statically Indeterminate Structures

As the simplest model of a statically indeterminate structure, two parallel springs are shown in Fig.2.

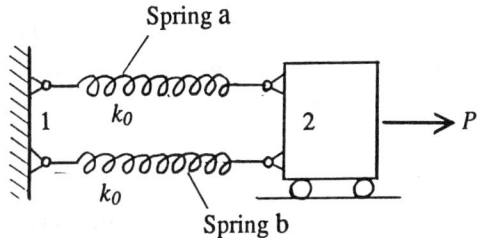

Fig. 2 Simple Statically Indeterminate Structure

D_2 can be expressed as $P/2k_0$ from basic technical knowledge. I figure out D_2 by means of the same method of statically determinate structures.

The axial forces of spring a, b caused by P can be represented by $P/2$, $N_{La} = N_{Lb} = P/2$. When a unit load is applied to point 2, the axial forces are expressed as, $N_{Ua} = N_{Ub} = 1/2$. D_2 can be obtained with these values using Eq.1

$$D_2 = \int_{l_0} \frac{\tfrac{1}{2}\tfrac{P}{2}}{EA_0}dx \; + \; \int_{l_0} \frac{\tfrac{1}{2}\tfrac{P}{2}}{EA_0}dx \;\; = \frac{P\,l_0}{4EA_0} + \frac{P\,l_0}{4EA_0} \qquad \text{Eq. 5}$$

By putting, $k_0 = EA_0 / l_0$

$$D_2 = \frac{P}{4k_0} + \frac{P}{4k_0} = \frac{P}{2k_0} \qquad \text{Eq. 6}$$

By setting $d_a = P / 4k_0$, $d_b = P / 4k_0$, this equation becomes

$$D_2 = d_a + d_b \qquad \text{Eq. 7}$$

We can think of d_a, d_b as constituents of D_2.

According to Eq.4, assuming that the quantity of each spring's deformation changes in proportion to the change of each spring constant respectively, I propose a new simplified calculation method. A new equation can be obtained from Eq.7, by dividing d_a by α_a and dividing d_b by α_b and adding both of them:

$$D_2' = \frac{d_a}{\alpha_a} + \frac{d_b}{\alpha_b} \qquad \text{Eq. 8}$$

I propose quite a daring assumption for structural deformation as follows.

"The quantity of each member's deformation which causes the displacement of certain point of a structure changes in proportion to the change of its member's stiffness, and this quantity is not affected by the change of any other member's stiffness."

This assumption is perfectly right for statically determinate structures. If there are no big changes in stress distribution, this assumption is almost right for statically indeterminate structures.

3. Constituents of drift of the top of a building caused by lateral force

3.1 A New Calculation Method

428

Fig.3.a indicates the stress resultants of a building caused by design lateral load. Fig.3.b indicates the stress resultants of the same building when Unit load is applied to the top of the building. In these figures, i means the sequence number of each member and l_i, w_i mean the length and weight of member i.

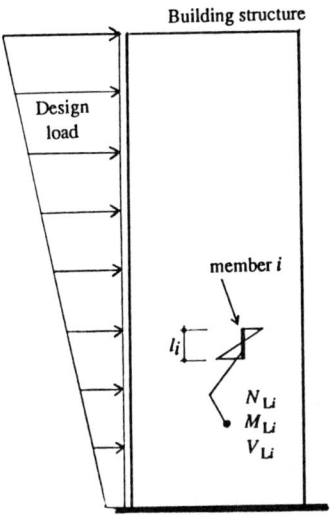

Fig. 3.a   Stress Resultants
Caused by Design Load

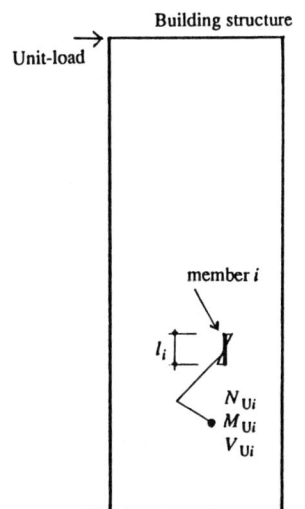

Fig. 3.b   Stress Resultants
Caused by Unit Load

By using the stress resultants of Fig.3.a, and Fig.3.b, and integrating Eq.1, the lateral displacement of the top of this building can be expressed as

$$D_{top} = \sum_{i=1}^{m} d_i \qquad \text{Eq. 9}$$

where

$$d_i = \frac{N_{Ui}\, N_{Li}}{EA_i}\, l_i + \int_{l_i} \frac{M_{Ui}\, M_{Li}}{E\, I_i}\, dx + \frac{\alpha\, V_{Ui}\, V_{Li}}{GA_i}\, l_i \qquad \text{Eq. 10}$$

Total weight of the structure can be given as

$$W_{total} = \sum_{i=1}^{m} w_i \qquad \text{Eq. 11}$$

The second term of Eq.12 can be easily obtained from Fig.4 as follows.

$$\int_l \frac{M_U M_L}{E I} dx = \frac{l}{6E I} \left(M_{Ua} (2M_{La} + M_{Lb}) + M_{Ub} (M_{La} + 2M_{Lb}) \right) \quad \text{Eq. 12}$$

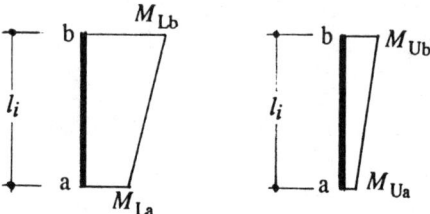

Fig. 4   Moment Distribution

3.2 Practical application to big buildings

When the structure is small, we can easily check and decompose each member's deformation to examine the source of the displacement of the top of a building. In practical design, generally a building has more than ten thousand members. Besides, the decomposition of each member's deformation shall significantly increase the number of data.

Consequently, The number shall become so great that we can hardly grasp all of them. Spread Sheet Programs for personal computers such as *Excel* and *Wingz*, will be quite useful to analyze these data. For example, they can sum up the deformations of columns, beams, braces and shear walls, separately.

4.   The way to minimize the displacement without changing total weight

The displacement of the top of a building can be expressed as

$$D_{top} = \sum_{i=1}^{m} d_i \qquad \text{Eq. 13}$$

where d_i = the deformation of member i

Total weight of steel can be expressed as

$$W_{total} = \sum_{i=1}^{m} w_i \qquad \text{Eq. 14}$$

where w_i = the weight of member i

On the assumption that section of members can be changed only by controlling the thickness of the steel plate without changing its type of figure, the Area and Moment Inertia and Weight of member i, A_i, I_i, w_i, shall be changed in proportion to the coefficient α_i which controls the thickness of the plate.

According to 2.3, the deformation of changed member i can be written as d_i / α_i

Therefore, the displacement of the top of the changed building is expressed as

$$D'_{top} = \sum_{i=1}^{m} \frac{d_i}{\alpha_i} \qquad \text{Eq. 15}$$

From the assumption, total weight of steel is constant. Therefore this is given by

$$W_{total} = W'_{total} = \sum_{i=1}^{m} \alpha_i \, w_i \qquad \text{Eq. 16}$$

The problem to be solved here is finding the minimum of Eq.15 under the subsidiary condition Eq.16.

By applying Lagrange multiplier λ, Eq.15 can be rewritten as

$$D'_{top} = \sum_{i=1}^{m} \frac{d_i}{\alpha_i} + \lambda \left(\sum_{i=1}^{m} \left(\alpha_i \, w_i \right) - W_{total} \right) \qquad \text{Eq. 17}$$

By partially differentiating Eq.17 by α_i ($i = 1,2,.... m$) and λ, and then setting these equations to be equal zero, we can find the values of α_i ($i = 1,2,.... m$) and λ which minimize D'_{top}. These equations become

$$\frac{\partial D'_{top}}{\partial \alpha_i} = -\frac{d_i}{\alpha_i^2} + \lambda \, w_i = 0 \qquad (i = 1, 2, \dots m) \qquad \text{Eq. 18}$$

$$\frac{\partial D'_{top}}{\partial \lambda} = \sum_{i=1}^{m} \left(\alpha_i \, w_i \right) - W_{total} = 0 \qquad \text{Eq. 19}$$

From Eq.18, α_i can be written as

$$\alpha_i = \sqrt{\frac{d_i}{\lambda \, w_i}} \qquad (i = 1, 2, \dots m) \qquad \text{Eq. 20}$$

By substituting Eq.20 into Eq.19, consequently we obtain the weight of changed member i as follows

$$w'_i = \alpha_i w_i = \frac{\sqrt{d_i \ w_i}}{\displaystyle\sum_{j=1}^{m} \sqrt{d_j \ w_j}} W_{total} \qquad (i = 1, 2, \ldots m) \qquad \text{Eq. 21}$$

where the solutions of this problem α_i ($i = 1,2,\ldots m$), λ can be easily obtained.

In conclusion, the weight of member i whose section was changed to minimize the displacement of the top, can be obtained by allocating total weight of steel in proportion to the square root of the product of d_i and w_i. This simple formula could be understood intuitively and helpful to structural designers.

5. Application to Tall Building

For example, this method is applied to a steel structure as schematized in Figure 1. Although the structure is fifteen-story high and is not actually a high-rise building, this example is suitable for proving the usefulness of the theory. The sections of columns and girders and braces in original design are determined by simplified calculation method based on Japanese seismic design standard.

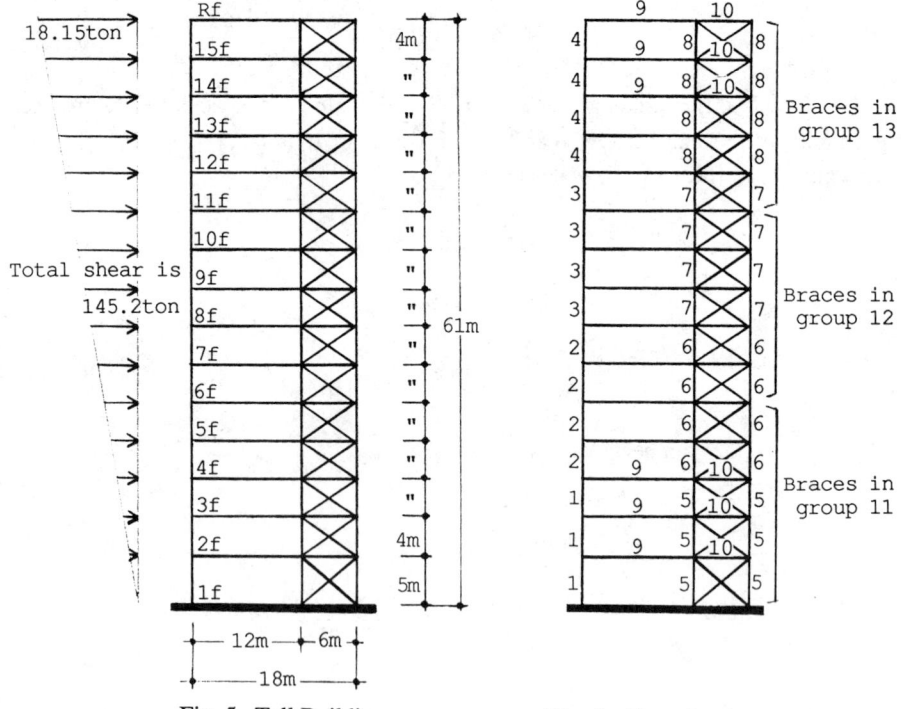

Fig. 5 Tall Building

Fig. 6 Classification to 13 groups

The shapes of these sections are listed on Table 1. When the distributed horizontal force as shown in Figure 5 is subjected to the structure, the horizontal displacement of the top is figured out to be 39.8 cm. As mentioned in section 3.2, this structure has so many members that 105 members are classified into 13 groups as shown in Fig.6 and the calculation for optimum design is performed. In section 4, d_i and w_i represent the deformation and weight of member i respectively. In this example, d_i and w_i can be defined as the sum of the deformations and weights of all members which belong to group i.

The calculation process is shown in Table 2. d_i and w_i are the calculation results of original structure. α_i is the coefficient to change the section of group i members.

The case that α_i is more than 1.0, indicates the structural deformation shall be decreased by changing the section of group i members for larger one. The case that α_i is less than 1.0, indicates the structural deformation shall be increased a little even if smaller sectional members are replaced in group i. Accordingly, using smaller sectional members for reducing weight could be efficient.

d'_i indicates the sum of deformations of all changed members which belong to group i. The summation of d'_i (i=1~13) represents the estimated horizontal displacement at the top of the structure (=31.3 cm).

At the top of changed building, the horizontal displacement caused by the previous force, becomes 29.9 cm. The calculation result is smaller than the estimated value by 1.4 cm, and consequently the method is proved to be quite efficient.   The reason why the

Table 1   Sections of Columns, Girders and Braces

Columns		Girders		Braces	
12f-15f	Box-450x450x9			11f-15f	H-150x150x7x10
8f-11f	Box-450x450x12	2f-Rf	H-600x200x11x17	6f-10f	H-175x175x7.5x11
4f-7f	Box-450x450x16			1f-5f	H-200x200x8x12
1f-3f	Box-450x450x19				

Table 2   Calculation Procedure

group	w_i (tons)	d_i (cm)	$\sqrt{d_i w_i}$	w'_i (tons)	α_i	d'_i (cm)
1	3.16	.42	1.149	1.72	.545	.77
2	3.35	.56	1.371	2.05	.613	.91
3	2.56	.64	1.285	1.93	.751	.86
4	1.95	.76	1.216	1.82	.935	.81
5	6.32	12.90	9.030	13.54	2.141	6.02
6	6.70	6.71	6.706	10.05	1.500	4.47
7	5.13	1.84	3.068	4.60	.897	2.05
8	3.90	.45	1.334	2.00	.512	.89
9	18.99	11.60	14.844	22.25	1.172	9.90
10	9.50	.17	1.265	1.90	.200	.85
11	3.66	2.00	2.705	4.06	1.109	1.80
12	2.90	1.33	1.964	2.94	1.016	1.31
13	2.27	.46	1.020	1.53	.673	.68
total	70.39	39.84	46.957	70.39		31.32

calculation result is not equal to the estimated value, is that this structure is a statically indeterminate one and the stress distribution varies by changing sections.

The original and changed structure have the same total weight (70.4 ton). The displacement at the top can be reduced to be 75 % of original design by this method.

6.   Conclusion

Design is the product of human creativity. Even if Optimum Design Method is completed by computers, I think very few structural engineers will use computer design directly without any consideration. This paper considers mainly structural deformation and proposes the new method to improve structural design.

In Computer age, technical intuitions are more difficult to be obtained as compared with Hand-calculation age. Therefore, some conservative engineers are afraid that frequent use of computers might be ruining the quality of engineers. Although I do not dare to oppose it, actually computer has been making great progress day by day and I think the important thing is how to use computers more efficiently. We should try to find out new methods which enhance engineers' technique and creativity by using computers. I believe that providing the process of computer calculation could be helpful, and I try to perform it in this paper.

This paper has developed a method on the assumption that the stiffness of each member is proportional to its weight. Next time, I would like to proceed with the method considering the change of each member's stiffness corresponding to the change of its depth and width. Furthermore, I will try to study deformation analysis which includes plastic deformation caused by large external loads.

7.   Acknowledgement

I appreciate the help of Mrs. Masako Yoneda, in making of English version of the paper.

8.   References

[1] S.J.Arora and E.J.Hang, "Methods of Design Sensitively Analysis in Structural Optimization", AIAA Journal, Vol.17, No.19, pp.970-974, 1979
[2] S.P.Timoshenko and James M. Gore, "Mechanics of Materials", Van Nostrand Reinhold, 1977
[3] Edward J.Teal, "Seismic Design Practice for Steel Buildings", Engineering Jounal/ American Institute of Steel Construction/ 4th quarter 1975
[4] Roy Becker, Farzad Naeim and Edward J. Teal, "Seismic Design Practice for Steel Buildings", Steel Committee of California, June 1988
[5] Akira Wada, "How to Reduce Drift of Buildings", Fourth U.S.-Japan Workshop on the Improvement of Building Structural Design Practices, Applied Technology Council, August 1990

Proceedings of the Second Conference on Tall Buildings in Seismic Regions
55th Regional Conference
May 16 and 17, 1991, Los Angeles, California

The Role of Active Structural Control Systems
in Super-Tall Building Structures

125-Story Tower: A Case Study

Nabih Youssef, S.E.[1]
T. Jeff Guh, Ph.D., S.E.[2]

In Collaboration With

Sami Masri, Ph.D.[3]

An important characteristic associated with **Super-Tall Building Structures** (STBS) is the relative importance of their lateral load resisting and stabilizing systems to counter the effect of dynamic environmental loadings such as earthquake ground motions, hurricanes and high winds. Unlike low-rise or mid-rise structures, of which the structural analysis and design with respect to dynamic forces have generally been a process of adapting the gravity load resisting systems to provide the needed lateral resistance and stability, super-tall buildings require specific lateral structural systems devoted to the optimization of the building's dynamic performance and serviceability, calling for close evaluation of the inherently conflicting constraints of economics, structural reliability, and the considerable uncertainties in quantifying dynamic loadings.

Given a prescribed dynamic loading condition, a conventional lateral load-resisting system of a building structure is designed to limit its responses within the allowable serviceability and damageability limits by balancing the structural supply and demand of stiffness, strength, stability, and energy dissipation/absorption capacities. However, due to the high uncertainties in predicting wind or earthquake loads, an optimal conventional structural system with sufficient performance confidence is often difficult to achieve for a super-tall building without incurring substantial cost.

Since its introduction to the structural engineering profession in the early '70s, the concept of **active structural control** has been demonstrated by researchers and engineers to be a viable solution to enhance the performance and serviceability of super-tall building structures. By implementing appropriate automatic control mechanisms in conventional structural systems, the dynamic behavior of a super-tall building can be actively monitored with real-time response measurement, and modified through the use of **feedback/feedforward** technology as the dynamic load condition changes. This strategy can significantly improve the dynamic structural response of a super-tall building and, at the same time, control and minimize the intrinsic uncertainties in dynamic environmental loadings.

---

[1] President, Nabih Youssef & Associates.
[2] Senior Engineer, Nabih Youssef & Associates, 660 South Figueroa Street, Suite 1660, Los Angeles, California 90017.
[3] Professor, Department of Civil Engineering, University of Southern California, Los Angeles, California 90007.

Despite an abundance of recent research activities in the field of active structural control, there remain several important issues that need be investigated from a practical design and construction point of view, such as the effect of model and loading uncertainty, system nonlinearity, response time lag, and construction feasibility; just to name a few. In order to identify the special problems associated with active structural control system in the structural design of super-tall buildings, a series of analytical studies involving actual design projects have been conducted through a joint effort of NYA and USC.

For the studies reported herein, the dynamic performance of a proposed 125-story super-tall building in Los Angeles is analyzed parametrically to evaluate the effectiveness of active structural control systems. Based on the system characteristics of the active structural control system and the structural system configuration and dimensions, a 3-dimensional analytical model is formulated for dynamic time history analysis under earthquake and wind loadings. Actual historic records of earthquakes and storms are modified in amplitudes and frequency contents to provide the input excitations. The modeling and analysis are conducted in such a way as to take the material and system nonlinearities into proper perspective.

Attention is given to the building's interstory drift response for optimal damage mitigation, and the floor diaphragm accelerations for human perceptibility control. The analytical results obtained from this study indicate that the introduction of the active structural control system contributes significant improvement to the building's dynamic performance and results in favorable cost savings as compared with alternative conventional structural systems. Areas that need further research are also pointed out and summarized.